Active Biomedical Materials and Their Applications

Active Biomedical Materials and Their Applications

Editor

Christie Ying Kei Lung

MDPI

Basel • Beijing • Wuhan • Barcelona • Belgrade • Novi Sad • Cluj • Manchester

Editor
Christie Ying Kei Lung
University of Hong Kong
Hong Kong
China

Editorial Office
MDPI AG
Grosspeteranlage 5
4052 Basel, Switzerland

This is a reprint of articles from the Special Issue published online in the open access journal *Journal of Functional Biomaterials* (ISSN 2079-4983) (available at: https://www.mdpi.com/journal/jfb/special_issues/W55G16Y2HX).

For citation purposes, cite each article independently as indicated on the article page online and as indicated below:

Lastname, A.A.; Lastname, B.B. Article Title. *Journal Name* **Year**, *Volume Number*, Page Range.

ISBN 978-3-7258-2133-4 (Hbk)
ISBN 978-3-7258-2134-1 (PDF)
doi.org/10.3390/books978-3-7258-2134-1

Cover image courtesy of Christie Ying Kei Lung

Contents

Journal of
Functional Biomaterials

Editorial

Active Biomedical Materials and Their Applications

Christie Ying Kei Lung

Restorative Dental Sciences, Faculty of Dentistry, The University of Hong Kong, Hong Kong, China; cyklung@hku.hk

Active biomedical materials are designed to heal and restore the functions of people recovering after injuries or diseases. They are used to repair or replace a specific biological function of dysfunctional tissues (Figure 1). They are synthetic or natural materials, such as metals and their alloys, ceramics, composites, or polymers. They must meet several criteria, such as having good mechanical properties, excellent biocompatibility, high corrosion resistance, good tribological properties, and non-toxicity. The modification of existing biomedical materials and the development of new biomedical materials are effective strategies for improving material performance.

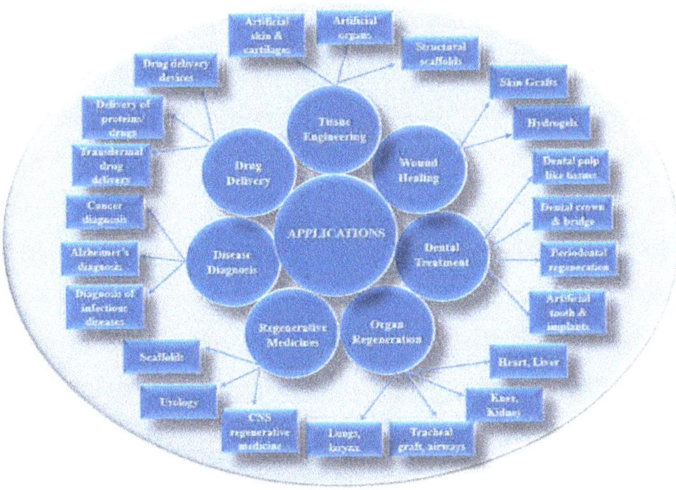

Figure 1. Applications of biomedical materials in different areas. Reprinted from Ref. [1].

Citation: Lung, C.Y.K. Active Biomedical Materials and Their Applications. *J. Funct. Biomater.* **2024**, *15*, 250. https://doi.org/10.3390/jfb15090250

Received: 15 August 2024
Accepted: 29 August 2024
Published: 30 August 2024

This Special Issue, "Active Biomedical Materials and Their Applications", provides updated information on the synthesis and properties of different materials, namely ceramics, polymers, nanoparticles, and composites, and their applications. It collects seven novel studies and four review articles. Experts from different research fields share their ideas and report their findings on these topics.

In the study by Tian, X. et al. (contribution 1), an injectable bioactive and a degradable calcium sulfate/hydroxyapatite (CaS/HA) cement was developed for the augmentation of fenestrated pedicle screws in an osteoporotic spine [2]. The results showed that CaS/HA augmentation was effective in enhancing the pull-out force of the fenestrated pedicle screws in osteoporotic vertebrae in comparison with non-augmented screws. This technique allowed for adequate cement distribution and the interdigitation ability of the CaS/HA cement in the trabecular bone.

In the review by Liu, D. et al. (contribution 2), an overview of currently used biodegrac able bone cements, such as calcium phosphates, calcium sulfates, and organic–inorgani composites, was described [3]. The degradation mechanism and the clinical performanc of the biodegradable cements were discussed in detail.

In the review by Itzhaki, E. et al. (contribution 3), the synthesis and characterizatio of proteinoid polymers and the nanocapsules containing synergistic drugs, cannabinoid and TRAIL for cancer theranostics were summarized [4]. Proteinoids are random polymer composed of amino acids synthesized using stepwise thermal polymerization. The in vitr and in vivo studies of the latest findings in this area were discussed.

In the review by Sousa, A.B. et al. (contribution 4), the authors discussed the potentia applications of specialized pro-resolving mediators (SPMs) in the development of new im munomodulatory biomaterials [5]. These mediators are a family of endogenous molecule that include lipoxins, resolvins, protectins, maresins, Cysteinyl-SPMs, and n-3 docosapen taenoic acid-derived SPMs. SPMs have important anti-inflammatory and pro-resolutiv actions, such as decreasing the recruitment of polymorphonuclear leukocytes, inducing th recruitment of anti-inflammatory macrophages, and increasing macrophage clearance o apoptotic cells through a process known as efferocytosis.

In the study by de Araujo, M.M. et al. (contribution 5), the authors develop solid lipid polymer hybrid nanoparticles (SLPHNs) as topical delivery systems for small interfering RNA (siRNA) molecules [6]. Their findings showed that the developed SLPHN–0.25% polyethyleneimine was a promising nanoplatform for the cutaneous delivery of siRNA, as it favored the ability of siRNA to penetrate the skin through the stratum corneum barrie and kept it in the epidermis.

In the study by Tărăboanță, I. et al. (contribution 6), a light-activated microhybric composite resin incorporated with green silver nanoparticles was developed to improve the antibacterial activity and surface hardness of the resin composite [7]. The authors reportec that the resin composite loaded with silver nanoparticles inhibited the Streptococcus mutans activity, but there was no improvement in surface hardness.

In the study by Lin, C.J. et al. (contribution 7), an ultrasonic-assisted digestion of a formic acid decellularized extracellular matrix (UdECM) hydrogel was developed for dia betic wound treatment [8]. The wound-healing effect of the UdECM hydrogel incorporated with rat platelet-rich plasma (R-PRP) and sacchachitin nanofibers (SCNFs) on diabetic wounds was examined using an animal model. The results showed that the UdECM/SC/R PRP group improved diabetic wound healing and the recovery of diabetic wounds in normal tissue.

In the study by Chauvin, A. et al. (contribution 8), a hydroxyapatite-based coating on zirconia was designed [9]. The zirconia was coated with polyethyleneimine-hydroxyapatite (PEI-HAp) followed by thermal treatment. The results showed that the adhesion of hydrox yapatite on zirconia was improved.

In the study by Sung, P. et al. (contribution 9), a network structure of anatase- and rutile-type TiO_2 with CuO fine particles was formed on titanium when it was subjected to H_2O_2-$Cu(OAc)_2$-heat treatments [10]. The results showed that the coated layer exhibited good antibacterial activity against E. coli and S. aureus when subjected to visible-light irradiation. This was due to the generation of reactive oxygen species and hydroxyl free radicals (·OH) under visible-light irradiation.

In the study by Soto ER et al. (contribution 10), they developed a new drug delivery system, yeast particles for the encapsulation of fungicides, tetraconazole (TET), and prothioconazole (PRO), with a high payload-loading capacity, encapsulation efficiency, and encapsulation stability [11]. The activity of the yeast particles in encapsulated PRO on azole-resistant C. albicans strains was enhanced compared to unencapsulated PRO.

In the review by Abdalla, M.M. et al. (contribution 11), the authors provided a comprehensive review of the applications of strontium compounds in dentistry [12]. Their comprehensive summary elucidated the uses and benefits of strontium, the mechanisms of its biological interactions, and its applications in dentistry.

In summary, the articles included in this Special Issue, "Active Biomedical Materials and Their Applications", report new data and contribute to recent developments in biomedical materials. Hopefully, these contributions will both practically benefit the readership and encourage further research in the field of biomedical materials.

I extend my gratitude to the authors, the reviewers, and the MDPI *JFB* editors and their team for their invaluable contributions.

Conflicts of Interest: The author declares no conflicts of interest.

References

1. Nikita, N.; Kamlesh, W.; Milind, U. An Overview on Biomaterials: Pharmaceutical and Biomedical Applications. *J. Drug Deliv. Ther.* **2021**, *11*, 154–161.
2. Tian, X.; Raina, D.B.; Vater, C.; Kilian, D.; Ahlfeld, T.; Platzek, I.; Nimtschke, U.; Tägil, M.; Lidgren, L.; Thomas, A.; et al. Evaluation of an Injectable Biphasic Calcium Sulfate/Hydroxyapatite Cement for the Augmentation of Fenestrated Pedicle Screws in Osteoporotic Vertebrae: A Biomechanical Cadaver Study. *J. Funct. Biomater.* **2022**, *13*, 269. [CrossRef] [PubMed]
3. Liu, D.; Cui, C.; Chen, W.; Shi, J.; Li, B.; Chen, S. Biodegradable Cements for Bone Regeneration. *J. Funct. Biomater.* **2023**, *14*, 134. [CrossRef] [PubMed]
4. Itzhaki, E.; Elias, Y.; Moskovits, N.; Stemmer, S.M.; Margel, S. Proteinoid Polymers and Nanocapsules for Cancer Diagnostics, Therapy and Theranostics: In Vitro and In Vivo Studies. *J. Funct. Biomater.* **2023**, *14*, 215. [CrossRef] [PubMed]
5. Sousa, A.B.; Barbosa, J.N. The Use of Specialized Pro-Resolving Mediators in Biomaterial-Based Immunomodulation. *J. Funct. Biomater.* **2023**, *14*, 223. [CrossRef] [PubMed]
6. de Araujo, M.M.; Borgheti-Cardoso, L.N.; Praça, F.G.; Marcato, P.D.; Bentley, M.V.L.B. Solid Lipid–Polymer Hybrid Nanoplatform for Topical Delivery of siRNA: In Vitro Biological Activity and Permeation Studies. *J. Funct. Biomater.* **2023**, *14*, 374. [CrossRef] [PubMed]
7. Tărăboanță, I.; Burlec, A.F.; Stoleriu, S.; Corciovă, A.; Fifere, A.; Batir-Marin, D.; Hăncianu, M.; Mircea, C.; Nica, I.; Tărăboanță-Gamen, A.C.; et al. Influence of the Loading with Newly Green Silver Nanoparticles Synthesized Using Equisetum sylvaticum on the Antibacterial Activity and Surface Hardness of a Composite Resin. *J. Funct. Biomater.* **2023**, *14*, 402. [CrossRef] [PubMed]
8. Lin, C.-J.; Lin, H.-L.; You, W.-C.; Ho, H.-O.; Sheu, M.-T.; Chen, L.-C.; Cheng, W.-J. Composite Hydrogels of Ultrasound-Assisted-Digested Formic Acid-Decellularized Extracellular Matrix and Sacchachitin Nanofibers Incorporated with Platelet-Rich Plasma for Diabetic Wound Treatment. *J. Funct. Biomater.* **2023**, *14*, 423. [CrossRef] [PubMed]
9. Chauvin, A.; Garda, M.-R.; Snyder, N.; Cui, B.; Delpouve, N.; Tan, L. Hydroxyapatite-Based Coatings on Silicon Wafers and Printed Zirconia. *J. Funct. Biomater.* **2024**, *15*, 11. [CrossRef] [PubMed]
10. Sung, P.-C.; Yokoi, T.; Shimabukuro, M.; Mokudai, T.; Kawashita, M. Apatite-Forming Ability and Visible Light-Enhanced Antibacterial Activity of CuO-Supported TiO_2 Formed on Titanium by Chemical and Thermal Treatments. *J. Funct. Biomater.* **2024**, *15*, 114. [CrossRef] [PubMed]
11. Soto, E.R.; Rus, F.; Ostroff, G.R. Yeast Particle Encapsulation of Azole Fungicides for Enhanced Treatment of Azole-Resistant Candida albicans. *J. Funct. Biomater.* **2024**, *15*, 203. [CrossRef] [PubMed]
12. Abdalla, M.M.; Sayed, O.; Lung, C.Y.K.; Rajasekar, V.; Yiu, C.K.Y. Applications of Bioactive Strontium Compounds in Dentistry. *J. Funct. Biomater.* **2024**, *15*, 216. [CrossRef] [PubMed]

Journal of *Functional Biomaterials*

MDPI

Article

Evaluation of an Injectable Biphasic Calcium Sulfate/Hydroxyapatite Cement for the Augmentation of Fenestrated Pedicle Screws in Osteoporotic Vertebrae: A Biomechanical Cadaver Study

Xinggui Tian [1,2], Deepak B. Raina [3], Corina Vater [1,2], David Kilian [2], Tilman Ahlfeld [2], Ivan Platzek [4], Ute Nimtschke [5], Magnus Tägil [3], Lars Lidgren [3], Alexander Thomas [1], Uwe Platz [1], Klaus-Dieter Schaser [1], Alexander C. Disch [1] and Stefan Zwingenberger [1,2,*]

1. University Center of Orthopaedic, Trauma and Plastic Surgery, University Hospital Carl Gustav Carus at Technische Universität Dresden, 01307 Dresden, Germany
2. Center for Translational Bone, Joint and Soft Tissue Research, Faculty of Medicine, University Hospital Carl Gustav Carus at Technische Universität Dresden, 01307 Dresden, Germany
3. Department of Clinical Sciences Lund, Faculty of Medicine, Lund University, 22184 Lund, Sweden
4. Department of Radiology, University Hospital Carl Gustav Carus at Technische Universität Dresden, 01307 Dresden, Germany
5. Institute of Anatomy, University Hospital Carl Gustav Carus at Technische Universität Dresden, 01307 Dresden, Germany

* Correspondence: stefan.zwingenberger@uniklinikum-dresden.de

Citation: Tian, X.; Raina, D.B.; Vater, C.; Kilian, D.; Ahlfeld, T.; Platzek, I.; Nimtschke, U.; Tägil, M.; Lidgren, L.; Thomas, A.; et al. Evaluation of an Injectable Biphasic Calcium Sulfate/Hydroxyapatite Cement for the Augmentation of Fenestrated Pedicle Screws in Osteoporotic Vertebrae: A Biomechanical Cadaver Study. *J. Funct. Biomater.* **2022**, *13*, 269. https://doi.org/10.3390/jfb13040269

Academic Editor: Christie Ying Kei Lung

Received: 23 October 2022
Accepted: 23 November 2022
Published: 1 December 2022

Abstract: Cement augmentation of pedicle screws is one of the most promising approaches to enhance the anchoring of screws in the osteoporotic spine. To date, there is no ideal cement for pedicle screw augmentation. The purpose of this study was to investigate whether an injectable, bioactive, and degradable calcium sulfate/hydroxyapatite (CaS/HA) cement could increase the maximum pull-out force of pedicle screws in osteoporotic vertebrae. Herein, 17 osteoporotic thoracic and lumbar vertebrae were obtained from a single fresh-frozen human cadaver and instrumented with fenestrated pedicle screws. The right screw in each vertebra was augmented with CaS/HA cement and the un-augmented left side served as a paired control. The cement distribution, interdigitation ability, and cement leakage were evaluated using radiographs. Furthermore, pull-out testing was used to evaluate the immediate mechanical effect of CaS/HA augmentation on the pedicle screws. The CaS/HA cement presented good distribution and interdigitation ability without leakage into the spinal canal. Augmentation significantly enhanced the maximum pull-out force of the pedicle screw in which the augmented side was 39.0% higher than the pedicle-screw-alone side. Therefore, the novel biodegradable biphasic CaS/HA cement could be a promising material for pedicle screw augmentation in the osteoporotic spine.

Keywords: biomaterial; biomechanical; calcium sulfate/hydroxyapatite; cement; osteoporosis; pedicle screw augmentation

1. Introduction

Osteoporosis is the most common age-related bone disease. Over 200 million people worldwide suffer from osteoporosis [1]. The prevalence of osteoporosis has been continuously escalating with increasingly elderly populations, with more than 70% of people over the age of 80 being affected [2]. With increased life expectancy, improved quality of life, and the desire to maintain physical activity in older adults, the need for spinal surgery in elderly patients suffering from osteoporosis has increased accordingly [3]. The prevalence of osteoporosis in men older than 50 and postmenopausal women when undergoing spinal surgery was reported as 14.5% and 51.3%, respectively, by Xie et al. [4]. Pedicle screws are the most commonly used implants to achieve posterior fixation of the lumbar and thoracic spine. The key to an adequate base for a stable fixation is the bony purchase of the pedicle screws in the pedicle and trabecular bone of the vertebral body [5]. The specific morphology of the trabecular meshwork affects the mechanical grasp of the screw and, therefore, influences the screw integration at the bone–metal interface [6]. In osteoporotic vertebrae, bone-metal osteointegration is significantly reduced which often leads to the loosening of the screw [6] with a rate of up to 62% [7].

Over the past few decades, numerous techniques have been developed to improve the anchorage of pedicle screws in order to reduce the risk of screw failure. These techniques primarily suggested larger outer diameters or an extended length of screws, screws with different thread profiles, a cylindrical or conical core shape, expanding screws, bi-cortical pedicle screws, coated pedicle screws, or cement augmentation screws [8,9]. The cement-augmented pedicle screw technique is considered one of the most promising approaches for the improvement of screw fixation. A recent meta-analysis showed that screw augmentation significantly reduced the loosening rate by a factor of 10 (non-augmentation screws vs. augmentation screws: 22.5% vs. 2.2%) [10]. To date, polymethyl methacrylate (PMMA) is the only material that has been clinically approved by the FDA to augment pedicle screw fixation. Due to its high viscosity and injectability, PMMA is advantageous for augmenting pedicle screws in the osteoporotic bone with a high stability. However, PMMA also has some drawbacks limiting its clinical application. PMMA is a rather stiff, high-strength epoxy resin with a higher mechanical strength than osteoporotic bone. This biomechanical mismatch leads to a shielding effect on the adjacent vertebrae or disc [11]. The strong exothermic reaction during PMMA polymerization can further damage the surrounding already-fragile bone, which weakens the fixation between the bone and cement and can finally lead to aseptic loosening [1]. Most importantly, due to its non-degradability and lack of bioactivity, PMMA neither forms a stable bio-bonding with the bone nor induces bone ingrowth or osseointegration [12]. Due to its non-resorbability, a chronic inflammatory response may occur, further compromising the strength and integrity of the implant [13]. PMMA excessively adheres to the screws, making it difficult to remove the screw during revision surgery [12]. In addition, its monomer component methyl methacrylate (MMA) is known to be toxic to healthy organs [9]. These PMMA-associated drawbacks have created a need for bioactive or bioresorbable functionalized cements to achieve better pedicle screw augmentation in osteoporotic bone. Calcium phosphate [9] as well as calcium sulfate cements [14] with high biodegradability were developed as PMMA-replacing alternative candidates for pedicle screw augmentation. However, the long curing time of calcium phosphate cements [15] and the rapid degradation rate of calcium sulfate cements [16] limit their clinical application.

In this study, we aim to show that a recently established injectable calcium sulfate/hydroxyapatite (CaS/HA) biomaterial, which has been approved for human use both in Europe and North America [17], may meet the requirements of pedicle screw augmentation for improved efficacy. The CaS/HA cement is a powder-based biomaterial that forms a paste when mixed with the liquid phase, with proper biocompatibility [18]. It can be injected for about 5 min, molded for up to 9 min after mixing, and is transformed into a solid mass approximately 15 min after initiating the mixing procedure [19]. The compressive strength of the material is higher compared to cancellous bone [20]. Clinical studies have shown that CaS/HA cement as a substitute for PMMA has achieved good clinical results in the treatment of osteoporotic vertebral fractures through percutaneous vertebroplasty [21,22], providing strong evidence that CaS/HA cement can be used to treat osteoporotic spinal diseases. Previous preclinical studies have shown that CaS/HA can enhance screw anchorage in native or synthetic bones [23,24]. Our latest study also showed that CaS/HA-enhanced lag-screw anchorage with high extraction forces is comparable to PMMA in cadaveric femoral heads as well as in patients undergoing treatment for intertrochanteric fractures [25]. In addition, CaS/HA cement is a degradable material that can serve as a scaffold for new bone ingrowth in order to achieve better osseointegration.

Its isothermal setting reaction will be compatible with surrounding tissues as it does nc lead to thermally induced necrosis [25]. Thus, the purpose of this study was to investigat whether degradable CaS/HA cement could be used for pedicle screw augmentation i osteoporotic vertebrae.

We hypothesized that CaS/HA cement could strengthen the interface between the ped cle screw and the vertebral trabecular bone with the potential to increase the pull-out forc of the pedicle screw in the osteoporotic vertebrae in a preliminary proof-of-principle stud

2. Materials and Methods

2.1. Methodology

A male human cadaver was used in this study as a model of osteoporotic vertebra to test the hypothesis that the CaS/HA cement could augment the pedicle screws wit a potential improvement of the immediate anchorage in the osteoporotic vertebrae. Th computed tomography (CT) scan was performed to measure the Hounsfield Units (HU The fenestrated pedicle screws were implanted into the osteoporotic vertebrae, in whic the right side of the vertebra was augmented with the CaS/HA cement and the left sid of the vertebra was not augmented. During the procedure, the C-arm X-ray was used t evaluate cement distribution and cement leakage. After the pedicle screw augmentatio the specimens were embedded in epoxy resin for pull-out testing to evaluate the pull-ou force. The maximum pull-out force was used to quantify the effect of cement augmentatio by comparison with the non-augmented side. (Figure 1).

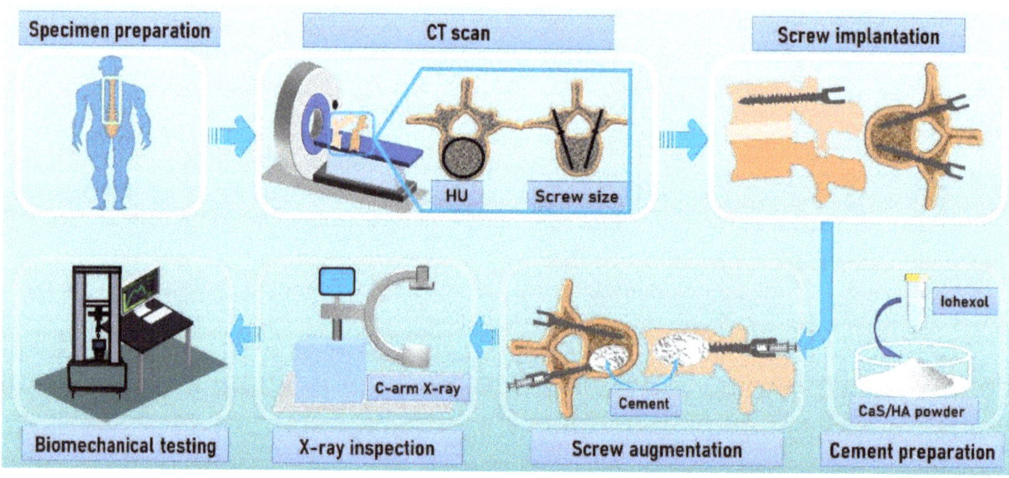

Figure 1. Flow chart summarizing the methodology. After the specimens' acquisition, computed tomography (CT) scan was performed to determine that the specimen met the experimental criteria and to determine the parameters of the pedicle screws. Pedicle screws were implanted and augmented with CaS/HA cement, and then a pull-out experiment was performed to evaluate the effect of the augmentation. Abbreviations: HU: Hounsfield units; CaS/HA: calcium sulfate/hydroxyapatite.

2.2. Specimen Preparation

A total of 17 vertebrae (12 thoracic and 5 lumbar) were obtained from a fresh-frozen cadaver of an 81-year-old male (Institute of Anatomy, University Hospital Carl Gustav Carus at Technische Universität Dresden, Dresden, Germany). All soft tissues on the vertebral body were completely removed and CT scans of all vertebral bodies were performed (Department of Radiology, University Hospital Carl Gustav Carus at Technische Universität Dresden). The diameter of the pedicle and the length of the vertebral body were measured to determine the respective size of each pedicle screw. The radiodensity in Hounsfield Units (HU) of the vertebral body, screw trajectory, and augmentation area were also measured using AGFA HealthCare software (AGFA; Mortsel, Belgium). The HU of vertebral bodies was measured according to a method described by Li et al. [26], and the HU of screw trajectory was measured using a method previously described by Zhang et al. [27]. In addition, a circular area with a diameter of 8 mm at the front end of the screw trajectory was defined as the ROI of the potentially augmented area in this study.

2.3. Pedicle Screw Implantation

The cortex and hyperplastic osteophytes of the lamina were removed with a rongeur, so that the polyaxial tulip-head of the pedicle screw could be rotated freely to prevent it from blocking the pedicle screw. A pilot hole was made by an awl oriented forward and inward along the pedicle axis. Then we checked the hole with a probe to ensure that the pedicle was not damaged. Each hole was tapped using a 4.5 mm tap, then each fenestrated pedicle screw (uCentum, Ulrich Medical, Ulm, Germany) was implanted with a ratchet screwdriver using a free-hand technique until the hub of the screw had been firmly fixed on the posterior cortex. Because the trajectory of the screw during implantation affects the required extraction force later, the classical trajectory of passing the center of the pedicle parallel to the superior and inferior endplates was adopted in this study [28]. All instrumentation was completed by one experienced spine surgeon. The specimens were stored in sealed vacuum bags at −20 °C until further preparation and testing.

2.4. CaS/HA Cement Preparation and Augmentation Procedure

According to the preparation protocol, the CaS/HA cement (Cerament, Bone Support AB, Lund, Sweden), which consists of a premixed powder of 60 weight% (wt%) CaS and 40 wt% HA, was mixed with iohexol (1 g CaS/HA:0.43 mL iohexol) in a 6-well plate for 30 s. The CaS/HA slurry was then transferred into the syringe used for injection. This process allowed the cement to cure for 4 min prior to injection [17,19]. During this period, the syringe was connected to the tail of the fenestrated pedicle screw through a special metal tube. Then the CaS/HA paste was injected into the right-side screws. A total of 1 mL CaS/HA paste was used for each thoracic vertebra and 2 mL was used for each lumbar vertebra. During injection, C-arm fluoroscopy radiographs were taken to evaluate cement distribution, interdigitation ability, and cement leakage. The CaS/HA cement was allowed to harden at room temperature for 24 h before mechanical testing.

2.5. Mechanical Testing

A customized metal container was used to hold the vertebra, while a band was used to fix the vertebra by insertion through the spinal canal. The vertebrae were then embedded in epoxy resin (Yachticon, Norderstedt, Germany) with the embedded line of each vertebra under the pedicle to avoid interference with mechanical testing (Figure 2A,B). This process required setting overnight at room temperature to obtain an exact, stable fixation. For the pull-out test, the customized container was fixed on the bottom of the testing machine (Z010, Zwick-Roell, Ulm, Germany) with a titanium rod. In this position, it could be freely adjusted in the sagittal plain of the pedicle to ensure that the implantation direction of each pedicle screw was consistent with the direction of the piston of the machine. In this case, the pedicle screw was securely attached to the piston with a nut. In addition to the pull-out force along the longitudinal axis of the screw, this procedure should avoid interference with additional forces in order to allow reliable quantification of the shear stability of the screw-bone interface (Figure 2C,D). The screws were pulled out at a displacement of 10 mm/min using a custom-made jig on the universal testing machine in displacement-controlled mode (Figure 2E). To avoid any biased tendency in the results by the test order, the tests were alternated between right (cement-augmented) and left (non-augmented) sides. The pull-out strength of a pedicle screw was defined as the point at which the force peaks in the force-displacement diagram in newtons (N). Stiffness was defined as the slope in the linear region of the force-displacement curve.

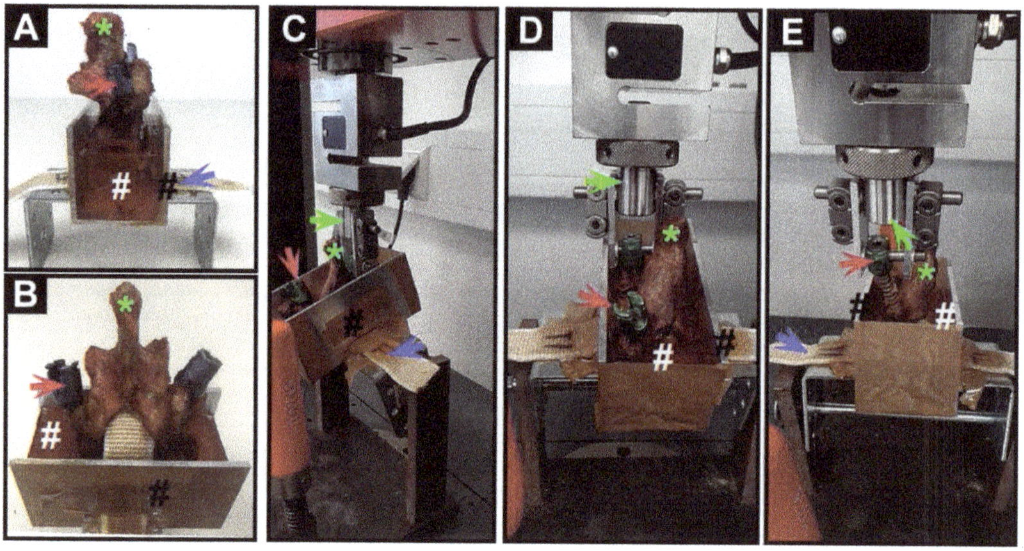

Figure 2. Photographs of vertebra fixation and biomechanical testing process. (**A**) Lateral view and (**B**) anterior view of a vertebra (*) fixed in the customized metal container (black #) filled with epoxy resin (white #); (**C**) lateral view; and (**D**) anterior view of the mechanical testing process of pedicle screws; and (**E**) the picture of pedicle screws being successfully pulled out from the vertebra. Red arrows indicate the pedicle screws; blue arrows indicate the fixed band; green arrows indicate the piston of the testing machine.

2.6. Statistics

Data were presented as mean ± standard deviation (SD). Statistical analysis was performed by using GraphPad Prism 8 (GraphPad Software, San Diego, CA, USA). Significant differences between the groups were evaluated using the paired t-test. The level of significance was chosen at $p < 0.05$ (*).

2.7. Ethical Approval

This study involved cadaveric specimens voluntarily donated by a donor to the Institute of Anatomy, University Hospital Carl Gustav Carus at Technische Universität Dresden for the advancement of science. This study has been carried out in accordance with the Code of Ethics of the World Medical Association (WMA Statement on Organ and Tissue Donation) and the local ethics policy of Dresden University Hospital for experiments.

3. Results

3.1. Specimen Preparation

For this cadaveric study, all thoracic and lumbar vertebrae from one cadaveric donor were collected. The T7 height was reduced which may have suffered a previous osteoporotic vertebral compression fracture, without previous cement-related surgical treatment. The applied individual, vertebra-specific pedicle screw sizes are listed in Table 1. The radiodensity of all vertebral bodies was below 150 HU while the HU values of all vertebral bodies except T3 (138) were also detected as lower than 135 (Figure 3A). No significant difference was detected comparing the HU of the pedicle screw trajectory between the potentially cement-augmented (right) side and the un-augmented (left) side (Figure 3B) and the HU between the potential augmented area and contralateral relative area (Figure 3C).

Table 1. The individual, vertebra-specific pedicle screw sizes defined according to CT scan measurement results.

Location	Size (d/l) in mm	Location	Size (d/l) in mm	Location	Size (d/l) in mm
T1	6/30	T7	6/45	L1	7/50
T2	5/35	T8	6/50	L2	7/50
T3	5/40	T9	6/50	L3	7/50
T4	5/40	T10	7/45	L4	7/45
T5	5/45	T11	7/45	L5	7/40
T6	6/45	T12	7/50		

The screw sizes were defined by diameter "d" and length "l" (in mm).

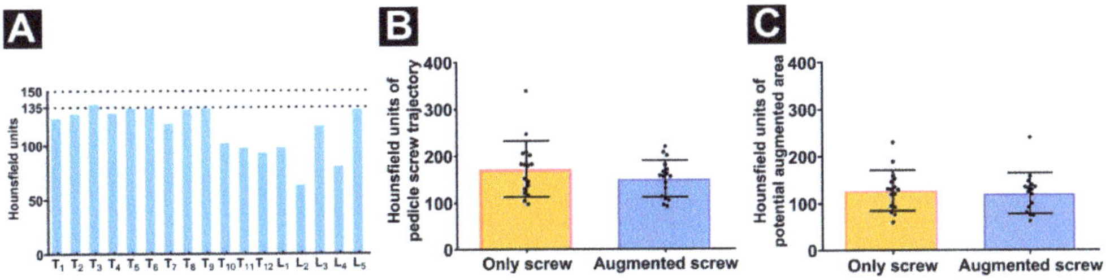

Figure 3. Quantification of radiodensity in CT scans: (**A**) Radiodensity in Hounsfield units (HU) for all vertebral bodies T1–T12 and L1–L5; (**B**) Radiodensity values (in HU) of pedicle screw trajectory comparing augmented screw side and only screw side; and (**C**) the HU between the potential augmented area and contralateral area. Data are presented as mean ± SD. The two horizontal dotted lines in panel A represent the critical values for diagnosing osteoporosis by HU in previous literature [29].

3.2. Pedicle Screw Implantation

All pedicle screws were successfully implanted into the vertebrae. After pedicle screw implantation, X-ray images were taken in order to demonstrate that the placement of all pedicle screws was accurate, and that no pedicle or vertebral body was damaged during the operation (Figure 4).

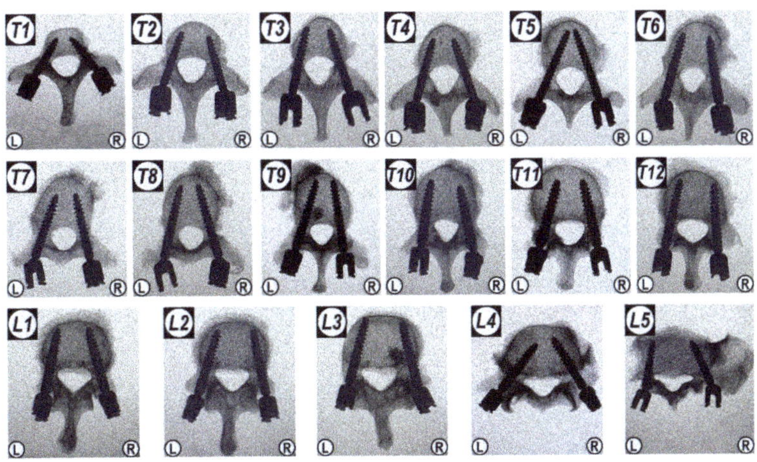

Figure 4. The anteroposterior X-ray images of all vertebrae after the fenestrated pedicle screw implantation prior to augmentation with CaS/HA material. "L" and "R" indicate the left and right side; T1–T12 and L1–L5 indicate the thoracic and lumbar position of the vertebrae.

3.3. Trajectories of Augmented and non-Augmented Pedicle Screws

The CaS/HA cement was completely and successfully injected in all vertebrae (1 mL in the thoracic vertebrae and 2 mL in the lumbar vertebrae) under fluoroscopic guidance. After cement injection, X-ray images demonstrated that the CaS/HA cement was evenly distributed around the anterior hole of the fenestrated pedicle screws both in the coronal plane from the anteroposterior radiograph and in the sagittal plane from the lateral radiograph in all vertebrae. No obvious cement leakage was seen into the spinal canal. In addition, the irregular shape of the cement edges also indicated the adequate cement-bone interdigitation ability in the trabecular bone (Figure 5).

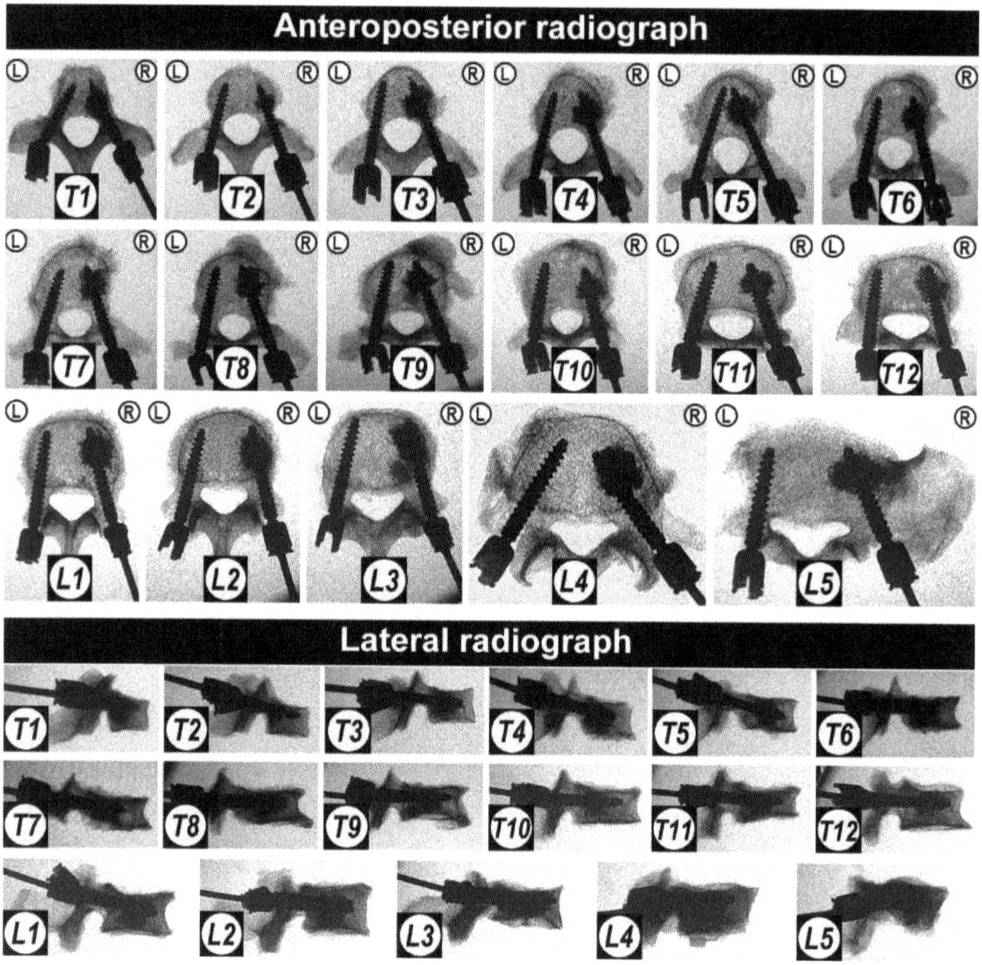

Figure 5. The anteroposterior radiograph and the lateral radiograph of the fenestrated pedicle screws augmented with CaS/HA cement. "L" and "R" indicate the left and right side; T1–T12 and L1–L5 indicate the thoracic and lumbar position of the vertebrae.

Further visual observation of the vertebral appearance after cement injection revealed that the cement was exposed in the superior endplate plane of four vertebrae (T1, T3, T5, and T9) and in the inferior endplate plane of three vertebrae (T7, L3, and L5) (Figure 6).

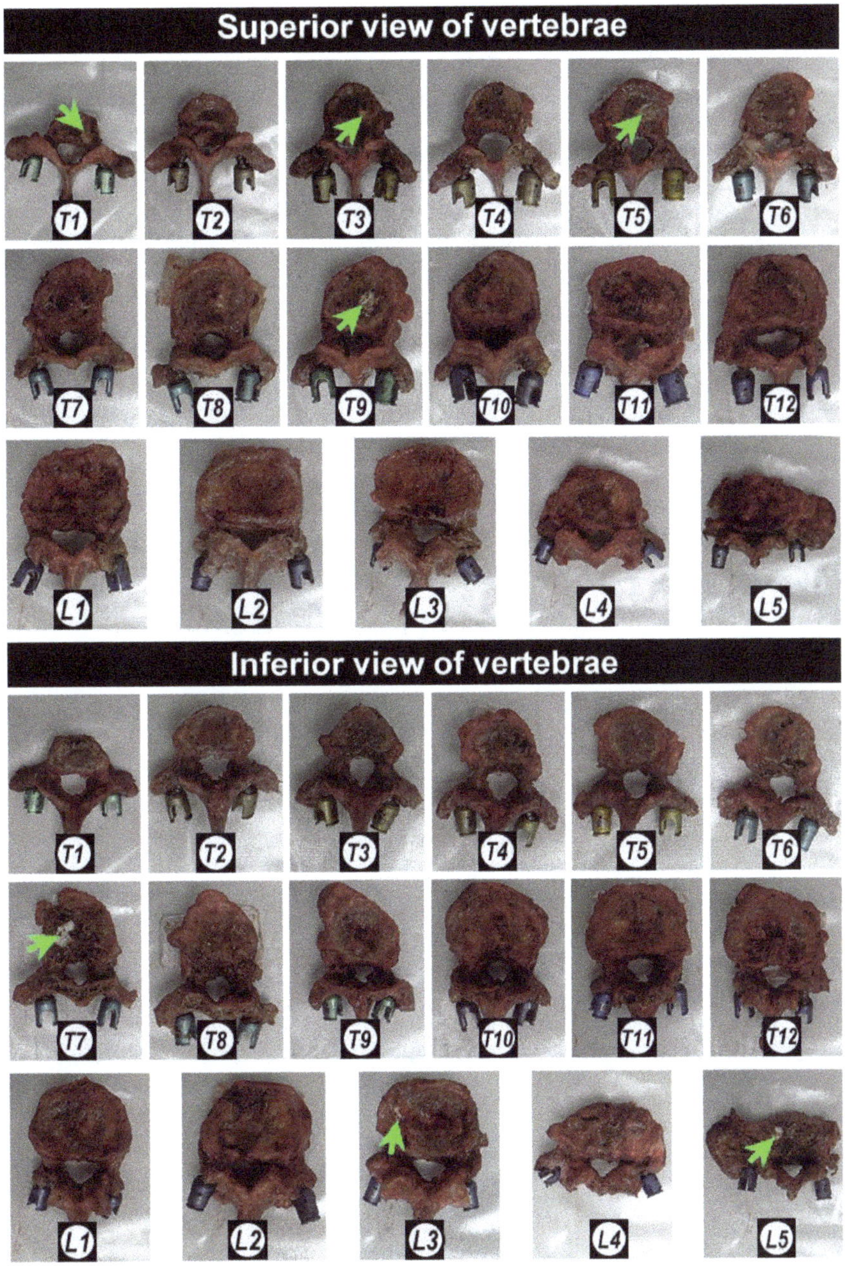

Figure 6. Photographs of the appearance of all the vertebrae after cement injection. Green arrows indicate the CaS/HA cement. T1–T12 and L1–L5 indicate the thoracic and lumbar position of the vertebrae.

3.4. CaS/HA Augmentation Leads to an Increased Pull-Out Force

All specimens were stabilized in the customized metal container to minimize variation. Regarding the maximum pull-out force, the determined force for the augmented screw was higher than the pull-out force of the pedicle screw without augmentation among 14 vertebrae. Whereas test outliers on the unenhanced side of T3 may be due to the destruction of the trabecular microstructure of the pedicle and vertebral body, cement augmentation significantly increased its pull-out resistance although the pedicle trabecular microarchitecture on the augmented side was also destroyed from the analysis of CT scans (Dragonfly, ORS Inc., Montreal, QC, Canada) (Supplementary Materials, Figure S1A). The maximum pull-out force of the screw on the contralateral side of three vertebrae was slightly higher than on the augmentation side. Each individual pull-out force of the augmentation side was lower than the corresponding pedicle screw-alone side occurring in the thoracic spine (T1, T7, and T11). From the analysis of the CT scans, it was found that the reason for these undesired results may be related to the microfracture of the right part of the vertebral body at T7 and T11 and the obvious cortical bone hyperplasia of the left pedicle at T1 detected by comparison with the right side (Supplementary Materials Figure S1B–D). In the lumbar spine, all the pull-out force figures of the augmentation sides were higher than the pedicle screw-alone side (Figure 7A). In general, the maximum pull-out force on the augmentation side was significantly higher compared to the screw-only side (non-augmentation screws vs. augmentation screws: 288.5 ± 105.6 N vs 401.1 ± 114.6 N) (Figure 7B). The maximal pull-out force was significantly increased by CaS/HA augmentation both in the thoracic (non-augmentation screws vs. augmentation screws: 278.8 ± 122.2 N vs. 359.5 ± 83.5 N) and lumbar (non-augmentation screws vs. augmentation screws: 311.8 ± 50.5 N vs. 500.8± 125.6 N) (Figure 7C,D). Overall, the maximum pull-out force of the augmented side was 39.0% higher than the pedicle screw-alone side (overall: 39.0%, thoracic: 28.9%, lumbar: 60.6%). The mechanical analysis of vertebral bone showed that the mean stiffness of the cement-augmented screw side was not statistically different to the screw-only side, including the thoracic and lumbar (Figure 7E–H).

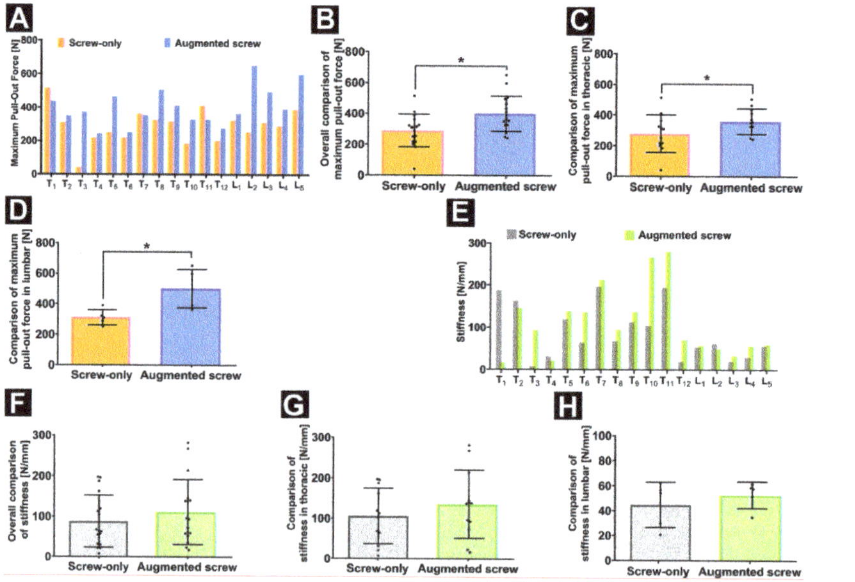

Figure 7. (**A–D**) The maximum pull-out force in N for screws pulled out from the cement-augmented screw side and screw-only side; (**A**,**B**) overall comparison; (**C**) comparison for thoracic vertebrae; and (**D**) lumbar comparison of screw only vs. augmented screw. (**E–H**) Stiffness in N/mm of vertebrae comparison between the cement-augmented screw side and screw-only side; (**E**,**F**) overall; (**G**) thoracic; and (**H**) lumbar comparison. Data are presented as mean ± SD, * $p < 0.05$.

Pedicle screw failure occurred at the screw-bone interface in both the screw-alone and the augmented screw sides. After testing, 10 specimens could be retrieved from the epoxy resin while the other vertebrae were damaged during removal of the osteoporotic vertebral body from the container due to the tight adhesion of the vertebral body and the epoxy resin. The vertebral cross-section photographs along the midline of the pedicle demonstrated that the cement was evenly distributed within the trabecular network in the vertebral body, which densified the loose architecture of the osteoporotic vertebral body. At the screw trajectory area, a local cavity was formed inside the vertebrae after screw extraction at the screw-only side; cement was still filled within the screw trajectory at the augmented side (Figure 8A,B). After testing, the threads of the augmented screws were filled with cement, but only little trabecular or bone marrow tissue filled the threads of the screws without augmentation (Figure 8C).

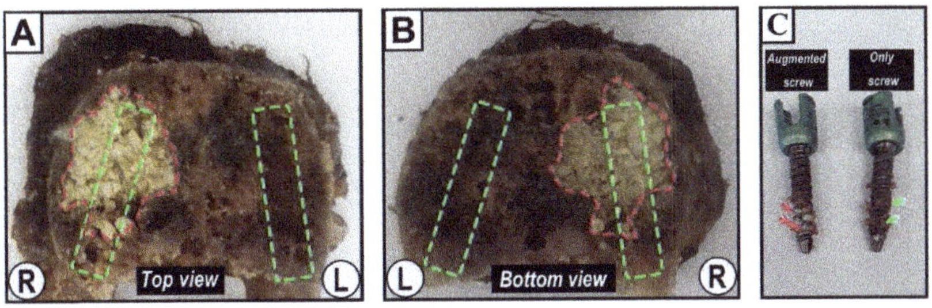

Figure 8. Representative specimens of vertebrae and fenestrated pedicle screws after pull-out testing. (**A**) The top section; and (**B**) the bottom section of vertebra L2 after axial cross-section along the midline of the pedicle. The red area of irregular shape indicates the CaS/HA cement-augmented area, while the rectangular area marked by the green dashed lines indicates the rough zone of the pedicle screw trajectory. "L" and "R" indicate the respective orientations. (**C**) A pair of pedicle screws (6.5/30 mm) from T1 after the pull-out experiment. Red arrows indicate remnants of CaS/HA cement; green arrows indicate remnants of the trabecular bone tissue.

4. Discussion

In this study, we described the novel application of a recently established CaS/HA biomaterial cement for the augmentation of pedicle screws in the osteoporotic spine. By injecting CaS/HA cement into the internal channel of the fenestrated pedicle screws, the cement can be evenly distributed into the screw threads and into the surrounding trabecular in order to strengthen the interlocking effect at the bone-screw interface (Figures 5 and 8). Biomechanical testing demonstrated that the described method of injectable CaS/HA cement improved the maximum pull-out force of fenestrated pedicle screws in osteoporotic vertebrae (Figure 7). The hypothesis that CaS/HA cement can enhance the interface between the pedicle screw and the osteoporotic bone and support the immediate anchoring of the pedicle screw in the osteoporotic vertebrae of the thoracic and lumbar spine was verified.

Anchoring of pedicle screws in the spine with poor bone quality remains challenging. The success of a pedicle screw instrumentation depends on the implant design as well as on the quality of the bone which could provide adequate compression between the threads and surrounding bone; this proves to be a critical aspect of fixation stability [30]. Previous studies found that there is a significant linear correlation between bone mineral density (BMD) and failure cycle and fatigue loads [31]; it is also suggested that for a BMD of thoracic and lumbar less than 80 mg/cm^3, the stability of the pedicle screws may be insufficient so that additional stabilization should be considered [31]. Dual energy X-ray absorptiometry (DEXA) is the current "gold standard" for BMD measurement and osteoporosis screening. Osteoporosis is defined by the World Health Organization as $\geq$2.5 standard deviations below the mean BMD, as measured by DEXA [1]. However, DEXA measurements of BMD are affected by lumbar degenerative changes, scoliosis, compression fractures, and cannot distinguish between cortical and cancellous bone [26]. Some studies have found that the detection of BMD based on the measured HU value not only has a certain linear correlation with the BMD assessed by DEXA and quantitative computed tomography, but can also be used for the auxiliary diagnosis of osteoporosis by passing the above shortcomings of the DEXA method [26]. Two recent studies suggested that using HU for the diagnosis of osteoporosis aiming for high sensitivity may set a cutoff value of 150 HU, while aiming for a more balanced sensitivity-specificity ratio may require a cutoff value of 135 HU [29,32]. The HU of all vertebral bodies in this study met the diagnostic criteria for osteoporosis. In addition, DEXA can only measure the overall BMD of the vertebrae without the ability to determine the BMD of the screw trajectory. The HU value of the screw trajectory can be used for pedicle screw stability prediction [27]. There were no significant differences in the HU values of the trajectories nor in the potential augmentation area of the augmented-screw side and the contralateral area of the screw-only side. This consolidates that the difference in the pull-out force was exclusively mediated by the augmentation effect of the cement.

At present, pedicle screw augmentation techniques mainly include prefilling with cement prior to solid screw insertion and fenestrated pedicle screws injected with cement through the inner channel, but the biomechanical testing results of PMMA augmentation are controversial [33]. Compared to traditional solid pedicle screws, the extraction force of the fenestrated pedicle screw with PMMA augmentation is either increased, equal, or even decreased [33]. However, here the cannulated screw has the advantages of a shorter surgery time and a reduced risk of cement leakage [9,33]. In this study, the commercially available, fenestrated standard pedicle screws with the same size for each vertebra were used for CaS/HA cement augmentation. The optimal amount of cement used for proper screw fixation is also controversial and numerous studies on PMMA recommend 1 mL in the thoracic spine and 1.5–3 mL in the lumbar spine [5,16,33]. With reference to the amount of PMMA, 1 mL of CaS/HA cement in the thoracic spine and 2 mL in the lumbar spine were used in this study. The CaS/HA cement was evenly distributed in the anterior end of fenestrated pedicle screws both in the coronal and sagittal plane and it significantly enhanced the pull-out force of the pedicle screw without spinal canal leakage. The exposure of the CaS/HA cement in the endplate is due to the tight junction between the annulus fibrosus and the cartilage endplate or the marginal bone of the vertebral body resulting in damage of the cartilage endplate during specimen preparation. These results proved that the CaS/HA cement augmented fenestrated pedicle screw is an effective technique and confirmed that a dosing volume of 1 mL cement in the thoracic spine and 2 mL in the lumbar spine is effective. However, the biomechanical evaluation and comparison of screw implantation between the two techniques as well as the correlation of cement volume and mechanical properties is worth further investigation. The CaS/HA cement overall showed a 39% increase in the maximum pull-out force, which was lower than that of PMMA (50–100% [16,34]), but the increase in the lumbar vertebrae was 60.6% in this range which may be related to vertebrae structure and bone cement amount. Due to the relatively small size of the thoracic vertebrae, the amount of cement that can be accommodated is limited, resulting in a limited increase in the pull-out resistance of the thoracic spine, which may also be related to the overall smaller force level of the thoracic spine [33]. The larger trabecular area of the lumbar vertebrae allows high doses of bone cement to be injected, which further leads to higher maximum pull-out forces. From a clinical point of view, the lumbar bears more stress than the thoracic because there is no sternum and ribs to share stress, and the higher pull-out resistance obtained may potentially reduce the risk of screw failure. The biomechanical relationship between cement volume and region needs to be further explored in another independent controlled trial. After the pull-out experiment, we observed that the CaS/HA cement was still evenly filled in the trabecular bone and between the threads of the screw, which confirms the good distribution and interdigitation ability of the CaS/HA cement [12]. This is another important reason why the cement enhances pull-out strength by filling the gap between the bony trabeculae and the thread of the screw in osteoporotic vertebrae [35]. This interface failure of CaS/HA cement occurs at the screw/bone interface, which is different from the failure modality of the PMMA composite/bone interface, indicating that the shear and tensile strength of CaS/HA, like calcium phosphate cement, are lower than those of cancellous bone [8,34].

Biphasic CaS/HA biomaterials have different rates of degradation, as the CaS phase is rapidly absorbed within 6–8 weeks in vivo while the HA phase persists for 9–12 months, integrating with new bone ingrowth over time [36]. The undesired degradation rates (too rapid or too slow) of single-phase degradable cement or the non-degradable nature of PMMA fail to match the bone ingrowth speed. Due to the lack of intrinsic osteoinductive cytokines, the functionalization of biomaterials with osteopromotive molecules is a promising option in bone tissue engineering [17], especially in the osteoporotic bone. Our previous studies have shown that CaS/HA materials can be used to locally co-deliver rhBMP-2 and zoledronic acid (ZA) to promote femoral defect repair [17] and bone regeneration in the osteoporotic femoral neck canal [37,38]. This is due to the fact that rhBMP-2 is physically entrapped in the resorbable CaS phase of the carrier for rapid release, whereas ZA is chemically bound to HA for long-term action [39]. Furthermore, Raina et al. showed that local co-delivery of rhBMP-2 and ZA by CaS/HA significantly enhanced screw anchoring [40]. Our recent data even indicates that the osteoporosis drug ZA seeks peri-implant HA particles and biologically activates them to induce significantly more new cancellous bone formation around the implant in an osteoporotic animal model of implant integration. All these data demonstrate that the suggested workflow is a promising option for the future usage of CaS/HA cement alone or after functionalization with osteo-regenerative molecules to enhance pedicle screw anchorage in patients with osteoporotic vertebrae. Further studies are warranted to elucidate whether this approach could aid in the reduction of pedicle screw failure and reduce the number of reoperations.

5. Limitations

Besides a successful outcome of the study, certain technical limitations need to be considered: the damaged cartilage endplate during specimen preparation due to the tight connection between the endplate and intervertebral disc in this study may affect the quantification of absolute pull-out force results. The anatomical variation of the pedicle in the available cadaveric vertebrae may also influence the biomechanical results. However, these limitations are not expected to affect a pairwise side-to-side comparison of individual vertebrae since specimen handling did not influence the morphology and macroscopic anatomy of the bone specimens [33]. Another limitation is that the axial pull-out test used in this study cannot fully simulate a physiological fatigue load situation, even though it is considered a standard test method for the fixing strength of screws [41]. The relationship between cement volume and biomechanics requires further exploration in another independent controlled trial. In addition, this experiment was a preliminary proof-of-concept study with a very limited sample size; a large sample size trial would be useful to further confirm our results.

6. Conclusions

This study demonstrates a novel application of the recently established injectable, biodegradable biphasic CaS/HA cement for augmentation of fenestrated pedicle screws in osteoporotic spine. In a cadaveric model, the CaS/HA augmentation was proven as an effective technique which significantly enhanced the required pull-out force of fenestrated pedicle screws in osteoporotic vertebrae in comparison with non-augmented screws. The technique allowed adequate cement distribution and interdigital ability to the surrounding trabecular bone. In addition, the CaS/HA material has been approved for clinical utilization, which could allow for an easier translation to the clinical application of this technique. In conclusion, this biodegradable biphasic CaS/HA cement could be a promising material for pedicle screw augmentation in the osteoporotic spine.

Supplementary Materials: The following supporting information can be downloaded at: https://www.mdpi.com/article/10.3390/jfb13040269/s1, Figure S1: CT scans of vertebrae with (A) outlier (T3) and (B-D) suboptimal pull-out test results (T1, T7, T11). "L" and "R" indicate the left and right side. The white rectangular areas indicate the potential cortical bone hyperplasia of the left pedicle in T1 and the potential microfracture areas inT3, T7 and T11.

Author Contributions: Conceptualization, X.T., D.B.R., M.T., L.L., A.C.D. and S.Z.; methodology, X.T., D.B.R., C.V., T.A., I.P., U.N., L.L., A.C.D. and S.Z.; software, X.T., T.A., I.P. and S.Z.; validation, X.T. and S.Z.; formal analysis, X.T. and S.Z.; investigation, X.T., D.B.R., C.V., T.A., I.P., U.N., A.T. and S.Z.; resources, D.B.R., L.L., U.N., K.-D.S., A.C.D. and S.Z.; data curation, X.T. and S.Z.; writing—original draft preparation, X.T. and D.K.; writing—review and editing, all authors; visualization, X.T.; supervision, U.P., K.-D.S., A.C.D. and S.Z.; project administration, S.Z. All authors have read and agreed to the published version of the manuscript.

Funding: This research received no external funding.

Institutional Review Board Statement: This study involved cadaveric specimens voluntarily donated by a donor. This study has been carried out in accordance with The Code of Ethics of the World Medical Association (WMA Statement on Organ and Tissue Donation) and the local ethics policy of Dresden University Hospital for experiments.

Informed Consent Statement: Not applicable.

Data Availability Statement: All data generated or analyzed during this study are included in this article.

Acknowledgments: We are very grateful to Ulrich Medical (Ulm, Germany) for the provided fenestrated pedicle screws and the donor for donating his remains in order to advance science.

Conflicts of Interest: L.L. is a board member of Ortho Cell, Australia and BONESUPPORT AB, Sweden. L.L., M.T., and D.B.R. hold stocks in Moroxite AB, Sweden. The authors declare that they have no other competing interests pertaining to this study.

References

1. Krenzlin, H.; Foelger, A.; Mailänder, V.; Blase, C.; Brockmann, M.; Düber, C.; Ringel, F.; Keric, N. Novel Biodegradable Composite of Calcium Phosphate Cement and the Collagen I Mimetic P-15 for Pedicle Screw Augmentation in Osteoporotic Bone. *Biomedicines* **2021**, *9*, 1392. [CrossRef] [PubMed]
2. Reginster, J.Y.; Burlet, N. Osteoporosis: A still increasing prevalence. *Bone* **2006**, *38* (Suppl. 1), 4–9. [CrossRef] [PubMed]
3. Park, S.B.; Chung, C.K. Strategies of spinal fusion on osteoporotic spine. *J. Korean Neurosurg. Soc.* **2011**, *49*, 317–322. [CrossRef]
4. Xie, F.; Zhou, B.; Wang, J.; Liu, T.; Wu, X.; Fang, R.; Kang, Y.; Dai, R. Microstructural properties of trabecular bone autografts: Comparison of men and women with and without osteoporosis. *Arch. Osteoporos.* **2018**, *13*, 18. [CrossRef] [PubMed]
5. Paré, P.E.; Chappuis, J.L.; Rampersaud, R.; Agarwala, A.O.; Perra, J.H.; Erkan, S.; Wu, C. Biomechanical evaluation of a novel fenestrated pedicle screw augmented with bone cement in osteoporotic spines. *Spine (Phila Pa 1976)* **2011**, *36*, E1210–E1214. [CrossRef] [PubMed]
6. Abousayed, M.; Boktor, J.G.; Sultan, A.M.; Koptan, W.; El-Miligui, Y. Augmentation of fenestrated pedicle screws with cement in patients with osteoporotic spine. *J. Craniovertebr. Junction Spine* **2018**, *9*, 20–25. [CrossRef]
7. El Saman, A.; Meier, S.; Sander, A.; Kelm, A.; Marzi, I.; Laurer, H. Reduced loosening rate and loss of correction following posterior stabilization with or without PMMA augmentation of pedicle screws in vertebral fractures in the elderly. *Eur. J. Trauma Emerg. Surg.* **2013**, *39*, 455–460. [CrossRef]
8. Tomé-Bermejo, F.; Piñera, A.R.; Alvarez-Galovich, L. Osteoporosis and the Management of Spinal Degenerative Disease (I). *Arch. Bone Jt. Surg.* **2017**, *5*, 272–282.
9. Shea, T.M.; Laun, J.; Gonzalez-Blohm, S.A.; Doulgeris, J.J.; Lee, W.E., 3rd; Aghayev, K.; Vrionis, F.D. Designs and techniques that improve the pullout strength of pedicle screws in osteoporotic vertebrae: Current status. *Biomed. Res. Int.* **2014**, *2014*, 748393. [CrossRef]
10. Rometsch, E.; Spruit, M.; Zigler, J.E.; Menon, V.K.; Ouellet, J.A.; Mazel, C.; Härtl, R.; Espinoza, K.; Kandziora, F. Screw-Related Complications after Instrumentation of the Osteoporotic Spine: A Systematic Literature Review with Meta-Analysis. *Glob. Spine J.* **2020**, *10*, 69–88. [CrossRef]
11. Sun, H.; Liu, C.; Liu, H.; Bai, Y.; Zhang, Z.; Li, X.; Li, C.; Yang, H.; Yang, L. A novel injectable calcium phosphate-based nanocomposite for the augmentation of cannulated pedicle-screw fixation. *Int. J. Nanomed.* **2017**, *12*, 3395–3406. [CrossRef] [PubMed]
12. Sun, H.; Liu, C.; Li, X.; Liu, H.; Zhang, W.; Yang, H.; Li, C.; Yang, L. A novel calcium phosphate-based nanocomposite for the augmentation of cement-injectable cannulated pedicle screws fixation: A cadaver and biomechanical study. *J. Orthop. Translat.* **2020**, *20*, 56–66. [CrossRef] [PubMed]
13. Frick, C.; Dietz, A.C.; Merritt, K.; Umbreit, T.H.; Tomazic-Jezic, V.J. Effects of prosthetic materials on the host immune response: Evaluation of polymethyl-methacrylate (PMMA), polyethylene (PE), and polystyrene (PS) particles. *J. Long Term Eff. Med. Implant.* **2006**, *16*, 423–433. [CrossRef] [PubMed]

14. Rohmiller, M.T.; Schwalm, D.; Glattes, R.C.; Elalayli, T.G.; Spengler, D.M. Evaluation of calcium sulfate paste for augmentation of lumbar pedicle screw pullout strength. *Spine J.* **2002**, *2*, 255–260. [CrossRef]
15. Cho, W.; Wu, C.; Erkan, S.; Kang, M.M.; Mehbod, A.A.; Transfeldt, E.E. The effect on the pullout strength by the timing of pedicle screw insertion after calcium phosphate cement injection. *J. Spinal Disord. Tech.* **2011**, *24*, 116–120. [CrossRef]
16. Hoppe, S.; Keel, M.J. Pedicle screw augmentation in osteoporotic spine: Indications, limitations and technical aspects. *Eur. J. Trauma Emerg. Surg.* **2017**, *43*, 3–8. [CrossRef]
17. Raina, D.B.; Matuszewski, L.M.; Vater, C.; Bolte, J.; Isaksson, H.; Lidgren, L.; Tägil, M.; Zwingenberger, S. A facile one-stage treatment of critical bone defects using a calcium sulfate/hydroxyapatite biomaterial providing spatiotemporal delivery of bone morphogenic protein-2 and zoledronic acid. *Sci. Adv.* **2020**, *6*, eabc1779. [CrossRef]
18. Alfotawi, R.; Naudi, K.; Dalby, M.J.; Tanner, K.E.; McMahon, J.D.; Ayoub, A. Assessment of cellular viability on calcium sulphate/hydroxyapatite injectable scaffolds. *J. Tissue Eng.* **2013**, *4*, 2041731413509645. [CrossRef]
19. Abramo, A.; Geijer, M.; Kopylov, P.; Tägil, M. Osteotomy of distal radius fracture malunion using a fast remodeling bone substitute consisting of calcium sulphate and calcium phosphate. *J. Biomed. Mater. Res. B Appl. Biomater.* **2010**, *92*, 281–286. [CrossRef]
20. Nilsson, M.; Wielanek, L.; Wang, J.S.; Tanner, K.E.; Lidgren, L. Factors influencing the compressive strength of an injectable calcium sulfate-hydroxyapatite cement. *J. Mater. Sci. Mater. Med.* **2003**, *14*, 399–404. [CrossRef]
21. Masala, S.; Nano, G.; Marcia, S.; Muto, M.; Fucci, F.P.; Simonetti, G. Osteoporotic vertebral compression fractures augmentation by injectable partly resorbable ceramic bone substitute (Cerament™ | SPINE SUPPORT): A prospective nonrandomized study. *Neuroradiology* **2012**, *54*, 589–596. [CrossRef] [PubMed]
22. Marcia, S.; Boi, C.; Dragani, M.; Marini, S.; Marras, M.; Piras, E.; Anselmetti, G.C.; Masala, S. Effectiveness of a bone substitute (CERAMENT™) as an alternative to PMMA in percutaneous vertebroplasty: 1-year follow-up on clinical outcome. *Eur. Spine J.* **2012**, *21* (Suppl. 1), S112–S118. [CrossRef] [PubMed]
23. Kok, J.; Širka, A.; Liu, Y.; Tarasevičius, Š.; Belickas, J.; Tägil, M.; Lidgren, L.; Isaksson, H.; Raina, D.B. Augmenting a dynamic hip screw with a calcium sulfate/hydroxyapatite biomaterial. *Med. Eng. Phys.* **2021**, *92*, 102–109. [CrossRef] [PubMed]
24. Zampelis, V.; Tägil, M.; Lidgren, L.; Isaksson, H.; Atroshi, I.; Wang, J.S. The effect of a biphasic injectable bone substitute on the interface strength in a rabbit knee prosthesis model. *J. Orthop. Surg. Res.* **2013**, *8*, 25. [CrossRef] [PubMed]
25. Raina, D.B.; Markevičiūtė, V.; Stravinskas, M.; Kok, J.; Jacobson, I.; Liu, Y.; Sezgin, E.A.; Isaksson, H.; Zwingenberger, S.; Tägil, M.; et al. A New Augmentation Method for Improved Screw Fixation in Fragile Bone. *Front. Bioeng. Biotechnol.* **2022**, *10*, 816250. [CrossRef]
26. Li, D.; Sun, C.; Jiang, J.; Lu, F.; Xia, X.; Wang, H.; Zou, F.; Ma, X. A study of screw placement to obtain the optimal pull-out resistance of lumbar pedicle screws-analysis of Hounsfield units measurements based on computed tomography. *BMC Musculoskelet. Disord.* **2022**, *23*, 124. [CrossRef]
27. Zhang, R.J.; Li, H.M.; Gao, H.; Jia, C.Y.; Xing, T.; Shen, C.L. Associations between the hounsfield unit values of different trajectories and bone mineral density of vertebrae: Cortical bone and traditional trajectories. *Am. J. Transl. Res.* **2020**, *12*, 3906–3916.
28. Kuhns, C.A.; Reiter, M.; Pfeiffer, F.; Choma, T.J. Surgical strategies to improve fixation in the osteoporotic spine: The effects of tapping, cement augmentation, and screw trajectory. *Global Spine J.* **2014**, *4*, 47–54. [CrossRef]
29. Ahern, D.P.; McDonnell, J.M.; Riffault, M.; Evans, S.; Wagner, S.C.; Vaccaro, A.R.; Hoey, D.A.; Butler, J.S. A meta-analysis of the diagnostic accuracy of Hounsfield units on computed topography relative to dual-energy X-ray absorptiometry for the diagnosis of osteoporosis in the spine surgery population. *Spine J.* **2021**, *21*, 1738–1749. [CrossRef]
30. Chevalier, Y.; Matsuura, M.; Krüger, S.; Traxler, H.; Fleege, C.; Rauschmann, M.; Schilling, C. The effect of cement augmentation on pedicle screw fixation under various load cases: Results from a combined experimental, micro-CT, and micro-finite element analysis. *Bone Jt. Res.* **2021**, *10*, 797–806. [CrossRef]
31. Weiser, L.; Huber, G.; Sellenschloh, K.; Viezens, L.; Püschel, K.; Morlock, M.M.; Lehmann, W. Insufficient stability of pedicle screws in osteoporotic vertebrae: Biomechanical correlation of bone mineral density and pedicle screw fixation strength. *Eur Spine J.* **2017**, *26*, 2891–2897. [CrossRef] [PubMed]
32. Zaidi, Q.; Danisa, O.A.; Cheng, W. Measurement Techniques and Utility of Hounsfield Unit Values for Assessment of Bone Quality Prior to Spinal Instrumentation: A Review of Current Literature. *Spine (Phila Pa 1976)* **2019**, *44*, E239–E244. [CrossRef] [PubMed]
33. Leichtle, C.I.; Lorenz, A.; Rothstock, S.; Happel, J.; Walter, F.; Shiozawa, T.; Leichtle, U.G. Pull-out strength of cemented solid versus fenestrated pedicle screws in osteoporotic vertebrae. *Bone Jt. Res.* **2016**, *5*, 419–426. [CrossRef] [PubMed]
34. Elder, B.D.; Lo, S.F.; Holmes, C.; Goodwin, C.R.; Kosztowski, T.A.; Lina, I.A.; Locke, J.E.; Witham, T.F. The biomechanics of pedicle screw augmentation with cement. *Spine J.* **2015**, *15*, 1432–1445. [CrossRef] [PubMed]
35. Son, H.J.; Choi, S.H.; Heo, D.R.; Kook, I.; Lee, M.K.; Ahn, H.S.; Kang, C.N. Outcomes of the use of cement-augmented cannulated pedicle screws in lumbar spinal fusion. *Spine J.* **2021**, *21*, 1857–1865. [CrossRef]
36. Wang, J.S.; Tägil, M.; Isaksson, H.; Boström, M.; Lidgren, L. Tissue reaction and material biodegradation of a calcium sulfate/apatite biphasic bone substitute in rat muscle. *J. Orthop. Translat.* **2016**, *6*, 10–17. [CrossRef]
37. Raina, D.B.; Širka, A.; Qayoom, I.; Teotia, A.K.; Liu, Y.; Tarasevicius, S.; Tanner, K.E.; Isaksson, H.; Kumar, A.; Tägil, M.; et al. Long-Term Response to a Bioactive Biphasic Biomaterial in the Femoral Neck of Osteoporotic Rats. *Tissue Eng. Part A* **2020**, *26*, 1042–1051. [CrossRef]

38. Širka, A.; Raina, D.B.; Isaksson, H.; Tanner, K.E.; Smailys, A.; Kumar, A.; Tarasevičius, Š.; Tägil, M.; Lidgren, L. Calciur Sulphate/Hydroxyapatite Carrier for Bone Formation in the Femoral Neck of Osteoporotic Rats. *Tissue Eng. Part A* **2018**, 2 1753–1764. [CrossRef]
39. Raina, D.B.; Isaksson, H.; Hettwer, W.; Kumar, A.; Lidgren, L.; Tägil, M. A Biphasic Calcium Sulphate/Hydroxyapatite Carrie Containing Bone Morphogenic Protein-2 and Zoledronic Acid Generates Bone. *Sci. Rep.* **2016**, *6*, 26033. [CrossRef]
40. Raina, D.B.; Larsson, D.; Sezgin, E.A.; Isaksson, H.; Tägil, M.; Lidgren, L. Biomodulation of an implant for enhanced bone-implan anchorage. *Acta Biomater.* **2019**, *96*, 619–630. [CrossRef]
41. Kueny, R.A.; Kolb, J.P.; Lehmann, W.; Püschel, K.; Morlock, M.M.; Huber, G. Influence of the screw augmentation techniqu and a diameter increase on pedicle screw fixation in the osteoporotic spine: Pullout versus fatigue testing. *Eur. Spine J.* **2014**, 2 2196–2202. [CrossRef] [PubMed]

Journal of
Functional Biomaterials

MDPI

Review

Biodegradable Cements for Bone Regeneration

Dachuan Liu [1,†], Chen Cui [2,†], Weicheng Chen [1], Jiaxu Shi [1], Bin Li [1,*] and Song Chen [1,*]

1 Department of Orthopaedic Surgery, Orthopaedic Institute, The First Affiliated Hospital, Suzhou Medical College, Soochow University, Suzhou 215006, China
2 Department of Orthopaedic Surgery, The Third Affiliated Hospital of Soochow University, 185 Ju Qian Road, Changzhou 213003, China
* Correspondence: binli@suda.edu.cn (B.L.); chensong@suda.edu.cn (S.C.)
† These authors contributed equally to this work.

Abstract: Bone cements such as polymethyl methacrylate and calcium phosphates have been widely used for the reconstruction of bone. Despite their remarkable clinical success, the low degradation rate of these materials hampers a broader clinical use. Matching the degradation rate of the materials with neo bone formation remains a challenge for bone-repairing materials. Moreover, questions such as the mechanism of degradation and how the composition of the materials contribute to the degradation property remain unanswered. Therefore, the review provides an overview of currently used biodegradable bone cements such as calcium phosphates (CaP), calcium sulfates and organic-inorganic composites. The possible degradation mechanism and clinical performance of the biodegradable cements are summarized. This paper reviews up-to-date research and applications of biodegradable cements, hoping to provide researchers in the field with inspirations and references.

Keywords: biomaterials; biodegradable bone cements; calcium phosphates; calcium sulfates; bone regeneration

Citation: Liu, D.; Cui, C.; Chen, W.; Shi, J.; Li, B.; Chen, S. Biodegradable Cements for Bone Regeneration. *J. Funct. Biomater.* **2023**, *14*, 134. https://doi.org/10.3390/jfb14030134

Academic Editors: Christie Ying Kei Lung and Daniele Botticelli

Received: 23 January 2023
Revised: 21 February 2023
Accepted: 24 February 2023
Published: 27 February 2023

1. Introduction

Bone defects may be caused by severe trauma, tumors and systemic disease [1–3]. Every year, numerous patients in the world undergo bone reconstruction treatment because of congenital defects, tumor resection and bone defects caused by fracture [4]. Currently, autologous bone graft is the gold standard for the treatment of bone defects [5]. However, its main disadvantages include a lack of supply, pain, infection and other complications [6]. Allogeneic bone transplantation is an alternative method, but there are many postoperative complications [7,8]. Therefore, artificial bone-repairing materials are promising for the repair of bone defects. Currently, most implants are made of metallic biomaterials. The most commonly used metals for medical applications are titanium and titanium alloys as well as stainless steel and cobalt–chromium–molybdenum alloys [9]. For example, Ti6Al4V and NiTi alloys are known for their high wear resistance, ductility and high hardness. Some researchers have incorporated new elements in the alloys to address their biological toxicity and excessive modulus of elasticity [10]. However, metal implants still cannot be used for filling irregular orthopedic wounds and are hardly degradable, requiring secondary surgical removal. In addition, the composition of metal implants differs greatly from that of the human body, so there is still a necessity to develop implant materials that are more compatible with human tissue and degradable.

Bone cement, as a kind of bone-filling material, has been applied in the orthopedic field. It is of great significance for filling and repairing the irregular orthopedic trauma site and is widely used in various orthopedic and dental implant fixations [11,12]. However, the most commonly used bone cements such as polymethylmethacrylate (PMMA) are nondegradable. The poor biodegradability may cause many complications, such as intramedullary hypertension and pulmonary embolism [13,14]. Therefore, the development of bone cement with good biodegradability is essential for their biomedical applications.

Biodegradable materials are the second generation of biomaterials, which play an indispensable role in tissue repair and regeneration [15–17]. From 2015 to 2024, the global orthopedic biomedical materials market increased from $4.3 billion to 46.5 billion [18]. Orthopedic biodegradable materials provide support, osteoconductivity and osteoinduction at the implantation site, and some of them also assist in the completion of bone tissue repair. They are gradually degraded in the organism through dissolution, enzymatic digestion and cellular phagocytosis at a certain degradation rate during the formation of new bone. The degradation products of the materials produce either no host response or a slight host response to the organism, and in some cases, promote the bone repair. The degradation products are continuously absorbed or excreted from the tissue as it grows in, and the repaired bone tissue completely replaces the implanted material, leaving no residual material in the body when the bone repair is completed.

According to the development status of biomaterials, biodegradable materials mainly consist of biodegradable polymers, biodegradable metal materials and biodegradable cements/ceramics (Figure 1). After implanting the materials into the defect site, the newly formed tissue grows into the implant's interior, and some biomolecules and ions produced during the degradation can interact with the bone injury microenvironment to promote the regeneration of bone tissue. Simultaneously, the mechanical properties of the implant gradually decrease, and the biological stress of the body transfers to the new bone tissue, which prevents the stress-shielding effect and stimulates bone tissue regeneration [19,20]. However, there are some shortcomings in biodegradable bone cements such as the mechanical strength of bone cements weakened during biodegradation, high price and inconvenient production, which all need to be continuously improved in the future [21].

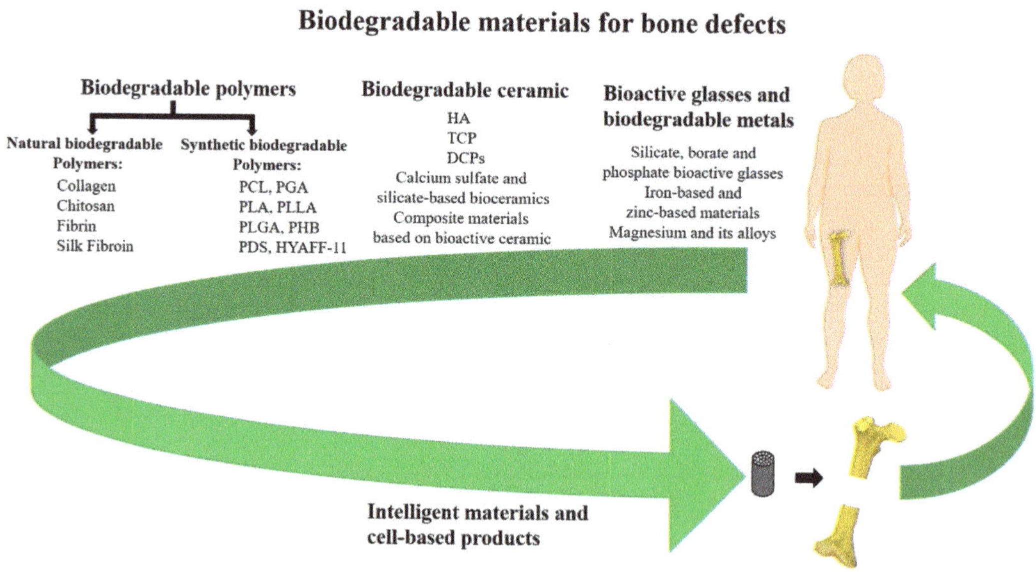

Figure 1. The classification of biodegradable materials used for bone defect repair [19].

A large number of examples have revealed the application prospect of degradable bone cements in the biomedical field for many years [11,19]. The subject itself has matured into an important research topic, and has achieved promising results in vitro and in vivo, which has aroused more and more interest from researchers [22,23]. In this review, we summarize the recent progress in biodegradable bone cements. Different types of biodegradable bone cements were introduced, and the mechanism of bone cements' degradation was analyzed. Finally, the clinical application of biodegradable bone cements was summarized.

Most of the previous studies summarized the mechanical properties and osteogenic ability of biological bone cement, but we focused on another important property of biological bone cement—degradability. In this review, we put forward our views on possible future directions for biodegradable bone cement research, including translating these products from laboratory to clinical. We believe that understanding the degradation process of different bone cements and optimizing the controllability of their biodegradation are the development trends of biological bone cements. This is because the biological bone cement can achieve its maximum effectiveness only when the rate of degradation matches the rate of new bone formation. It is expected that this timely review will provide a comprehensive and strong knowledge base for degradable bone cements [24].

2. Categories of Biodegradable Bone Cements

Bone cements are self-setting bone substitute materials, which are generally composed of powder and liquid phases [25]. In the last couple of decades, bone cements have been extensively applied in the repair and regeneration of tissue engineering due to their good biocompatibility, biodegradability and excellent osteogenesis [19,26]. To date, the most commonly used biodegradable bone cements mainly include calcium phosphate cements (CPCs), calcium sulfate cements and other composite bone cements [11]. The biggest advantage of biodegradable bone cements is that they are gradually degraded by chemical dissolution and cell absorption after implantation in the bone defect and finally replaced by the newly formed bone tissue [27].

2.1. Calcium Phosphate Cements

Calcium phosphate cements consist of an aqueous solution and one or several calcium phosphates [28]. Despite many formulations and compositions proposed, they can be classified into three groups according to their end products: apatite (HA, $Ca_5(PO_4)_3OH$), brushite (DCPD, $CaHPO_4 \cdot 2H_2O$) and monetite (DCPA, $CaHPO_4$) [29,30]. The degradability of these three bone cements will be described below.

2.1.1. Apatite Cements

There are two different reaction paths for the setting of apatite cements [31]. The first is based on an acid–base reaction, where one acidic and one alkaline calcium phosphate source react to form a neutral product. Tetracalcium phosphate (TTCP) is the most frequently applied alkaline calcium phosphate source. The mixture of TTCP with an acidic calcium phosphate source (DCPA or DCPD) produces the precipitation of HA (Equation (1)) (Figure 2) [32].

$$2\,CaHPO_4 + 2\,Ca_4(PO_4)_2O \rightarrow Ca_{10}(PO_4)_6(OH)_2 \quad (1)$$

The second type of reaction is a simple hydration reaction. Calcium-deficient HA (CDHA) is prepared via the hydrolysis (Equation (2)) of a single CaP compound (Figure 2), such as α-$Ca_3(PO_4)_2$ (α-TCP), which is applied in most clinical products.

$$3\,\alpha\text{-}Ca_3(PO_4)_2 + H_2O \rightarrow Ca_9(HPO_4)(PO_4)_5(OH) \quad (2)$$

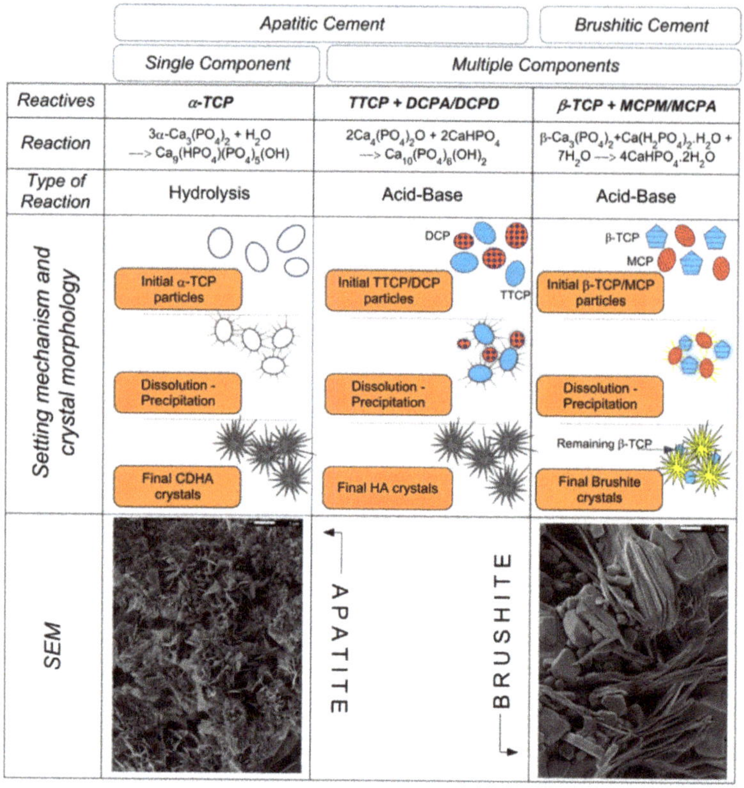

Figure 2. Classification of calcium phosphate cements, with examples of the most common formulations. Reprinted (adapted) with permission from [33]. 2012, Elsevier Ltd.

Apatite cements are generally degraded through chemical dissolution and active degradation mediated by cellular activity, and the active degradation is dominant. Due to the limited degradation of cellular activity, HA is the least soluble and most stable of the different CPCs [34,35]. Many experiments indicated that apatite cements showed only slight degradation and new bone regeneration after being implanted in vivo for several months or even several years. The slow degradation of apatite not only hinders the regeneration of new bone, but also severely limits its clinical application.

2.1.2. Brushite Cements

Brushite cements are prepared using an acid–base reaction. The components of acid–base reactions are usually composed of an acidic phosphorus source and basic calcium source [36]. Many formulations have been reported in the research. The most commonly applied formulation is monocalcium phosphate monohydrate (MCPM) (Figure 2) and β-tricalcium phosphate (β-TCP) (Equation (3)) [37,38].

$$Ca(H_2PO_4)_2 \cdot H_2O + \beta\text{-}Ca_3(PO_4)_2 + 7\,H_2O \rightarrow 4\,CaHPO_4 \cdot 2\,H_2O \tag{3}$$

Under physiological conditions, the dissolution rate of brushite cements is almost three times that of HA cements [31]. However, compared with the regeneration rate of bone, the degradation rate of brushite cements is still slightly slower. Flautre et al. studied a brushite cement applied in non-load-bearing defects [39]. After 6 months, the newly formed bone was not obviously increased and approximately half of the bone defect was still occupied by undegraded cement. Britta et al. prevented the formation of HA precipitate by adjusting the liquid–solid ratio, but the cements with two different powder-to-liquid ratios displayed only slight degradation after 10 months [20]. Maenz et al. used poly(l-lactide-co-glycolide) acid (PLGA) nanofibers to enhance the mechanical property of brushite cements. Although the addition of PLGA fiber promoted bone regeneration, the brushite cement still retained a large volume after 3 months. When brushite cements are used in bone defect repair operations, its good degradation and replacement via regenerative bone is preferred. Ideally, the biodegradation rate of brushite cements should be in line with the newly formed bone to allow the newly formed bone to gradually restore its mechanical properties.

Therefore, an urgent problem to be solved for brushite cement is to improve the degradation rate. At present, there are several methods to improve the degradation rate of brushite cements: (i) changing the powder-to-liquid ratio, (ii) improving porosity and (iii) inhibiting phase conversion [11]. These parameters could be adjusted in such a way that the biodegradation rate is line with the rate of bone formation. The most commonly used method is to introduce different pore sizes into bone cement to improve the local metabolism and the degradation rate of cement. Real et al. used an acid-based reaction between NaH_2PO_4 and $NaHCO_3$ to fabricate CO_2 bubbles into CPCs [40]. At 3 months, while the control CPCs still maintained their integrity, macroporous CPCs were almost completely degraded [41]. Additionally, Félix-Lanao et al. studied the effect of PLGA porogens on CPCs degradation [42]. They found that relatively low PLGA porogens were applied (e.g., 10–20 wt.%), being able still to promote CPC degradation and accelerate bone regeneration [43]. These two studies demonstrate that increasing porosity is effective to promote CPC degradation. Modulating the porosity can adjust the degradation rate of CPCs to match the rate of new bone formation, which will greatly contribute to the expansion of cements' use. Besides improving porosity, preventing brushite cements from recrystallizing to form apatite is also an important way to increase the degradation rate [44]. Apelt et al. and Theiss et al. incorporate Mg^{2+} into the brushite cement to prevent phase conversion [30,45]. In a study, Grover et al. used brushite CPCs modified with pyrophosphate to inhibit phase conversion, causing a broader resorption and promoting new bone regeneration compared with the control group [46]. They proposed an innovative method to optimize the degradation performance of brushite cement. These two studies improved the degradation performance of brushite cement from the source of its slow degradation, enabling brushite cement to enhance its degradation performance while retaining its osteogenic ability and biocompatibility.

2.1.3. Monetite Cements

Monetite is the anhydrous form of brushite. Monetite cements can be obtained by adjusting the reaction conditions of brushite cements. For example, setting brushite cements in excessively low pH conditions, in water-deficient environments, or in the presence of metallic ions, favoring monetite formation [47]. Another way of fabricating monetite cements is via the thermal dehydration of already-set brushite cements (Equation (4)):

$$CaHPO_4\cdot 2\,H_2O \rightarrow CaHPO_4 + 2\,H_2O \tag{4}$$

The degradation mechanism of monetite cements is the same as that of brushite, mainly divided into active absorption and via passive dissolution [48]. In a comparison between monetite cements and autologous grafts on grounds of bone healing efficiency, histomorphometry results indicated that 42% of the monetite cements were resorbed and that the newly formed bone within the implant occupied 43% of its volume [49]. This indicated that monetite cement resorption can offer a good balance between implant degradation and new bone regeneration on the premise of maintaining mechanical stability, an advantage that separated monetite from brushite and HA.

As a biodegradable bone cement with the potential to replace brushite cements, monetite cements have a very broad application prospect. Compared with brushite cements, it was found that monetite cements resorb at a faster rate than brushite in vivo. The main reason responsible for the higher resorption rate lies in that monetite is not transformed into apatite as readily as brushite [50]. Sheikh et al. designed monetite grafts of differing physical form by autoclaving and dry heating (under vacuum). Upon implantation for 3 months, both types of monetite cements showed complete resorption and no monetite converted into apatite [51]. Moreover, Tamimi et al. implanted monetite granules synthesized via autoclaving into alveolar bone defects. After six months, the amount of resorbed monetite (74%) and regenerated bone (60%) was obviously higher in comparison with the hydroxyapatite group and no apatite formation was observed in patients [52]. The monetite is an excellent calcium phosphate cement, and its advantage is that it will not recrystallize to form apatite and has a fast degradation rate. It has great research value as a substitute material for brushite.

2.2. Calcium Sulfate Cements

Calcium sulfate ($CaSO_4$), also referred to as gypsum, is the first synthetic scaffold used for bone regeneration [53–55]. The application of calcium sulfate as a filler in bone defects was reported in 1892 [56]. In 1996, Wright Medical Company optimized the form and size of calcium sulfate cements and developed the bone repair product of Osteoset® to promote the application of calcium sulfate in the orthopedic field. It has excellent biocompatibility and biodegradability; and the degradative products even cause no tissue inflammatory reaction [57].

The promotion of bone repair via calcium sulfate cements was first confirmed in animal experiments. Lillo and Peltier applied calcium sulfate cements to fill bone defects in a canine model and reported that calcium sulfate cements obviously promote bone regeneration covered by periosteum [58]. Radentz and Collings used calcium sulfate as plugging agents to promote bone defect repair in intrabony defects of a canine model [59]. Histological results showed that bone in the calcium sulfate-filled defects had a denser trabecular pattern. In 2001, Kelly et al. used calcium sulfate cements to repair bone defects caused by benign bone tumors, trauma and periprosthetic bone loss [60]. The percentage of bone regeneration and replacement of the calcium sulfate cements was 80% of the original defect size after 2 months. Calcium sulfate cements have also been applied in the management of acute traumatic defects. Yu et al. used injectable calcium sulfate cements to fill impacted metaphyseal defects in 31 patients with tibial plateau fractures [53]. The result showed that calcium sulfate cements improved the safety and stability of early knee motion. The advantage of calcium sulfate cement is that it is biocompatible and safe because it does not cause inflammation when implanted in the body.

However, the rapid degradation of calcium sulfate does not match the bone formation process, which limits its clinical application [61]. It is reported that calcium sulfate cement generally takes 4–6 weeks to completely degrade under the condition of good blood vessels. However, the degradation time of calcium sulfate was prolonged to 6–10 weeks in the poorly vascularized bone defect. In addition, it has been reported that osteoporosis can accelerate the degradation of calcium sulfate [62]. In normal rat caudal vertebra defects, calcium sulfate degraded about 80% in 4 weeks and almost completely dissolved in 8 weeks [63]. In the case of osteoporosis, calcium sulfate in the defects of coccygeal vertebra is completely degraded after 4 weeks. Rechenberg et al. found that calcium sulfate cements were already completely resorbed at 8 weeks in a drill hole model of sheep [64]. In order to match the rate of calcium sulfate degradation with the rate of new bone growth, many researchers combine calcium sulfate with other substances to reduce the porosity of calcium sulfate or steam treatment [65]. In short, reducing the degradation rate of calcium sulfate to be in line with the newly formed bone will be the next development direction of calcium sulfate cements.

2.3. Organic-Inorganic Composites

In order to further improve the properties of bone cement in various aspects, researchers often add organic components to the inorganic phase. Because of the addition of organic components, the degradation performance of composite bone cement becomes more complicated. The degradation rate of the composites mainly depends on the properties of the inorganic components in the composites [66]. However, the addition of organic components can accelerate or delay the degradation of bone cement to some extent. Therefore, adjusting the degradation rate of bone cement by adding organic components will be the future development direction for the personalized treatment of bone defects.

At present, the organic components added in composite bone cement can be divided into natural polymers and synthetic polymers [67]. Natural polymers mainly include gelatin, sodium alginate, silk fibroin and chitosan; the synthetic polymer is mainly composed of Polycaprolactone (PCL), Poly (Lactic Acid) (PLA), PLGA, and so on [68]. These polymers are often added into bone cement in different ways (polymer chain, microspheres) to regulate the degradation rate to meet the corresponding clinical needs. Habraken et al. added gelatin microspheres into CPCs to study their effect on the degradation of bone cement in vitro [69]. The results showed that the degradation curve of CPCs incorporated with gelatin microspheres decreased linearly, and the mass loss was about 5% at 3 months. Guo et al. prepared CPC/PLGA composites to explore in vivo degradation and mechanical property changes. They found that CPC/PLGA composites degraded significantly faster than the CPC group and lost 10% more weight at 12 weeks than CPC group [70].

In general, the addition of polymers enriches the types of biodegradable bone cement and improves its degradability and provides a good method for the clinical customization of personalized biodegradable bone cement.

3. The Mechanism of Degradability

It is known that degradable bone cements can fill and repair bone defects to promote bone regeneration [71]. However, they will be gradually degraded as a foreign matter in implanted bones after implantation into the human body. The degradation modes of bone cements mainly include chemical dissolution and resorbing by the action of osteoclasts. The chemical dissolution belongs to passive degradation (Figure 3), and resorbing to active degradation [72,73].

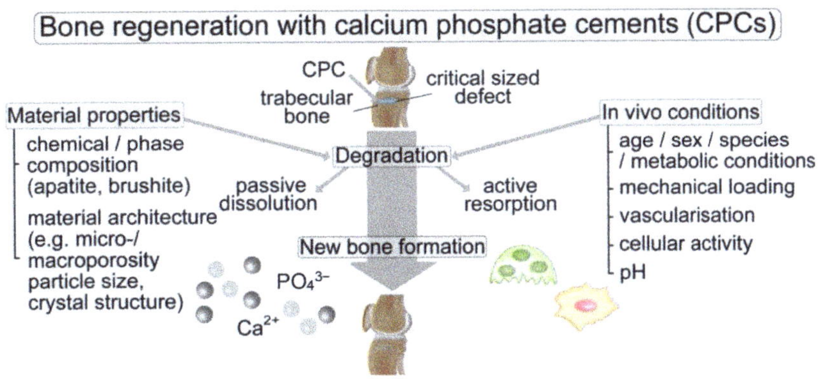

Figure 3. Natural process of bone repair in the fracture zone [27].

3.1. Chemical Dissolution

The whole process of chemical dissolution involves physical and chemical changes without the involvement of cells [74]. The degree and speed of chemical dissolution depended on the composition, structures and the microenvironment of implant locations [37].

Among all biodegradable bone cements, the calcium sulfate cement is one of the most completely and fastest degraded cements [75]. Despite various differences between calcium sulfate hemihydrate (CSH) and calcium sulfate dihydrate (CSD), all calcium sulfates are freely soluble in physiological conditions. As the degradation of calcium sulfates, the calcium phosphate, a new phase, is formed after degradation. This phenomenon is not only the competition of two processes of precipitation and dissolution, but also a good sign for cell adhesion [57]. It is precisely because the local calcium ion concentration increase through the chemical dissolution of calcium sulfate that new bone formation is induced.

Compared with the apatite cement cements, brushite cements have been shown to degrade to a much greater extent in vivo [76]. This is mainly due to their excellent solubility at a physiological pH, which leads to degradation by chemical dissolution and active resorption by cells. The rate of chemical dissolution of CPCs is determined by the surface area, Ca/P ratio, the crystallinity and pH or perfusion with bodily liquids [11 Dissolution occurs when brushite is placed in an environment that is under-saturated in calcium and phosphate ions, and where calcium and phosphate ions would dissolve directly. The process of dissolution stops once the solubility limit has been reached. Brushite dissolution supersaturates the environment with respect to HA, ultimately resulting in HA precipitation [37]. Because of the slow degradation rate of brushite cements, brushite cements would recrystallize to form HA during the long degradation process, which hinders its further dissolution. The resorption mechanism of monetite is the same as brushite cement implants, which is mainly mediated via chemical dissolution and cell absorption [24]. The only difference is that monetite cements dos not recrystallize to form HA, which makes monetite cements degrade faster than brushite cements.

Hydroxyapatite is insoluble in water, but can be slightly dissolved after being soaked in water for a long time [77]. Because of poorly solubility, apatite cements are primarily resorbed via active resorption through osteoclasts [61]. As for some organic–inorganic composites, they may be chemically degraded through the joint action of the active and passive methods in vivo.

3.2. Resorbed by the Action of Osteoclasts

Another important degradation type is active degradation mediated by cells such as osteoclasts, giant cells and macrophages [78]. The cements as implantation materials cause the inflammatory response of body's immune system, which can induce the immune cells migration to the cements [79]. It has been reported that the degraded granules or fragments of part-degradable bone cements may stimulate the immune response, promote the secretion of inflammatory factors, such as IL-1 β, IL-6 and TNF-α, activate osteoclasts further increase the release of inflammatory factors and activate macrophages and giant cells in bone marrow. Then the activated macrophages and giant cells will endocytose and exocytose the dissolved substances into the microenvironment between granules and cells to maintain the homeostasis of the bone environment [72]. Therefore, the cells of the mononuclear phagocytic system play a critical role in the degradation of bone implantations.

Calcium sulfate cements do not cause inflammatory reactions in vivo, so the degradable type mediated is not the main degradation type of calcium sulfate cements [57,80]. In addition, histological staining revealed that no cells are involved in implant degradation in calcium sulfate cement implants.

It is reported that the degradation of CPCs is mainly involved in giant cells and osteoclasts (Figure 4). But the results in vivo showed that macrophages, not osteoclasts, play an important role in the active resorption of brushite cements [45]. It has been suggested that cement disintegration and phagocytosis play a critical role, showing that brushite cement degradation may be a complicated process. If the larger cement particles were produced in the degradation process and cannot be swallowed by macrophages, the foreign matter giant cells would be responsible for degradation [81]. Sheraly et al. proved that CPC implants can be resorbed by osteoclasts, and they successfully used gastrointestinal proton pump inhibitors (PPIs) to delay osteoclast-mediated resorption of CPCs [82]. On the basis of confirming that osteoclasts participate in CPC degradation, Christian et al. quantitatively analyzed the rate of bone cement degradation involved by osteoclasts. They analyzed the active absorption efficiency of bone cements by measuring tartrate-resistant acid phosphatase (TRAP) activity and calcitonin receptor (CT-R) expression [76,83,84]. In addition, the researchers indirectly affected the degradation of calcium phosphate cement by adding different ions to the cement. Bernhardt et al. found an increased osteoclastic resorption of Cr^{3+}-doped brushite cements compared to the non-incorporated brushite cement [85,86]. They also found that Co^{2+} ions promoted the formation of osteoclasts and accelerated the degradation of bone cement in vivo. Another study showed that Sr^{2+} substitution in apatite cement can effectively attenuate osteoclastic absorption and delayed the absorption of the bone cement [87]. Some studies tried to use Mg^{2+} ions and Cu^{2+} to regulate the formation or viability of osteoclasts to indirectly regulate the degradation of bone cement [84].

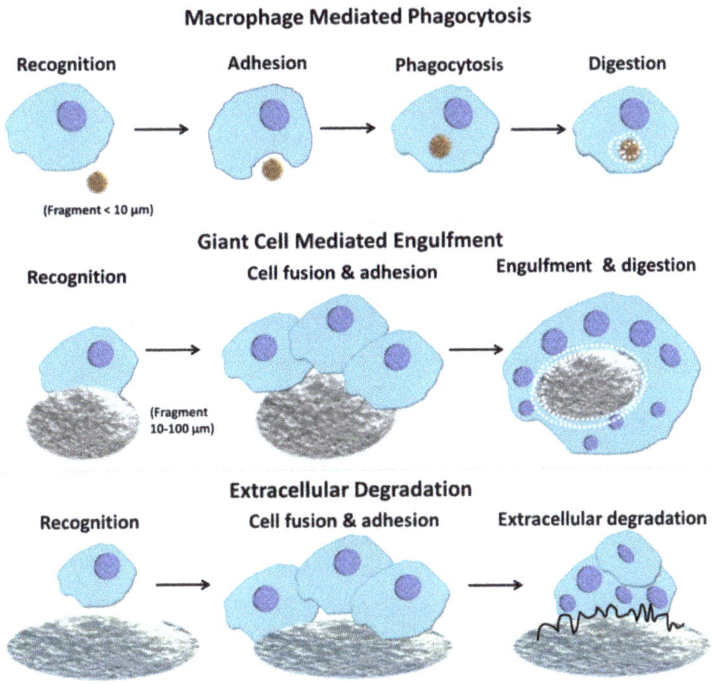

Figure 4. Macrophage response to biomaterials depending on the size of the implanted materials [44].

The active degradation of bone cements by osteoclasts plays a key role in bone cements degradation. The resorption activity of osteoclasts may be regulated by various active substances, hormones and other factors, such as interleukin, the parathyroidal hormone, calcitonin, nerve growth factor, etc. [88].

3.3. *Recrystallized to Form Apatite*

In addition to the above two degradation modes, a small part of the calcium phosphate cement also has another degradation mode, recrystallizing, to form apatite (Figure 5). The resorption of brushite cements has been displayed to be highly unpredictable with strong dependence on many conditions [89]. One of the major conditions is phase transformation, transforming to less soluble phases such as HA, affecting the rate of resorption [50]. After the initial fast degradation of the cements, the remaining brushite is converted into insoluble apatite. It has been found that after 6 months of in vivo implantation in sheep, brushite cements completely transform into poor crystalline-carbonated apatite [90]. Many studies on the degradation of brushite cements had found that certain organic matters (i.e., citrate ions) could cause an obvious reduction in this energy barrier, and is probably involved in the process of brushite to HA conversion in vivo. In other words, physiological inhibitors of HA mineralization, such as pyrophosphate (in pyrophosphoric acid-based brushite cements), acidic amino acids and magnesium ions, can be used to prevent brushite reprecipitation as HA. For instance, Apelt et al. and Theiss et al. prevented phase transformation of brushite into insoluble phases by adding magnesium ions into the bone cement [45]. Another method to avoid the transformation into HA in vivo is to reduce the free calcium ions which are essential for phase transformation. Schroter et al. increased the powder-to-liquid rates of cements to reduce the porosity and avoided the transformation of brushite into insoluble HA [27]. In brief, inhibiting brushite cements from transforming into insoluble the HA phase can accelerate the degradation rate of brushite cements to some extent.

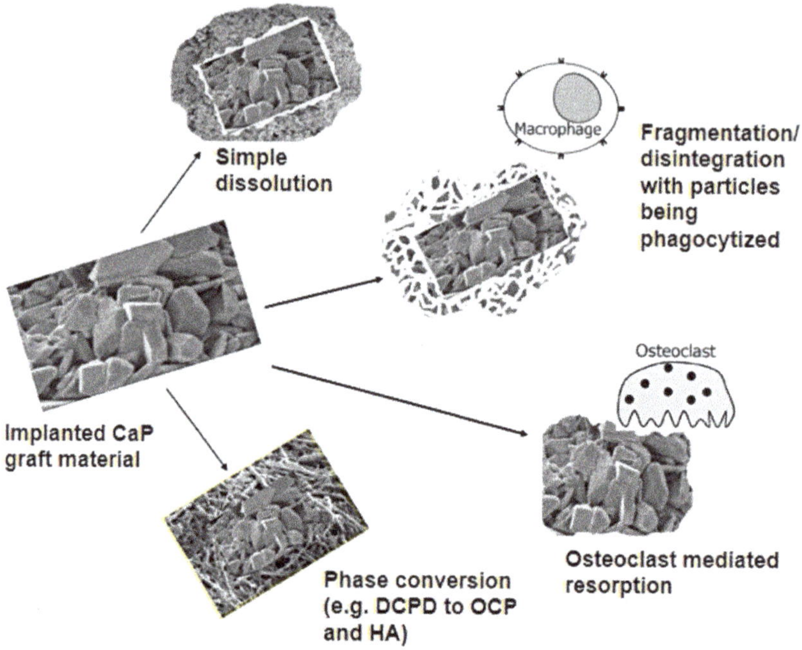

Figure 5. The fate of CaP biomaterials after implantation. (OCP: Octacalcium phosphate) [44].

4. Evaluation of Degradability

Ideally, CPCs are replaced with new bone formation in good time without a temporary reduction in mechanical properties [27]. However, the degradation of CPCs varies with their components and products. Therefore, different methods are required to evaluate the degradation of cement from various aspects.

In vitro, immersion and weight loss methods are commonly used to predict the degradation behavior of cements. In the measurements, the geometry and mass of the cements are detected before and after immersion [22]. Chang et al. investigated the degradation of magnesium phosphate/calcium silicate composite (MPC/CS) bone cements by submerging the samples in a Tris–HCl buffer solution. After a period of time, the weight loss of each sample was measured to assess the degradation of the bone cements in vitro [91]. The addition of CS to MPC resulted in an increased rate of degradation. The critical aspect of this method is to choose an immersion solution that better simulates the environment in vivo. For example, simulated body fluids (SBF) are widely used because this composition is similar to that of body fluids. However, minerals will deposit on the surface of the sample in SBF and thus affect the experiment results [23]. Tris–HCl, a pH 7 buffer solution, can avoid the mineralization due to its pH-balancing ability [22]. Moreover, the study of Rohanová et al. showed that unbuffered the Dulbecco's Modified Eagle's Medium (DMEM) could be an appropriate solution for degradation experiment [92].

Evaluation of degradability in vivo is commonly performed on tissue sections. The residual materials were observed on the stained sections, and the degradation of the materials are evaluated by comparing the residuals with the initial implanted materials.

In addition, materials and surrounding tissues are assessed qualitatively and quantitatively via image analysis systems such as Micro-CT (micro-computed tomography) and SEM (scanning electron microscopy). Choll et al. established a model of osteoporosis using rodents suffering from osteoporosis [62]. Each group was packed with $CaSO_4$ or $CaSO_4/CaPO_4$ cement to fill the tail defects. After 8 weeks, cement resorption profiles, bone mineral density, average cortical thickness, average trabecular thickness, average trabecular spacing and cements resorption profiles were evaluated via Micro-CT. Histologic sections were performed on spines obtained after surgery and 8 weeks. The study concluded that the pro-inflammatory and pro-osteolytic bone environments of osteoporotic disease may lead to the altered osteoconductivity and accelerated resorption of both materials.

The Norian SRS®, an apatite-forming commercial cement, was studied in sheep with non-loading defects in the humerus and femur [30,64]. Six months after implantation, the tissue sections showed only material degradation and marginal new bone formation. Penel et al. reported the degradation of brushite-forming cement in the femoral bone defects of sheep [90]. In the analysis of the phase of cement specimens, there was no sign of brushite detectable at 6 months or 12 months. However, a mixture of β-TCP and carbonated apatite was found, suggesting that the DCPD phase converted into slow-degrading apatite in vivo. However, the DCPD phase was transformed to slowly degraded apatite in vivo because a mixture of β-TCP and carbonated apatite was found. The phase analysis verified the unpredictability of the degradation of brushite cements.

In clinical settings, imaging observations and histopathological observations also play important roles in the study of degradability to learn the patients' prognosis. Hillmeier et al. treated 33 patients with vertebral compression fractures via augmentation with calcium phosphate bone cement in vertebroplasty [93]. Radiomorphometric evaluation showed no significant difference in degeneration at 6 months postoperatively in comparison to patients treated with PMMA. Libicher et al. investigated the degradation of calcium phosphate bone cement in vertebral compression fractures in patients with osteoporosis after vertebral osteoplasty [94]. After 1 year, the cement implant was evaluated via CT for volume and the average degradation rate was 2 vol.%. This method exhibited high accuracy for quantifying the amount of bone cements in the vertebral body. However, given the low volume of patients in the research, the reliability of this approach necessitates more study. Heo et al. studied 14 patients with osteoporotic compressive fractures managed with vertebroplasty with brushite cements [95]. The injected CPCs exhibited unpredictable and variable morphological changes observed via radiological evaluation. Half of the implantation resulted in resorption, coagulation and fracturing. In addition, 78.6% of participants showed post-treatment advancements in vertebral compression over time. The compression has continued to progress for over 2 years. The results of this clinical study suggested that this bone cement is not mechanically strong enough to be used in vertebroplasty.

Furthermore, the degradation of CPCs cements in vivo can be evaluated by the change in mechanical properties before and after cement implantation. The degradation of cements accompanied with changes in mechanical properties [96]. The most widely studied mechanical performance of CPCs is the compressive strength [97]. Compared to cancellous bone, apatite cements exhibit similar or greater compressive strength, which is typically maintained in living organisms after implantation [27]. The new bone combined with the slow degradation of CPCs results in the formation of bone cement/bone complexes with sufficient mechanical strength and which may exceed that of intact bone trabeculae [98]. Nevertheless, the initial fast degradation of these cements may result in an unexpected reduction in mechanical properties, which may lead to failure in clinical treatment [95].

5. Clinical Performance of Biodegradable Cement

The CPCs were invented a century ago and have been widely researched as bone substitutes over the past 40 years [99]. A number of commercial products of calcium phosphates are available nowadays. In clinical applications, CPC is primarily used to treat patients with non-weight-bearing bone fractures to promote the recovery of joint functio around the fracture. Transtemporal approaches to the petrous apex and CP (cerebellopon tine) angle are standard procedures in the armamentarium of the neurotologist. Recentl calcium phosphate cement (hydroxyapatite) has been used to close cranial defects in severa medical centers, as part of an FDA-IDE (Food and Drug Administration-Investigationa Device Exemption) study in human subjects [100]. The results showed that apatite cemen had the potential to become a standard tool for the surgical management of the skull bas and temporal bone defects. Marcio et al. evaluated the efficacy and safety of HA as a highl inlaid graft on intact skulls. In 166 BFOA (bi-fronto-orbital advancement) and 19 secondar cranioplasty procedures with high inlay HA, only 1 patient had complications [101]. β-TC was also widely used clinically as a bone repair material for maxillary sinus elevatio Suba et al. investigated the therapeutic effect of maxillary sinus elevation using β-TCP a a bone graft [102]. The results showed no statistical difference in the new bone densit microstructure of the osteo–graft material interface in the graft area between the β-TCP an autologous bone groups after 6 months. Furthermore, β-TCP degradation was significantl lower than that of the grafted autologous bone.

Apatite CPCs are used for the treatment of vertebroplasty or vertebroplasty due t their high compressive strength in CPC bone cements [27]. Nakano et al. investigated th treatment of apatite-forming CPC Biopex® and Biopex-R® in the clinical management o osteoporotic vertebral compression fractures after vertebroplasty [103]. After follow-u over 2 years, the clinical outcome was satisfactory, with maintained bone union and pai relief in patients. The CT images displayed that new bone was forming at the bone–cemen interface and Biopex appears to be gradually degrade. Nevertheless, the degradation o cements and newly formed new bone were unquantified. In a study by Gumpert et al. [100 with the application of apatite cement KyphOS® in osteoporosis, a considerable slow degradation was observed on CT scans [104]. The researchers indicated that the cemen was biocompatible and osseointegrated well, as demonstrated via biopsies performed on and a half years later. In addition, the cement appeared to be gradually absorbed by th multinucleated cells, while osteoblasts formed new bone. However, the authors noted tha the degradation of KyphOS was slow and may take several years. To our knowledge, ther are few clinical studies on brushite cement for osteoporotic compression fractures.

Although CPCs were promising materials for bone repair, their low mechanica strength and high brittleness still limit their application in load-bearing situations [105] Degradation that is too slow or too fast can prevent recovery, and adjusting the degradatio behavior to different clinical requirements remains a major challenge [106,107]. Whil solving the above disadvantages, new commercial products need to be improved for mor availability in the following directions: (i) improving the injectability and porosity of CPC to optimize their osteoconductivity, (ii) using CPCs as drug carriers for the treatment o orthopedic diseases or (iii) using open macroporous CPC blocks for tissue engineering [31]

6. Conclusions and Outlook

Since the last century, more and more bone cements have been developed for bone regeneration. In addition to basic bone conduction and bone induction properties, these scaffolds should also be biocompatible and have a biodegradation rate similar to the rate of new bone formation to meet the time required for new bone to recover its mechanical properties. For example, researchers isolated an anhydrous form of brushite called monetite In an experiment on the efficiency of bone healing with monetite, it was measured that 42% of the monetite cements were resorbed and that the new bone formed within the implant occupied 43% of its volume. This suggested a good balance between phosphate cement reabsorption and new bone regeneration, which helped maintain the mechanical stability of the bone defect. Biodegradable cements can be roughly divided into ceramic (calcium phosphate and calcium sulfate), polymer and composite materials. The degradation of implants in vivo depends on complex physicochemical reactions and cellular mechanisms After the slow dissolution and particle formation of the bone cements, phagocytosis and absorption by the recruited macrophages and osteoclasts lead to the biodegradation and reabsorption of the bone cement scaffold in vivo. Although biodegradable bone cement has many clinical applications, future research is required to further understand the degradation process of different cements and optimize the controllability of biodegradation.

At present, there are still many urgent issues to be solved for biodegradable biomaterials. Basic research has been performed on the mechanical properties of biodegradable materials used for load bearing bone defects, but the control of their biodegradable properties and the mechanism of degradation are still not well understood. The ideal bone cement should provide an adhesion site for osteoprogenitor cells to deposit the bone matrix and gradually mineralize into bone. For this purpose, the cement should be slowly absorbed, allowing newly formed bone to penetrate and grow within the material. We think that the future development direction of biodegradable materials for load bearing bone defects should be the optimization of their controllability, which can be achieved by changing the porosity of material, ion doping or the preparation of composite materials. In the future, a variety of biomaterials that degrade at the same rate as new bone formation should be developed to meet the needs of patients with load-bearing bone defects.

Author Contributions: Conceptualization, S.C. and B.L.; writing–original draft preparation, D.L. and C.C.; writing–review and editing, S.C., W.C. and D.L.; visualization, J.S. and C.C.; supervision, B.L. and S.C.; All authors have read and agreed to the published version of the manuscript.

Funding: The authors are grateful for the funding support from the National Natural Science Foundation of China (82002275, 32271421, 81925027), and the Priority Academic Program Development of Jiangsu Higher Education Institutions.

Data Availability Statement: Not applicable.

Acknowledgments: Authors thank the Department of Orthopaedic Surgery, Orthopaedic Institute, The First Affiliated Hospital, Suzhou Medical College, Soochow University.

Conflicts of Interest: The authors declare no conflict of interest.

References

1. Xue, N.; Ding, X.; Huang, R.; Jiang, R.; Huang, H.; Pan, X.; Min, W.; Chen, J.; Duan, J.A.; Liu, P.; et al. Bone Tissue Engineering in the Treatment of Bone Defects. *Pharmaceuticals* **2022**, *15*, 879. [CrossRef] [PubMed]
2. Wubneh, A.; Tsekoura, E.K.; Ayranci, C.; Uludağ, H. Current state of fabrication technologies and materials for bone tissue engineering. *Acta Biomater.* **2018**, *80*, 1–30. [CrossRef] [PubMed]
3. Zhang, J.; Huang, Y.; Wang, Y.; Xu, J.; Huang, T.; Luo, X. Construction of biomimetic cell-sheet-engineered periosteum with a double cell sheet to repair calvarial defects of rats. *J. Orthop. Translat.* **2023**, *38*, 1–11. [CrossRef]
4. Cui, L.; Xiang, S.; Chen, D.; Fu, R.; Zhang, X.; Chen, J.; Wang, X. A novel tissue-engineered bone graft composed of silicon-substituted calcium phosphate, autogenous fine particulate bone powder and BMSCs promotes posterolateral spinal fusion in rabbits. *J. Orthop. Translat.* **2021**, *26*, 151–161. [CrossRef] [PubMed]
5. Li, J.; Cao, F.; Wu, B.; Yang, J.; Xu, W.; Wang, W.; Wei, X.; Liu, G.; Zhao, D. Immobilization of bioactive vascular endothelial growth factor onto Ca-deficient hydroxyapatite-coated Mg by covalent bonding using polydopamine. *J. Orthop. Translat.* **2021**, *30*, 82–92. [CrossRef]
6. Steijvers, E.; Ghei, A.; Xia, Z. Manufacturing artificial bone allografts: A perspective. *Biomater. Transl.* **2022**, *3*, 65–80. [CrossRef]
7. Koons, G.L.; Diba, M.; Mikos, A.G. Materials design for bone-tissue engineering. *Nat. Rev. Mater.* **2020**, *5*, 584–603. [CrossRef]
8. Dimitriou, R.; Jones, E.; McGonagle, D.; Giannoudis, P.V. Bone regeneration: Current concepts and future directions. *BMC Med.* **2011**, *9*, 1–10. [CrossRef]
9. Jimenez-Marcos, C.; Mirza-Rosca, J.C.; Baltatu, M.S.; Vizureanu, P. Experimental Research on New Developed Titanium Alloys for Biomedical Applications. *Bioengineering* **2022**, *9*, 686. [CrossRef]
10. Baltatu, M.S.; Spataru, M.C.; Verestiuc, L.; Balan, V.; Solcan, C.; Sandu, A.V.; Geanta, V.; Voiculescu, I.; Vizureanu, P. Design, Synthesis, and Preliminary Evaluation for Ti-Mo-Zr-Ta-Si Alloys for Potential Implant Applications. *Materials* **2021**, *14*, 6806. [CrossRef]
11. Lodoso-Torrecilla, I.; van den Beucken, J.; Jansen, J.A. Calcium phosphate cements: Optimization toward biodegradability. *Acta Biomater.* **2021**, *119*, 1–12. [CrossRef]
12. No, Y.J.; Xin, X.; Ramaswamy, Y.; Li, Y.; Roohaniesfahani, S.; Mustaffa, S.; Shi, J.; Jiang, X.; Zreiqat, H. Novel injectable strontium-hardystonite phosphate cement for cancellous bone filling applications. *Mater. Sci. Eng. C Mater. Biol. Appl.* **2019**, *97*, 103–115. [CrossRef]

13. Jaffe, J.D.; Edwards, C.J.; Hamzi, R.; Khanna, A.K.; Olsen, F. Bone Cement Implantation Syndrome: Incidence and Associated Factors in a United States Setting. *Cureus* **2022**, *14*, e31908. [CrossRef]
14. Kim, Y.J.; Lee, J.W.; Park, K.W.; Yeom, J.S.; Jeong, H.S.; Park, J.M.; Kang, H.S. Pulmonary cement embolism after percutaneous vertebroplasty in osteoporotic vertebral compression fractures: Incidence, characteristics, and risk factors. *Radiology* **2009**, *251*, 250–259. [CrossRef]
15. Li, C.; Guo, C.; Fitzpatrick, V.; Ibrahim, A.; Zwierstra, M.J.; Hanna, P.; Lechtig, A.; Nazarian, A.; Lin, S.J.; Kaplan, D.L. Design of biodegradable, implantable devices towards clinical translation. *Nat. Rev. Mater.* **2019**, *5*, 61–81. [CrossRef]
16. Pişkin, E. Biodegradable polymers as biomaterials. *J. Biomater. Sci. Polym. Ed.* **1995**, *6*, 775–795. [CrossRef]
17. Peng, Y.; Li, J.; Lin, H.; Tian, S.; Liu, S.; Pu, F.; Zhao, L.; Ma, K.; Qing, X.; Shao, Z.; et al. Endogenous repair theory enriches construction strategies for orthopaedic biomaterials: A narrative review. *Biomater. Transl.* **2021**, *2*, 343–360. [CrossRef]
18. Hasan, A.; Byambaa, B.; Morshed, M.; Cheikh, M.I.; Shakoor, R.A.; Mustafy, T.; Marei, H.E. Advances in osteobiologic materials for bone substitutes. *J. Tissue Eng. Regen. Med.* **2018**, *12*, 1448–1468. [CrossRef]
19. Wei, S.; Ma, J.X.; Xu, L.; Gu, X.S.; Ma, X.L. Biodegradable materials for bone defect repair. *Mil. Med. Res.* **2020**, *7*, 54. [CrossRef]
20. Kanter, B.; Geffers, M.; Ignatius, A.; Gbureck, U. Control of in vivo mineral bone cement degradation. *Acta Biomater.* **2014**, *10*, 3279–3287. [CrossRef]
21. Pisecky, L.; Luger, M.; Klasan, A.; Gotterbarm, T.; Klotz, M.C.; Hochgatterer, R. Bioabsorbable implants in forefoot surgery: A review of materials, possibilities and disadvantages. *EFORT Open Rev.* **2021**, *6*, 1132–1139. [CrossRef] [PubMed]
22. Jana, A.; Das, M.; Balla, V.K. In vitro and in vivo degradation assessment and preventive measures of biodegradable Mg alloys for biomedical applications. *J. Biomed. Mater. Res. A* **2022**, *110*, 462–487. [CrossRef] [PubMed]
23. López, H.Y.; Cortés-Hernández, D.A.; Escobedo, S.; Mantovani, D. In Vitro Bioactivity Assessment of Metallic Magnesium. *Key Eng. Mater.* **2006**, *310*, 453–456. [CrossRef]
24. Zhou, H.; Yang, L.; Gbureck, U.; Bhaduri, S.B.; Sikder, P. Monetite, an important calcium phosphate compound-Its synthesis, properties and applications in orthopedics. *Acta Biomater.* **2021**, *127*, 41–55. [CrossRef] [PubMed]
25. Hurle, K.; Oliveira, J.M.; Reis, R.L.; Pina, S.; Goetz-Neunhoeffer, F. Ion-doped Brushite Cements for Bone Regeneration. *Acta Biomater.* **2021**, *123*, 51–71. [CrossRef]
26. Sheikh, Z.; Najeeb, S.; Khurshid, Z.; Verma, V.; Rashid, H.; Glogauer, M. Biodegradable Materials for Bone Repair and Tissue Engineering Applications. *Materials* **2015**, *8*, 5744–5794. [CrossRef]
27. Schroter, L.; Kaiser, F.; Stein, S.; Gbureck, U.; Ignatius, A. Biological and mechanical performance and degradation characteristics of calcium phosphate cements in large animals and humans. *Acta Biomater.* **2020**, *117*, 1–20. [CrossRef]
28. Rajzer, I.; Castano, O.; Engel, E.; Planell, J.A. Injectable and fast resorbable calcium phosphate cement for body-setting bone grafts. *J. Mater. Sci. Mater. Med.* **2010**, *21*, 2049–2056. [CrossRef]
29. Xu, H.H.; Wang, P.; Wang, L.; Bao, C.; Chen, Q.; Weir, M.D.; Chow, L.C.; Zhao, L.; Zhou, X.; Reynolds, M.A. Calcium phosphate cements for bone engineering and their biological properties. *Bone Res.* **2017**, *5*, 17056. [CrossRef]
30. Apelt, D.; Theiss, F.; El-Warrak, A.O.; Zlinszky, K.; Bettschart-Wolfisberger, R.; Bohner, M.; Matter, S.; Auer, J.A.; von Rechenberg, B. In vivo behavior of three different injectable hydraulic calcium phosphate cements. *Biomaterials* **2004**, *25*, 1439–1451. [CrossRef]
31. Bohner, M. Calcium orthophosphates in medicine: From ceramics to calcium phosphate cements. *Injury* **2000**, *31*, 37–47. [CrossRef]
32. Miyamoto, Y.; Ishikawa, K.; Takechi, M.; Toh, T.; Yoshida, Y.; Nagayama, M.; Kon, M.; Asaoka, K. Tissue response to fast-setting calcium phosphate cement in bone. *J. Biomed. Mater. Res.* **1997**, *37*, 457–464. [CrossRef]
33. Ginebra, M.P.; Canal, C.; Espanol, M.; Pastorino, D.; Montufar, E.B. Calcium phosphate cements as drug delivery materials. *Adv. Drug Deliv. Rev.* **2012**, *64*, 1090–1110. [CrossRef]
34. Fernández, E.; Gil, F.J.; Ginebra, M.-P.; Driessens, F.C.M.; Planell, J.A.; Best, S.M. Calcium phosphate bone cements for clinical applications. *J. Mater. Sci. Mater. Med.* **1999**, *10*, 169–176. [CrossRef]
35. An, J.; Liao, H.; Kucko, N.W.; Herber, R.P.; Wolke, J.G.; van den Beucken, J.J.; Jansen, J.A.; Leeuwenburgh, S.C. Long-term evaluation of the degradation behavior of three apatite-forming calcium phosphate cements. *J. Biomed. Mater. Res. A* **2016**, *104*, 1072–1081. [CrossRef]
36. Tamimi, F.; Sheikh, Z.; Barralet, J. Dicalcium phosphate cements: Brushite and monetite. *Acta Biomater.* **2012**, *8*, 474–487. [CrossRef]
37. Gallo, M.; Tadier, S.; Meille, S.; Gremillard, L.; Chevalier, J. The in vitro evolution of resorbable brushite cements: A physico-chemical, micro-structural and mechanical study. *Acta Biomater.* **2017**, *53*, 515–525. [CrossRef]
38. Dorozhkin, S.V.; Epple, M. Biological and Medical Significance of Calcium Phosphates. *Angew. Chem. Int. Ed. Engl.* **2002**, *41*, 3130–3146. [CrossRef]
39. Flautre, B.; Delecourt, C.; Blary, M.-C.; Van Landuyt, P.; Lemaître, J.; Hardouin, P. Volume effect on biological properties of a calcium phosphate hydraulic cement: Experimental study in sheep. *Bone* **1999**, *25*, 35–39. [CrossRef]
40. Del Real, R.P.; Wolke, J.; Vallet-Regí, M.; Jansen, J.A. A new method to produce macropores in calcium phosphate cements. *Biomaterials* **2002**, *23*, 3673–3680. [CrossRef]
41. Del Real, R.P.; Ooms, E.; Wolke, J.G.C.; Jansen, J.A. In vivo bone response to porous calcium phosphate cement. *J. Biomed. Mater. Res. A* **2002**, *65*, 30–36. [CrossRef]
42. Felix Lanao, R.P.; Leeuwenburgh, S.C.; Wolke, J.G.; Jansen, J.A. In vitro degradation rate of apatitic calcium phosphate cement with incorporated PLGA microspheres. *Acta Biomater.* **2011**, *7*, 3459–3468. [CrossRef] [PubMed]

43. Felix Lanao, R.P.; Leeuwenburgh, S.C.; Wolke, J.G.; Jansen, J.A. Bone response to fast-degrading, injectable calcium phosphate cements containing PLGA microparticles. *Biomaterials* **2011**, *32*, 8839–8847. [CrossRef] [PubMed]
44. Sheikh, Z.; Brooks, P.J.; Barzilay, O.; Fine, N.; Glogauer, M. Macrophages, Foreign Body Giant Cells and Their Response to Implantable Biomaterials. *Materials* **2015**, *8*, 5671–5701. [CrossRef] [PubMed]
45. Theiss, F.; Apelt, D.; Brand, B.; Kutter, A.; Zlinszky, K.; Bohner, M.; Matter, S.; Frei, C.; Auer, J.A.; von Rechenberg, B. Biocompatibility and resorption of a brushite calcium phosphate cement. *Biomaterials* **2005**, *26*, 4383–4394. [CrossRef]
46. Grover, L.M.; Wright, A.J.; Gbureck, U.; Bolarinwa, A.; Song, J.; Liu, Y.; Farrar, D.F.; Howling, G.; Rose, J.; Barralet, J.E. The effect of amorphous pyrophosphate on calcium phosphate cement resorption and bone generation. *Biomaterials* **2013**, *34*, 6631–6637. [CrossRef]
47. Tamimi, F.; Le Nihouannen, D.; Eimar, H.; Sheikh, Z.; Komarova, S.; Barralet, J. The effect of autoclaving on the physical and biological properties of dicalcium phosphate dihydrate bioceramics: Brushite vs. monetite. *Acta Biomater.* **2012**, *8*, 3161–3169. [CrossRef]
48. Tamimi, F.; Torres, J.; Bassett, D.; Barralet, J.; Cabarcos, E.L. Resorption of monetite granules in alveolar bone defects in human patients. *Biomaterials* **2010**, *31*, 2762–2769. [CrossRef]
49. Torres, J.; Tamimi, I.; Cabrejos-Azama, J.; Tresguerres, I.; Alkhraisat, M.; Lopez-Cabarcos, E.; Hernandez, G.; Tamimi, F. Monetite granules versus particulate autologous bone in bone regeneration. *Ann. Anat.* **2015**, *200*, 126–133. [CrossRef]
50. Sheikh, Z.; Zhang, Y.L.; Grover, L.; Merle, G.E.; Tamimi, F.; Barralet, J. In vitro degradation and in vivo resorption of dicalcium phosphate cement based grafts. *Acta Biomater.* **2015**, *26*, 338–346. [CrossRef]
51. Sheikh, Z.; Zhang, Y.L.; Tamimi, F.; Barralet, J. Effect of processing conditions of dicalcium phosphate cements on graft resorption and bone formation. *Acta Biomater.* **2017**, *53*, 526–535. [CrossRef]
52. Lewin, S.; Kihlstrom Burenstam Linder, L.; Birgersson, U.; Gallinetti, S.; Aberg, J.; Engqvist, H.; Persson, C.; Ohman-Magi, C. Monetite-based composite cranial implants demonstrate long-term clinical volumetric balance by concomitant bone formation and degradation. *Acta Biomater.* **2021**, *128*, 502–513. [CrossRef]
53. Thomas, M.V.; Puleo, D.A. Calcium sulfate: Properties and clinical applications. *J. Biomed. Mater. Res. B Appl. Biomater.* **2009**, *88*, 597–610. [CrossRef]
54. Bohner, M. Bioresorbable ceramics. In *Degradation Rate of Bioresorbable Materials*; Woodhead Publishing: Sawston, UK, 2008; pp. 95–114.
55. Huan, Z.; Chang, J. Self-setting properties and in vitro bioactivity of calcium sulfate hemihydrate-tricalcium silicate composite bone cements. *Acta Biomater.* **2007**, *3*, 952–960. [CrossRef]
56. Kutkut, A.; Andreana, S. Medical-grade calcium sulfate hemihydrate in clinical implant dentistry: A review. *J. Long Term Eff. Med. Implants* **2010**, *20*, 295–301. [CrossRef]
57. Thomas, M.V.; Puleo, D.A.; Al-Sabbagh, M. Calcium sulfate: A review. *J. Long Term Eff. Med. Implants* **2005**, *15*, 599–607. [CrossRef]
58. Beuerlein, M.J.S.; McKee, M.D. Calcium sulfates: What is the evidence? *J. Orthop. Trauma* **2010**, *24*, 46–51. [CrossRef]
59. Radentz, W.H.; Collings, C.K. The implantation of plaster of paris in the alveolar process of the dog. *J. Periodontol.* **1965**, *36*, 357–364. [CrossRef]
60. Kelly, C.M.; Wilkins, R.M. Treatment of benign bone lesions with an injectable calcium sulfate-based bone graft substitute. *Orthopedics* **2004**, *27*, 131–135. [CrossRef]
61. Hu, G.; Xiao, L.; Fu, H.; Bi, D.; Ma, H.; Tong, P. Study on injectable and degradable cement of calcium sulphate and calcium phosphate for bone repair. *J. Mater. Sci. Mater. Med.* **2010**, *21*, 627–634. [CrossRef]
62. Wang, M.L.; Massie, J.; Allen, R.T.; Lee, Y.P.; Kim, C.W. Altered bioreactivity and limited osteoconductivity of calcium sulfate-based bone cements in the osteoporotic rat spine. *Spine J.* **2008**, *8*, 340–350. [CrossRef] [PubMed]
63. Ferguson, J.; Diefenbeck, M.; McNally, M. Ceramic Biocomposites as Biodegradable Antibiotic Carriers in the Treatment of Bone Infections. *J. Bone Jt. Infect.* **2017**, *2*, 38–51. [CrossRef] [PubMed]
64. Von Rechenberg, B.; Genot, O.R.; Nuss, K.; Galuppo, L.; Fulmer, M.; Jacobson, E.; Kronen, P.; Zlinszky, K.; Auer, J.A. Evaluation of four biodegradable, injectable bone cements in an experimental drill hole model in sheep. *Eur. J. Pharm. Biopharm.* **2013**, *85*, 130–138. [CrossRef] [PubMed]
65. Tsai, Y.Y.; Wang, S.F.; Kuo, S.T.; Tuan, W.H. Improving biodegradation behavior of calcium sulfate bone graft tablet by using water vapor treatment. *Mater. Sci. Eng. C Mater. Biol. Appl.* **2013**, *33*, 121–126. [CrossRef]
66. Miyazaki, T.; Sugawara-Narutaki, A.; Ohtsuki, C. Organic-Inorganic Composites Toward Biomaterial Application. *Front. Oral. Biol.* **2015**, *17*, 33–38. [CrossRef]
67. Chung, J.J.; Fujita, Y.; Li, S.; Stevens, M.M.; Kasuga, T.; Georgiou, T.K.; Jones, J.R. Biodegradable inorganic-organic hybrids of methacrylate star polymers for bone regeneration. *Acta Biomater.* **2017**, *54*, 411–418. [CrossRef]
68. Salgado, A.J.; Coutinho, O.P.; Reis, R.L. Bone tissue engineering: State of the art and future trends. *Macromol. Biosci.* **2004**, *4*, 743–765. [CrossRef]
69. Habraken, W.; Wolke, J.G.C.; Mikos, A.G.; Jansen, J.A. Porcine gelatin microsphere/calcium phosphate cement composites: An in vitro degradation study. *J. Biomed. Mater. Res. B Appl. Biomater.* **2009**, *91*, 555–561. [CrossRef]
70. Dagang, G.; Haoliang, S.; Kewei, X.; Yong, H. Long-term variations in mechanical properties and in vivo degradability of CPC/PLGA composite. *J. Biomed. Mater. Res. B Appl. Biomater.* **2007**, *82*, 533–544. [CrossRef]
71. Ouyang, Y.; Zhang, R.; Chen, H.; Chen, L.; Xi, W.; Li, X.; Zhang, Q.; Yan, Y. Novel, degradable, and cytoactive bone cements based on magnesium polyphosphate and calcium citrate. *New J. Chem.* **2022**, *46*, 13137–13148. [CrossRef]

72. Lu, J.; Descamps, M.; Dejou, J.; Koubi, G.; Hardouin, P.; Lemaitre, J.; Proust, J.P. The biodegradation mechanism of calcium phosphate biomaterials in bone. *J. Biomed. Mater. Res.* **2002**, *63*, 408–412. [CrossRef] [PubMed]
73. Ambard, A.J.; Mueninghoff, L. Calcium phosphate cement: Review of mechanical and biological properties. *J. Prosthodont.* **2006**, *15*, 321–328. [CrossRef] [PubMed]
74. Hsu, P.Y.; Kuo, H.C.; Syu, M.L.; Tuan, W.H.; Lai, P.L. A head-to-head comparison of the degradation rate of resorbable bioceramics. *Mater. Sci. Eng. C Mater. Biol. Appl.* **2020**, *106*, 110175. [CrossRef] [PubMed]
75. Kumar, Y.; Nalini, K.B.; Menon, J.; Patro, D.K.; Banerji, B.H. Calcium sulfate as bone graft substitute in the treatment of osseous bone defects, a prospective study. *J. Clin. Diagn. Res.* **2013**, *7*, 2926–2928. [CrossRef]
76. Xia, Z.; Grover, L.M.; Huang, Y.; Adamopoulos, I.E.; Gbureck, U.; Triffitt, J.T.; Shelton, R.M.; Barralet, J.E. In vitro biodegradation of three brushite calcium phosphate cements by a macrophage cell-line. *Biomaterials* **2006**, *27*, 4557–4565. [CrossRef]
77. Ramesh, N.; Moratti, S.C.; Dias, G.J. Hydroxyapatite-polymer biocomposites for bone regeneration: A review of current trends. *J. Biomed. Mater. Res. B Appl. Biomater.* **2018**, *106*, 2046–2057. [CrossRef]
78. Hannink, G.; Arts, J.J. Bioresorbability, porosity and mechanical strength of bone substitutes: What is optimal for bone regeneration? *Injury* **2011**, *42*, 22–25. [CrossRef]
79. Ingham, E.; Green. T.R.; Stone. M.H.; Kowalski. R.; Watkins. N.; Fisher, J. Production of TNF-alpha and bone resorbing activity by macrophages in response to different types of bone cement particles. *Biomaterials* **2000**, *99*, 1005–1013. [CrossRef]
80. Hammouche, S.; McNicholas, M. Biodegradable Bone Regeneration Synthetic Scaffolds: In Tissue Engineering. *Curr. Stem Cell Res. Ther.* **2012**, *7*, 134–142. [CrossRef]
81. Ohura, K.; Bohner, M.; Hardouin, P.; Lemaître, J.; Pasquier, G.; Flautre, B. Resorption of, and bone formation from, new beta-tricalcium phosphate-monocalcium phosphate cements: An in vivo study. *J. Biomed. Mater. Res.* **1996**, *30*, 193–200. [CrossRef]
82. Sheraly, A.; Lickorish, D.; Sarraf, F.; Davies, J.E. Use of Gastrointestinal Proton Pump Inhibitors to Regulate Osteoclast Mediated Resorption of Calcium Phosphate Cements In Vivo. *Curr. Drug Deliv.* **2009**, *6*, 192–198. [CrossRef] [PubMed]
83. Winkler, T.; Hoenig, E.; Huber, G.; Janssen, R.; Fritsch, D.; Gildenhaar, R.; Berger, G.; Morlock, M.M.; Schilling, A.F. Osteoclastic bioresorption of biomaterials: Two- and three-dimensional imaging and quantification. *Int. J. Artif. Organs.* **2010**, *33*, 198–203. [CrossRef] [PubMed]
84. Grossardt, C.; Ewald, A.; Grover, L.M.; Barralet, J.E.; Gbureck, U. Passive and active in vitro resorption of calcium and magnesium phosphate cements by osteoclastic cells. *Tissue Eng. Part A* **2010**, *16*, 3687–3695. [CrossRef] [PubMed]
85. Bernhardt, A.; Schamel, M.; Gbureck, U.; Gelinsky, M. Osteoclastic differentiation and resorption is modulated by bioactive metal ions Co^{2+}, Cu^{2+} and Cr^{3+} incorporated into calcium phosphate bone cements. *PLoS ONE* **2017**, *12*, e0182109. [CrossRef]
86. Rentsch, B.; Bernhardt, A.; Henss, A.; Ray, S.; Rentsch, C.; Schamel, M.; Gbureck, U.; Gelinsky, M.; Rammelt, S.; Lode, A. Trivalent chromium incorporated in a crystalline calcium phosphate matrix accelerates materials degradation and bone formation in vivo. *Acta Biomater.* **2018**, *69*, 332–341. [CrossRef]
87. Schumacher, M.; Wagner, A.S.; Kokesch-Himmelreich, J.; Bernhardt, A.; Rohnke, M.; Wenisch, S.; Gelinsky, M. Strontium substitution in apatitic CaP cements effectively attenuates osteoclastic resorption but does not inhibit osteoclastogenesis. *Acta Biomater.* **2016**, *37*, 184–194. [CrossRef]
88. Qixin, Z.; Jingyuan, D.; Xia, Z.; Zeng, H.; Li, S.; Yan, Y.; Chen, F. Biodegradation of Tricalcium Phosphate Ceramics by Osteoclasts. *J. Tongji Med. Univ.* **1998**, *18*, 257–262. [CrossRef]
89. Bannerman, A.; Williams, R.L.; Cox, S.C.; Grover, L.M. Visualising phase change in a brushite-based calcium phosphate ceramic. *Sci. Rep.* **2016**, *6*, 32671. [CrossRef]
90. Penel, G.; Leroy, N.; Van Landuyt, P.; Flautre, B.; Hardouin, P.; Lemaître, J.; Leroy, G. Raman microspectrometry studies of brushite cement: In vivo evolution in a sheep model. *Bone* **1999**, *25*, 81–84. [CrossRef]
91. Liu, W.; Huan, Z.; Wu, C.; Zhou, Z.; Chang, J. High-strength calcium silicate-incorporated magnesium phosphate bone cement with osteogenic potential for orthopedic application. *Compos. B Eng.* **2022**, *247*, 110324. [CrossRef]
92. Kokubo, T.; Takadama, H. How useful is SBF in predicting in vivo bone bioactivity? *Biomaterials* **2006**, *27*, 2907–2915. [CrossRef]
93. Hillmeier, J.; Meeder, P.J.; Noldge, G.; Kock, H.J.; Da Fonseca, K.; Kasperk, H.C. Balloon kyphoplasty of vertebral compression fractures with a new calcium phosphate cement. *Orthopade* **2004**, *33*, 31–39. [CrossRef]
94. Libicher, M.; Vetter, M.; Wolf, I.; Noeldge, G.; Kasperk, C.; Grafe, I.; Da Fonseca, K.; Hillmeier, J.; Meeder, P.J.; Meinzer, H.P.; et al. CT volumetry of intravertebral cement after kyphoplasty. Comparison of polymethylmethacrylate and calcium phosphate in a 12-month follow-up. *Eur. Radiol.* **2005**, *15*, 1544–1549. [CrossRef]
95. Heo, H.D.; Cho, Y.J.; Sheen, S.H.; Kuh, S.U.; Cho, S.M.; Oh, S.M. Morphological changes of injected calcium phosphate cement in osteoporotic compressed vertebral bodies. *Osteoporos. Int.* **2009**, *20*, 2063–2070. [CrossRef]
96. Welch, R.D.; Berry, B.H.; Crawford, K.; Zhang, H.; Zobitz, M.; Bronson, D.; Krishnan, S. Subchondral defects in caprine femora augmented with in situ setting hydroxyapatite cement, polymethylmethacrylate, or autogenous bone graft: Biomechanical and histomorphological analysis after two-years. *J. Orthop. Res.* **2002**, *20*, 464–472. [CrossRef]
97. Zhang, J.; Liu, W.; Schnitzler, V.; Tancret, F.; Bouler, J.M. Calcium phosphate cements for bone substitution: Chemistry, handling and mechanical properties. *Acta Biomater.* **2014**, *10*, 1035–1049. [CrossRef]
98. Kobayashi, N.; Ong, K.; Villarraga, M.; Schwardt, J.; Wenz, R.; Togawa, D.; Fujishiro, T.; Turner, A.S.; Seim, H.B., 3rd; Bauer, T.W. Histological and mechanical evaluation of self-setting calcium phosphate cements in a sheep vertebral bone void model. *J. Biomed. Mater. Res. A* **2007**, *81*, 838–846. [CrossRef]

99. Roberts, S.C.; Brilliant, J.D. Tricalcium phosphate as an adjunct to apical closure in pulpless permanent teeth. *J. Endod.* **1975**, *1*, 263–269. [CrossRef]
100. Kamerer, D.B.; Hirsch, B.E.; Snyderman, C.H.; Costantino, P.; Friedman, C.D. Hydroxyapatite cement: A new method for achieving watertight closure in transtemporal surgery. *Am. J. Otol.* **1994**, *15*, 47–49.
101. Brussius Coelho, M.; Rtshiladze, M.; Aggarwala, S.; Hunt, J.; Peltz, T.; Gardner, D.; Gianoutsos, M. Twenty-Year Experience of Use of Onlay Hydroxyapatite Cement for Secondary Cranioplasty. *J. Craniofac. Surg.* **2021**, *32*, 300–304. [CrossRef]
102. Suba, Z.; Takacs, D.; Matusovits, D.; Barabas, J.; Fazekas, A.; Szabo, G. Maxillary sinus floor grafting with beta-tricalcium phosphate in humans: Density and microarchitecture of the newly formed bone. *Clin. Oral Implants Res.* **2006**, *17*, 102–108. [CrossRef] [PubMed]
103. Nakano, M.; Hirano, N.; Zukawa, M.; Suzuki, K.; Hirose, J.; Kimura, T.; Kawaguchi, Y. Vertebroplasty using calcium phosphate cement for osteoporotic vertebral fractures study of outcomes at a minimum follow-up of two years. *Asian Spine J.* **2012**, *6*, 34–42. [CrossRef] [PubMed]
104. Gumpert, R.; Bodo, K.; Spuller, E.; Poglitsch, T.; Bindl, R.; Ignatius, A.; Puchwein, P. Demineralization after balloon kyphoplasty with calcium phosphate cement: A histological evaluation in ten patients. *Eur. Spine J.* **2014**, *23*, 1361–1368. [CrossRef]
105. Kruger, R.; Groll, J. Fiber reinforced calcium phosphate cements-on the way to degradable load bearing bone substitutes? *Biomaterials* **2012**, *33*, 5887–5900. [CrossRef] [PubMed]
106. Habraken, W.; Habibovic, P.; Epple, M.; Bohner, M. Calcium phosphates in biomedical applications: Materials for the future? *Mater. Today* **2016**, *19*, 69–87. [CrossRef]
107. Chow, L.C. Next generation calcium phosphate-based biomaterials. *Dent. Mater. J.* **2009**, *28*, 1–10. [CrossRef]

Journal of
Functional
Biomaterials

MDPI

Review

Proteinoid Polymers and Nanocapsules for Cancer Diagnostics, Therapy and Theranostics: In Vitro and In Vivo Studies

Ella Itzhaki [1], Yuval Elias [1], Neta Moskovits [2], Salomon M. Stemmer [2,3] and Shlomo Margel [1,*]

[1] Department of Chemistry and Institute of Nanotechnology and Advanced Materials, Bar-Ilan University, Ramat-Gan 5290002, Israel; elaeli3543@gmail.com (E.I.)
[2] Felsenstein Medical Research Center, Petah Tikva 49100, Israel
[3] Sackler Faculty of Medicine, Tel Aviv University, Tel Aviv 6997801, Israel
* Correspondence: shlomo.margel@biu.ac.il; Tel.: +972-3-5318861

Abstract: Proteinoids—simple polymers composed of amino acids—were suggested decades ago by Fox and coworkers to form spontaneously by heat. These special polymers may self-assemble in micrometer structures called proteinoid microspheres, presented as the protocells of life on earth Interest in proteinoids increased in recent years, in particular for nano-biomedicine. They were produced by stepwise polymerization of 3–4 amino acids. Proteinoids based on the RGD motif were prepared for targeting tumors. Nanocapsules form by heating proteinoids in an aqueous solution and slowly cooling to room temperature. Proteinoid polymers and nanocapsules suit many biomedical applications owing to their non-toxicity, biocompatibility and immune safety Drugs and/or imaging reagents for cancer diagnostic, therapeutic and theranostic applications were encapsulated by dissolving them in aqueous proteinoid solutions. Here, recent in vitro and in vivo studies are reviewed.

Keywords: proteinoid polymers; proteinoid nanocapsules; fluorescent nanocarriers; cancer theranostics; RGD nanoparticles

Citation: Itzhaki, E.; Elias, Y.; Moskovits, N.; Stemmer, S.M.; Margel, S. Proteinoid Polymers and Nanocapsules for Cancer Diagnostics, Therapy and Theranostics: In Vitro and In Vivo Studies. *J. Funct. Biomater.* **2023**, *14*, 215. https://doi.org/10.3390/jfb14040215

Academic Editor: Christie Ying Kei Lung

Received: 19 February 2023
Revised: 29 March 2023
Accepted: 6 April 2023
Published: 11 April 2023

1. Introduction

Proteinoids are random polymers composed of amino acids synthesized by stepwise thermal polymerization. They were discovered and studied in the 1950s by Fox and coworkers, who suggested that they formed spontaneously by high heat at the beginning of life on Earth. Fox et al. demonstrated that proteinoids may self-assemble in spherical micrometer structures, proteinoid microspheres, presented as the protocells of life [1–3]. This process occurs at rather high temperature (170–180 °C, for instance) in an inert atmosphere without a catalyst or a solvent. Lysine or aspartic/glutamic acid are critical, as they form cyclic products which act as solvents [4]. Various proteinoids may be prepared by using natural and synthetic amino acids at different ratios. The special features of each proteinoid influence the character of particles that are composed from it [5,6].

In early work by Rao and coworkers [7,8], an acidic proteinoid was prepared by thermal condensation of seven natural amino acids, mainly Glu, Asp and Gly [7]. The proteinoid was expected to be non-antigenic due to its low molecular weight (<10 kDa). The acidic gastric irritating drug methotrexate was encapsulated in self-assembled microspheres rather efficiently (~50% with ~10% loading) [8], conferring gastric (pH~1) stability (<10% released in two hours) and thus improving the prospects of oral delivery. The spherical microspheres were of uniform diameter (1–3 μm). Complete solubility in neutral blood pH allowed release of most of the encapsulated drug within one hour.

During the last decade, Margel and coworkers produced proteinoids by bulk stepwise polymerization of 3–4 amino acids without or with a biopolymer such as poly(L-lactic acid) (PLLA) [9–14]. Nanocapsules (NCs) form spontaneously by heating proteinoids to about 70 °C in an aqueous solution to completely dissolve the polymers, followed by slow cooling to room temperature. Different drugs and/or imaging reagents were encapsulated by dissolution in the proteinoid solution, and covalent binding to the surface of the NCs was achieved with or without a spacer arm. Proteinoid NCs suit many applications owing to their non-toxicity, biocompatibility and immune safety [10,11]. Incorporation of the tri-amino acid sequence arginine–glycine–aspartic acid (RGD) in the random proteinoid backbone enables targeting of tumors.

Kwon, Park & Kim [15] demonstrated that proteinoids could also act as carriers of a certain kind of drugs and be programmed to release them depending on the external conditions. This is made possible by introducing a disulfide bond in the proteinoid that lets it assemble in an aqueous phase. In an external environment which is reducing in nature, the disulfide bond is cleaved, and the micelle is likely to be loosened. As a result, the payload (drug) inside the micelle would be released.

Recent work by Adamatzky investigated the potential application of proteinoids in the field of computing systems [16]. Proteinoids have the potential to be utilized in unconventional computing owing to their unique electrical properties. Such proteinoids were also prepared by step-growth polymerization of amino acids in an aqueous environment. These proteinoids swell into hollow microspheres that produce an endogenous burst of electrical potential spikes and exhibit the capacity to alter patterns of their electrical activity in response to illumination. By forming interconnected networks through pores and tubes, proteinoid microspheres can enable programmable growth and have potential for creation of intricate computing systems. This capacity for novel growth patterns and electrical activity allows for the possibility of developing computing architectures that are more versatile and efficient than traditional systems.

The present review article builds upon the foundation established by Kolitz-Domb and Margel in 2018 [11] that focused primarily on the chemistry behind the preparation of a few proteinoids and their nanocapsules (e.g., P(KF), P(EF) in the absence or presence of PLLA), as well as their application in select fields such as cosmetics, agriculture and theranostics using the cancer drug doxorubicin. This review describes mainly the synthesis and characterization of P(RGD) polymers and nanocapsules containing synergistic drugs, cannabinoids and TRAIL for cancer theranostics. The main focus is on P(RGD) application in the realm of cancer research. Our aim is to provide an up-to-date in vitro and in vivo survey of the latest findings in this rapidly evolving field, shedding light on new developments and offering fresh insights into the potential of proteinoid nanocapsules for cancer diagnostic, therapeutic and theranostic applications, including targeted drug delivery systems.

2. Synthesis and Characterization of Proteinoids and NCs

2.1. Preparation of Proteinoids

Proteinoid chains were prepared by step-growth polymerization of amino acids at high temperature (depending on the amino acids, e.g., 180 °C) in an inert atmosphere with no solvent, initiator or catalyst [11,12]. A tri-functional amino acid—Glu/Asp/Lys—is an essential component, providing a solvent by cyclization and serving as an initiator (see Figure 1) [11]. Proteinoid preparation was recently reviewed [17]. Different proteinoid polymers can be obtained due to the extensive range of both natural and synthetic amino acids available (in most cases, the ratio between the monomers was 1:1). This review presents several examples of proteinoid polymers that have been studied for the purpose of cancer diagnostics, therapy and theranostics (see Table 1).

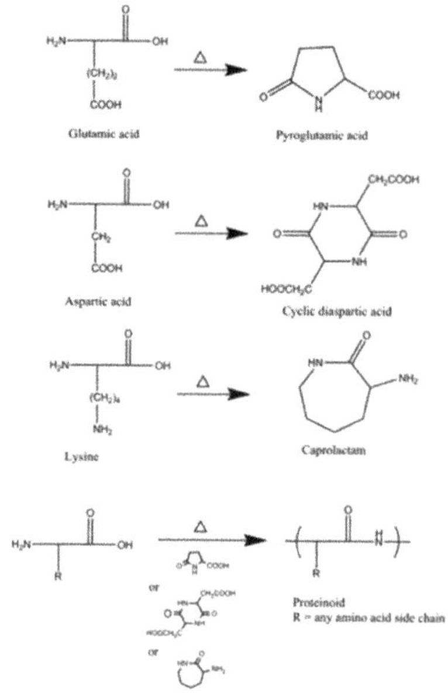

Figure 1. Proteinoid formation by polymerization with heat in a cyclic form of Lys/Asp/Glu as solvent [11].

Table 1. Various proteinoid polymers and their amino acid content [a].

Proteinoid Polymer	Amino Acid Content	Main Component of Amino Acids [a]	Segment
P(EF-PLLA)	L-glutamic acid L-phenylalanine	L-glutamic acid	Poly-L-lactic acid (PLLA)
P(KRHF)	L-lysine L-arginine L-histidine L-phenylalanine	L-lysine	PLLA
P(RGD)	D-arginine L-glycine L-aspartic acid	L-aspartic acid	—

[a] A tri-functional amino acid—Glu/Asp/Lys—is a main component, providing a solvent by cyclization and serving as an initiator.

2.2. Preparation of Nanocapsules (NCs)

Hollow proteinoid NCs were produced by a self-assembly process in an aqueous continuous phase [17]. In this process, the first step was to heat the aqueous phase containing the proteinoid to about 70–80 °C until a full dissolution of the polymer was observed, followed by slow cooling to room temperature for precipitation and formation of proteinoid NCs. The NCs form biocompatible carriers with a hydrophobic core and hydrophilic groups on the surface. Various compounds were encapsulated during self-assembly, as presented in Figure 2. Near infrared (NIR) fluorescent dyes were used for cancer diagnosis [12,17,18]. Recently, a synergistic combination of anti-cancer drugs was encapsulated [19] for personalized therapy.

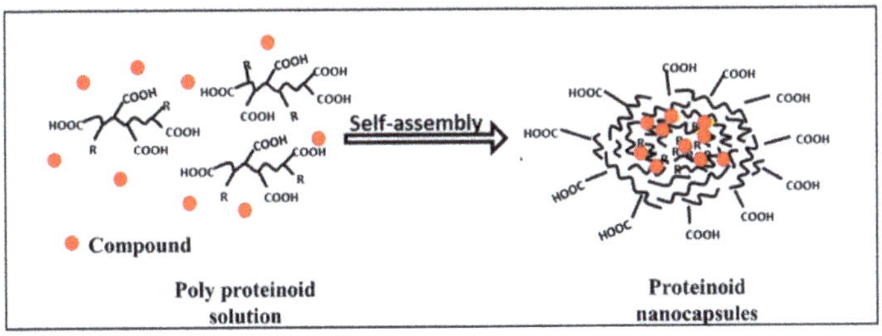

Figure 2. Self-assembly of proteinoid polymers; scribbled lines—polymer chains, dots—encapsulated compound (dyes/drugs) [11].

2.3. Characterization of Proteinoids and NCs

Unexpectedly, the molecular weight of all proteinoids made in the Margel lab was high (25–195 kDa) with a low polydispersity index (PDI of 1.01–1.27) [5]. These very unusual results for stepwise polymerization may be explained by the high temperature, which provides uniform long chains that resemble natural biopolymers. Self-assembled NCs were spherical with a uniform distribution, as shown by dynamic light scattering (DLS) and scanning electron microscopy (SEM) in Figure 3.

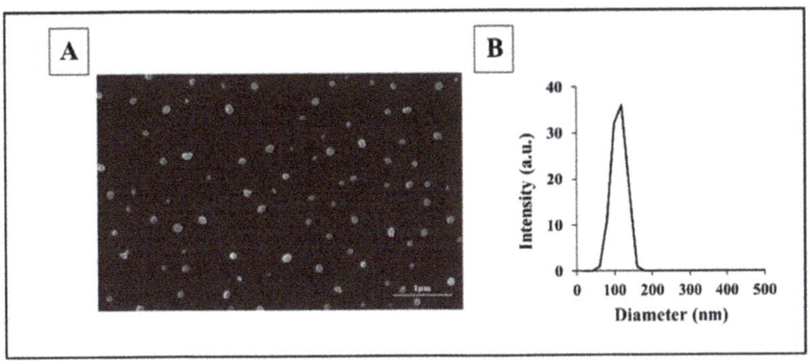

Figure 3. SEM image (**A**) and diameter histogram (**B**) of proteinoid nanocapsules (NCs) [19]. The diameters of more than 200 NCs were measured with Analysis Auto image analysis software version 3.2 (Soft Imaging System GmbH, Münster, Germany).

3. Cancer Diagnostics and Therapy, towards Theranostic Applications

Over the past decade, various proteinoids and nanocapsules (NCs) were synthesized in the Margel lab. In this section, we provide a brief overview of early applications, which were previously discussed in Kolitz-Domb's 2018 review [11] that focused on detection of colorectal cancer (CRC). We then present new developments in cancer diagnostics and therapy towards theranostic applications. Recently, theranostics, a field that combines therapy and diagnostics, has gained considerable attention. Compared to targeted conjugates of peptides with drugs/agents, theranostics offers significant advantages [20].

3.1. Colorectal Cancer (CRC) Diagnostics

The first application of proteinoids studied by Margel and coworkers aimed at early specific detection of colorectal cancer (CRC) [14,17]. Indocyanine green (ICG), a fluorescent dye in the NIR region, 700–1000 nm [21,22], that is FDA-approved was encapsulated in biodegradable proteinoid NCs composed of Glu and Phe with PLLA [10,14]. Peanut agglutinin (providing targeting) and anti-caci-nonembryonic antigen antibodies (termed anti-CEA) were attached [23]. These as well as hollow P(EF-PLLA) NCs allowed specific detection of CRC tumors in chicken embryo tumor implants. The high signal in tumors was attributed to receptor upregulation [14].

ICG encapsulation was optimized, and spherical particles were obtained with wet/dry diameters of 145 ± 20/70 ± 15 nm. The encapsulation increased photo-stability significantly, protecting the dye from light-induced bleaching. The NCs were non-toxic as expected, also at a high concentration of 2.5 mg/mL. Interestingly, biodistribution in mice (after injection) showed complete clearance after 24 h from various locations including the brain and bones (Figure 4). This is useful for targeted biological applications. Specific detection of colon tumors by the fluorescent NCs was demonstrated in chicken embryos and mice (Figure 5), offering a significant advantage over invasive colonoscopy. The anti-CEA-conjugated NCs specifically detected tumors in mice with a specific signal (reporting signal to noise ratios) (Figure 5A). On the other hand, no signal was obtained with NCs attached to an anti-rabbit non-selective antibody (see Figure 5B) [14].

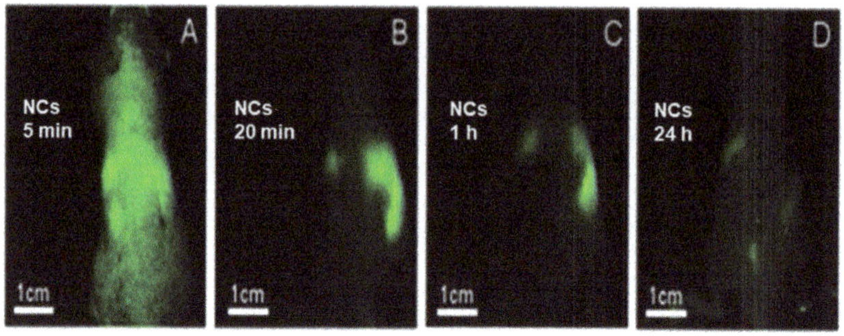

Figure 4. Whole-body images of NIR fluorescent P(EF-PLLA) NCs after intravenous (IV) injection to mice at (**A**) 5 min, (**B**) 20 min, (**C**) 1 h and (**D**) 24 h [14].

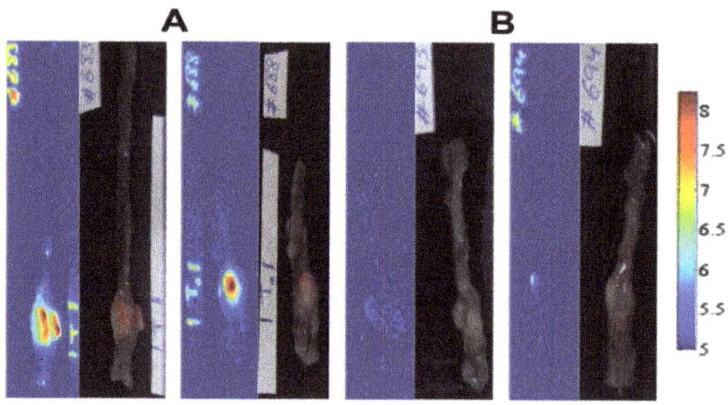

Figure 5. Specific targeting of colon tumors—P(EF-PLLA) NCs with anti-CEA (**A**) vs. IgG (**B**). Fluorescent (**left**) and grayscale (**right**) images [14].

3.2. Cancer Therapy with Doxorubicin-Loaded NCs

Kiel et al. used proteinoid NCs to encapsulate the anticancer compound doxorubicin (Dox) [10]. The NCs showed similar activity as the anticancer drug Doxil, comprising another nanocarrier, a polyethylene glycol (PEG) lipid surface that prolongs blood circulation and improves tumor uptake [24]. The side effects of Dox were avoided by sufficient encapsulation. Four basic L-amino acids were selected, along with PLLA; lysine was the main monomer, enabling the synthesis. A series of P(KRHF-PLLA) proteinoids with varying His/Phe ratios were prepared with very low PDI (1.01–1.03) and high molecular weight (122–149 kDa).

Particle diameter was optimized. A 2:3 His/Phe weight ratio produced the smallest hollow particles (36.2 ± 6.9 nm). The corresponding Dox-loaded particles were 112 ± 15 nm (by DLS). The nanometric diameter is important for cancer therapeutics, allowing particles to cross biological barriers, penetrate cells and evade immunogenic systems [25]. The NCs exhibited a narrow diameter distribution (PDI~1), an important characteristic for biomedicine. Drug content was optimized at 15% Dox (compared to 12.5% in Doxil) [14]. Successful encapsulation was confirmed by FTIR (93 ± 12%) and XRD (amorphous vs. crystalline phase). The diffractogram of free Dox shows clear peaks, indicating a crystalline phase, while the proteinoid shows an amorphous pattern. The XRD pattern of Dox-loaded NPs shows a reduction in peak number and intensity compared to free Dox, indicating a phase transformation from crystalline to amorphous doxorubicin [10].

PEG was conjugated to enhance stability and prevent drug leakage [26]. PEGylated Dox-loaded NCs were shown to be more stable when changing the environment, and the release of the drug was confirmed to occur only in blood, while non-PEGylated NCs also release it in serum and phosphate-buffered saline (PBS). This selective release becomes important when considering long-term shelf life, ensuring release only upon exposure to proteolytic enzymes and other blood ingredients.

An in vitro colorimetric assay (Figure 6) reveals a similar cytotoxic activity as free Dox of NCs and even higher toxicity of PEGylated NCs toward glioblastoma and ductal and colon carcinoma cells. The higher dose of loaded NCs, i.e., 0.05 mg/mL (drug loading: Dox concentration of 8.3 µg/mL), exceeded free Dox as shown in Figure 6.

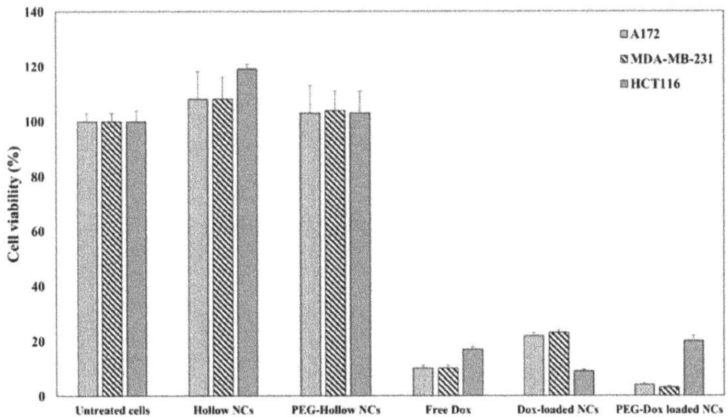

Figure 6. Viability of HCT116 colon carcinoma, ductal carcinoma and A172 glioblastoma cells 48 h after treatment with Doxil vs. Dox-loaded P(KGHF-PLLA) NCs, measured by XTT assay [10]. The cells were exposed to NCs at a concentration of 0.05 mg/mL, as well as to an equal quantity of free Dox.

4. Cancer Theranostics

Proteinoid NCs may encapsulate other drugs such as Taxol and Temozolomide and provide cancer theranostics by co-encapsulation of an anticancer drug and a NIR fluorescent dye. Moreover, conjugation of suitable targeting agents such as TRAIL to PEGylated Dox-loaded NCs is likely to enhance delivery and increase penetration rate [12].

4.1. RGD (ArgGlyAsp) for Specific Delivery to Tumors

The RGD motif was discovered in 1985 by Pierschbacher and Ruoslahti as the active component in fibronectin [27]. This peptide has high affinity to αvβ3 integrin, which is overexpressed in cancer cells and upregulated on the surface of growing blood vessels and is thus attracted to areas of angiogenesis, affording integrin-targeted nanodrugs for tumor imaging and treatment [28–30].

Both cyclic and linear RGD peptides were conjugated to NCs for targeted delivery [31 The optical activity, rotating polarized light to the left (L) or right (D), influences attachmen to cells. For example, replacing L-Asp with the D isomer yields an inactive peptide, whil D-Arg increases binding 10-fold [31,32]. However, recent studies of biomedical application do not emphasize this aspect [33–35].

Optimal RGD configuration offers significant improvement. Hadad and coworker recently prepared novel RGD proteinoid polymers and NCs [13]. Different configuration were used to randomly achieve the RGD sequence in about 13% [36] of the proteinoi backbone. Such P(RGD) proteinoid NCs may act both as a nanocarrier and as a targeting system due to the RGD motif. This work suggests a rapid method for economic synthesi of proteinoid NCs for therapeutic, diagnostic and theranostic cancer applications.

4.2. Preparation and Characterization of RGD Proteinoids

To determine optimal RGD configuration for targeted delivery, four proteinoid con figurations were synthesized: P(R^D/RGD^D/D). Asp serves as solvent and as a linker tha reduces the energy of polymerization. In contrast to previous publications, cross-linke products were not obtained, and all four configurations were obtained with 100% yield an were water soluble. The proteinoids had molecular weights of 67,660−69,066 Da, a ver low PDI [10–12] atypical of stepwise polymerization, which usually yields light polymer with high PDI [37].

4.3. Self-Assembly and Characterization of P(RGD) NCs

Hollow P(RGD) NCs were formed by self-assembly of the crude proteinoid by heating the solution and cooling slowly to room temperature. The diameters as determined by SEM were all similar at 55 ± 13, 48 ± 9 and 42 ± 9 nm for R^D and D^D, R^D, and D^D, respectively and 45 ± 11 nm for P(RGD). ICG was successfully encapsulated with dry diameters o 141 ± 24 nm for R^D and D^D, 95 ± 13/87 ± 12 nm for R^D/D^D and 86 ± 11 nm for P(RGD) The capsules thus swell considerably [38].

4.4. ICG Controlled Release

Drug release is commonly studied in vitro by incubation in serum or PBS [39]. ICG release was evaluated in physiological conditions by incubating the NCs for 2.5 h at 37 °C The kinetics of ICG release from P(R^DGD) NCs after treatment in serum/PBS is shown in Figure 7. As expected, there is little release to PBS, while in serum, peptide bonds are cleaved and ICG is significantly released.

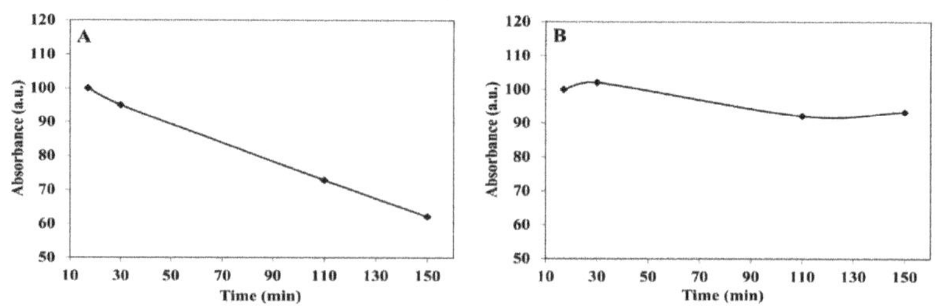

Figure 7. Absorbance of ICG-loaded P(R^DGD) NCs in human serum (**A**) vs. PBS (**B**) at 37 °C over 2.5 h [13].

4.5. Optimization of RGD Configuration

Angiogenesis, a cascade of cellular events in which new blood vessels are developed, is essential in many conditions including psoriasis and cancer [40]. Human vascular en- dothelial cells (HUVECs) interact with the environment using receptors known as integrins, which regulate growth/repair of blood vessels [40]. The high affinity of RGD to αvβ3 integrin attracts it to angiogenesis areas [27,29,41–43].

A scratch assay commonly employed for studying angiogenesis in vitro [39,44] was applied to find the optimal NC configuration. Confocal images of HUVECs after short (15 min) incubation with ICG-loaded NCs (Figure 8A–C) show clear accumulation in the injured scratch zone (red). The strongest signal is exhibited by $P(RGD^D)$ NCs. The fluorescence in the scratch zone (Figure 8D) clearly shows significantly higher intensity for $P(R^DGD)$-treated cells, in accordance with the high affinity of D-Arg to integrins expressed on the cell membrane. Hence, cell attachment appears to be negatively affected by D-Asp.

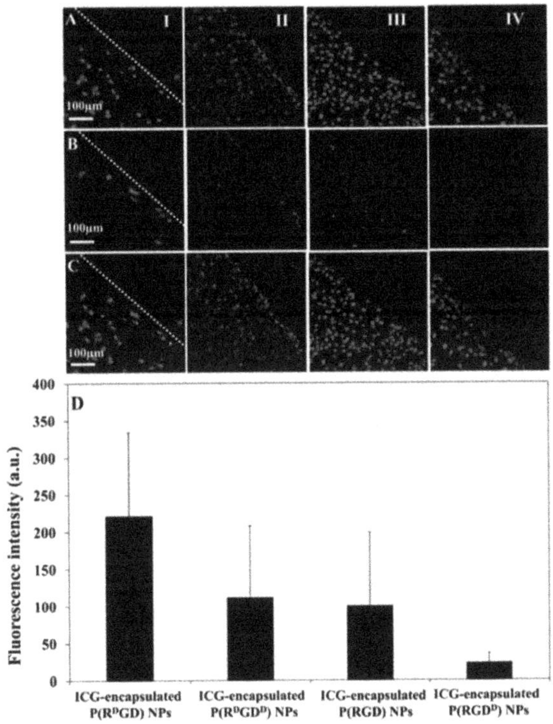

Figure 8. Scratch test assay [13]. HUVEC after treatment with ICG-loaded P(RGD) NCs show lines of nuclei with blue Hoechst dye (**A**) or red NCs (fluorescent, (**B**)) as well as in overlay images (**C**). Scale bar 100 μm (I–IV for R^DGD, R^DGD^D, RGD and RGD^D). Quantification of fluorescence intensity of scratch region images, analyzed by ImageJ 1.52a (**D**).

These results further suggest that the optical configuration directly influences the biological activity [29,31]. $P(R^DGD)$ NCs with the strongest signal in the scratch zone are hence the best candidates for targeted carrying of drugs to angiogenesis areas such as wounds or tumors.

4.6. Engineering of NIR Fluorescent PEGylated $P(R^DGD)$ NCs

The NIR dye ICG was loaded within $P(R^DGD)$ NCs during self-assembly of the proteinoid NCs. The dry diameter increased with ICG from 47 ± 9 to 95 ± 1 nm [37]. ICG-bearing NCs were PEGylated to improve stability. PEGylation was found to be important for drug delivery, resulting in improved blood circulation, [45] phagocytosis evasion [46] and better stability in serum [47]. The latter is critical for prevention of dye/drug leakage [48,49]. On the other hand, antibodies may form which specifically recognize and bind PEG [47]. The attachment of PEG (Mw 750 or 5000) to the ICG-encapsulated $P(R^DGD)$ NCs was done by conjugation of NHS ester groups (NHS-PEG) to the primary amine group of the NCs. The hydrodynamic diameter of the loaded NCs,

93 ± 20 nm, increased following PEGylation with NHS-PEG, 750 or 5000 Da, to 177 ± 30 and 216 ± 25, respectively.

To evaluate the release of ICG, the NCs were incubated for 2.5 h in physiological conditions (PBS or serum at 37 °C). The absorbance of ICG was measured by UV spectroscopy. The sustained release in serum probably results from proteases [12]. Clearly, after incubation in serum, the absorbance of the NCs without PEG decreased significantly, by 40%, while there was a significantly lower decrease in absorbance with 5000 and 750 Da PEG, 10% and 20%, respectively. This difference was attributed to the PEGylation, which prevents biodegradation; the longer the PEG chain, the higher the stability of encapsulated ICG [50]. The results clearly demonstrate the successful PEGylation as well as the major effect of PEG on the rate of ICG release from P(R^{D}GD) nanocapsules.

5. Application of P(RGD) NCs for Cancer Tumors

5.1. mCherry-Labeled Tumor in CAM Model

The well-known chick chorioallantois membrane (CAM) experiment is a simple, cost effective animal model that avoids animal suffering due to the use of embryonic tissue. The highly vascularized membrane around the embryo allows grafting of tumor explants, [51] enabling rapid in vivo studies with fewer mice required later.

A CAM window was exposed after incubation of fertile eggs, and implanted cells began to grow on a ring made of plastic (Figure 9A). ICG-loaded P(R^{D}GD) NCs with or without PEG were injected IV (Figure 9B), and mammary carcinoma explants were examined after 4/24 h (Figure 9C) [13].

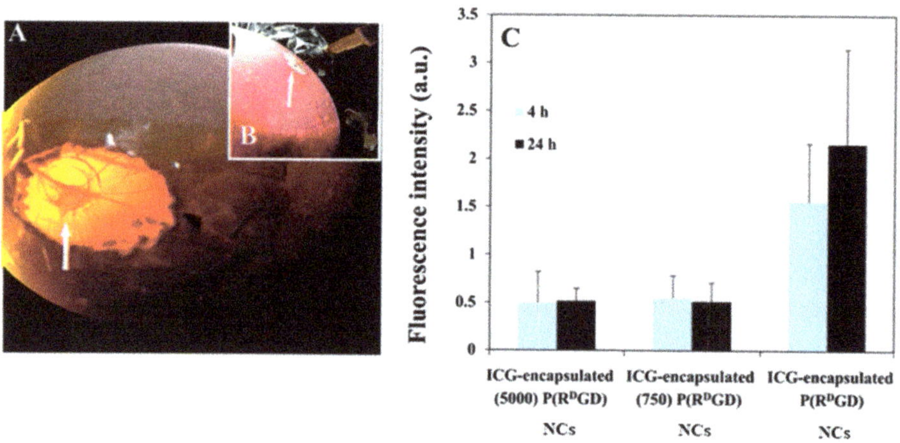

Figure 9. Tumor on CAM (top view, (**A**)) and P(R^{D}GD) NCs injected IV (**B**). Fluorescence from implanted carcinoma cells with encapsulated ICG 4/24 h after the injection (**C**) [36].

The non-PEGylated NCs showed significantly higher fluorescence compared to the PEGylated ones, indicating greater accumulation of ICG. This is in accord with the better packing with PEG with fewer bonds that may be bio-degraded, which significantly slows the release by preventing quick disassembly of the NCs at the tumor site.

5.2. mCherry-Labeled Tumor in Mouse Model

The targeting of the ICG-encapsulating P(R^{D}GD) NCs toward tumors for a prolonged duration (beyond 24 h) was evaluated in vivo using a xenograft model [36]. Balb/c mice were injected with mouse carcinoma cells subcutaneously, inducing a xenograft. After two weeks, the mice were treated IV with ICG or P(R^{D}GD) NCs loaded with ICG (with or without PEG) [13].

The fluorescence of tumors harvested after 24/48/120 h (Figure 10) shows that as time progresses, the NCs remain visible. This clearly indicates a targeted delivery (Figure 10A–C). For ICG, the fluorescence rapidly decreases and after 24 h is absent in the areas of the tumor (Figure 10D). The major role of PEG in stabilization is clearly seen; the fluorescence increases with time, while without PEG, it decreases (Figure 10E).

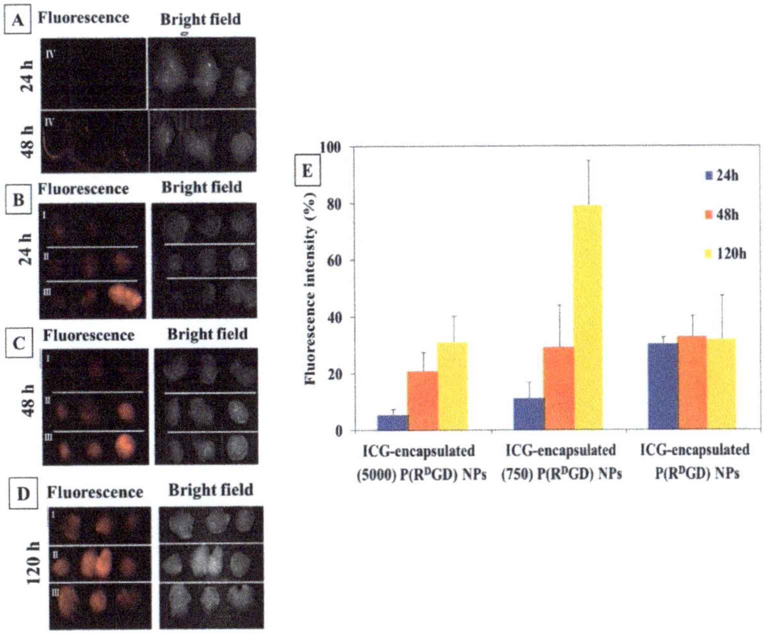

Figure 10. Images of P(R^{D}GD) NCs with encapsulated ICG with long (5000 Da, I) and short (750 Da, II) PEG chains and without PEG (III) toward carcinoma tumors after (**B–D**) 24/48/120 h and ICG (IV) after (**A**) 24/48 h. Tumor fluorescence intensity after NC treatment (**E**) [36].

The targeting of the NCs is affected by the PEG chain length. The migration toward the tumor is reduced with the longer chain, indicating enhanced shielding [52]. After 120 h, the short PEGylated NCs were the most efficient. Recent studies [53,54] suggest that small uncharged NCs may be ideal for targeted delivery, as they interact less with cells. Accordingly, the short PEGylated NCs are smaller than the long PEGylated ones (177 $\pm$ 30 vs. 216 $\pm$ 25 nm) and essentially uncharged (5 $\pm$ 3 mV); without PEG, the charge is quite significant (30 $\pm$ 7 mV).

5.3. In Vivo Anti-Tumor Therapy with Cannabidiol (CBD)

Polymeric and other NCs were developed to improve the antitumor activity of cannabinoids [55]. Lugasi and coworkers recently presented CBD-loaded NCs designed for targeting human tumors [9]. P(R^{D}GD) acidic proteinoids were prepared, and hollow/CBD-loaded NCs were formed with 97 $\pm$ 4/86 $\pm$ 3 nm diameter.

HPLC confirmed successful loading, and drug loading analysis found complete encapsulation. After lyophilization with trehalose and prolonged storage, redispersed particles retained their original diameter. Hollow NCs were found to be nontoxic, similar to CBD, and reached the cytoplasm. However, encapsulated CBD showed high toxicity, indicating efficient penetration and destruction of tumor cells.

A drug release study showed a high initial burst (about a third released in the initial 12 h) followed by gradual release (over 84 h). The acidic extracellular matrix in tumor cells promises faster release. NCs were found in all organs that were examined in vivo (Figure 11).

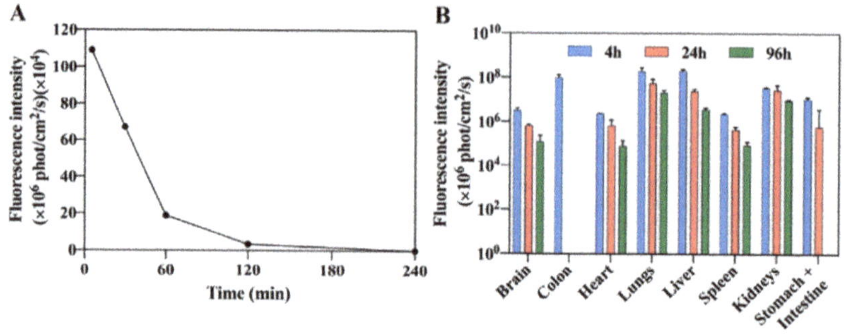

Figure 11. Blood fluorescence (**A**) and biodistribution (**B**) of Cy7-conjugated P(RGD) NCs [9]. Using the Maestro II in vivo imaging system, the fluorescence intensity of multiple organs was measured in mice at 4, 24 and 96 h after injection, following which the mice were sacrificed and the organs were harvested.

Figure 12 shows targeting, based on the high affinity of RGD to blood vessels and tumor receptors, and treatment of breast cancer and CRC. The targeted delivery significantly increased the efficacy, permitting lower CBD concentrations.

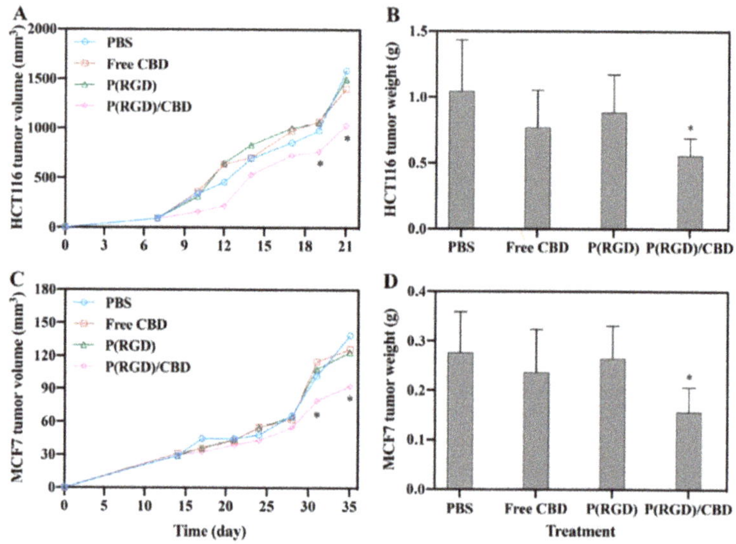

Figure 12. Tumor parameters in nude mouse xenograft model in HCT116 (**A**,**B**) and MCF7 (**C**,**D**) cells after 14 days of bi/triweekly IV injection of PBS, CBD and P(RGD) NCs [9]. * Student's t-test $p < 0.05$, error bars: standard deviations.

5.4. Conjugation of TRAIL

Tumor necrosis factor (TNF)-related apoptosis-inducing ligand (TRAIL) belongs to the TNF cytokine family, which induces apoptosis [56]. Hadad et al. recently conjugated TRAIL to the surface of hollow P(RGD) NCs [11]. Encapsulation of Dox thereby allows targeted dual cancer therapy.

Dox was encapsulated within P(RGD) NCs during the self-assembly process of the proteinoids. The encapsulation and targeted delivery of the drug should reduce its side effects such as cardiac damage and hair loss. TRAIL was protected from rapid enzyme degradation by covalent conjugation [56].

Polymers (Figure 13A) were self-assembled to form hollow (Figure 13D) and Dox-loaded (Figure 13B) and PEGylated NCs (Figure 13C). TRAIL was conjugated directly (Figure 13E) and indirectly using a spacer (Figure 13F). The NCs were characterized to determine the optimal method of conjugation. SEM confirmed the preservation of the NCs (shape and dimensions), and ELISA allowed quantitation of TRAIL binding.

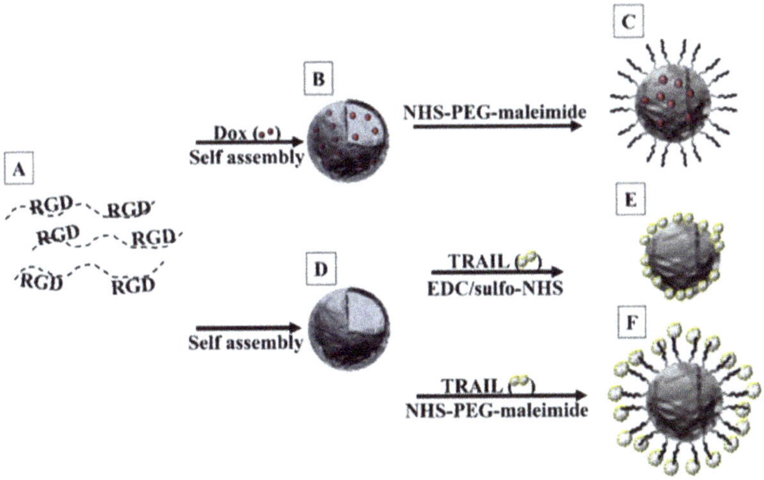

Figure 13. Formation of P(RGD) NCs. (**A**) Polymers are self-assembled (**B**) with Dox and (**C**) PEGylated or (**D**) conjugated to TRAIL (**E**) directly or (**F**) indirectly via PEG (3500 kDa) [12].

PEG was conjugated covalently to the NC surface, and its effect on Dox cytotoxicity and TRAIL release in ovarian cancer (CAOV-3) cells was studied. The hollow NCs were non-toxic, while all encapsulating/conjugated NPs were toxic and showed controlled release. Moreover, in vitro experiments demonstrated that Dox/P(RGD) and TRAIL-P(RGD) NCs were as effective as free Dox/TRAIL (viability of 2 and 9%), while PEGylation considerably reduced the activity (to 20 and 41 %), allowing extended release over several days.

The TRAIL-conjugated drug-loaded NCs are promising for treatment of ovarian cancer. Dox may be applied more safely, and TRAIL stability is increased, while retaining the efficiency of both agents.

5.5. Fluorescent NCs Containing Synergistic Drugs

Recently, efficacy/safety of P(RGD) NCs with synergistic drugs targeting tumors was assessed for cancer treatment [18]. Palbociclib (Pal), a CDK4/6 inhibitor, and Alpelisib (Alp), a P13K inhibitor, were co-encapsulated (25 w% of each) with yields of 72% and 95%, respectively. As the drugs have low solubility, different concentrations of Tween 80 were added to the mixed cloudy suspension to obtain a clear solution for drug loading into the NCs. Long-term stability, controlled release and cellular uptake, as well as synergistic cytotoxicity and induced cell death are evident from in vitro experiments.

The hydrodynamic diameter of the hollow NCs, 34 $\pm$ 5 nm, decreased following encapsulation of mixed drugs to 22 $\pm$ 3 nm. This may be explained by the presence of part of the drugs on the surface of the NCs, preventing water molecules due to their hydrophobic nature from being entrapped and adsorbed. To evaluate this hypothesis, zeta potentials were measured at a pH of 7.5, yielding -6 ± 2 and -0.3 ± 0.8 mV for hollow and drug-loaded NCs, respectively. The negative charge is attributed to aspartic acid on the NC surface. Thus, the increase in the zeta potentials is due to the drug loading and the physical interactions of the drugs with the NC surface, which mask the negative residue.

In vitro cell toxicity was determined by XTT assay on three human cancer cell lines—HCT116 (colon carcinoma), MCF-7 (breast adenocarcinoma) and A549 (lung carcinoma) treated for 24, 48, 72 and 96 h with Cy7-P(RGD) hollow and Pal,Alp-loaded NCs. Cell viability post treatment with the hollow NCs was not significantly different to untreated cells (above 100%), hence the NCs are non-toxic. Treatment with Cy7-P(RGD)/Pal,Alp NCs after 24, 48, 72 and 96 h demonstrated a gradual decrease in cell growth and cell death after 96 h. In contrast, free mixed drugs showed a decrease in cell viability already after 24 h. These results demonstrate the controlled release of the drugs from the NCs compared to the free drugs.

Very recently, NCs were examined in PDX models (colon, breast and gastric cancer) and showed similar results to free drugs with reduced toxicity [57]. One example is the RA-300 PDX model derived from a tumor biopsy of a 50-year-old male patient diagnosed with CRC (well-differentiated adenocarcinoma, stage IV).

For the efficacy experiment, the free mixed drugs were administered orally (per os marked as PO) 5 days a week, whereas the NCs were administered IV and/or intraperitoneally (IP) 2 days per week. As IV injection can lead to damage/blockage of veins it is advisable to limit the volume and frequency of such injections [58]. Therefore, to achieve the maximum amount of drug, three ways of administering the NCs were tested (IV, IP, or both). IV/IP administration entailed injection twice a week. The vehicle group received no drugs, whereas the free drug group received (PO) a total of 14 mg/week of Pal and 3.5 mg/week of Alp. The three P(RGD)/Pal,Alp (50%) groups included IV (0.72 and 0.95 mg/week of Pal and Alp, respectively), IP (1.3 and 1.7 mg/week, respectively) and IV + IP (2.0 and 2.6 mg/week, respectively), see Figure 14A.

The mice in the vehicle group were sacrificed on Day 14 post treatment due to excessive tumor growth (>1500 mm^3), while those treated with combined free drugs reached the tumor volume endpoint 2–3 weeks later. Tumor growth (compared to Day 0) is shown in Figure 14B. IV or IP administration led to a similar reduction (2.8-fold) in tumor growth compared to vehicle-treated mice, even though the mice received higher IP doses (2.6-fold). This observation is clinically important, as IP injections are simpler and can be self-administered. Alternating IV and IP administration resulted in a 4-fold reduction in tumor growth at Day 14 relative to the vehicle-treated group, whereas the free drugs demonstrated a 7-fold reduction in tumor growth (Figure 14B). Notably, P(RGD)/Pal,Alp (50%) administered IV + IP had a similar effect as the free drugs, although the doses were lower by 7-fold for Pal and 1.3-fold for Alp.

The high concentrations of free drugs required to achieve a significant inhibitory effect pose a main limitation due to treatment-related toxicity [59]. The most common side effects in patients treated with these drugs include neutropenia, diarrhea, leukopenia and fatigue [60,61]. We therefore evaluated the tolerability of the drug in each treatment group by monitoring the body weight and white/red blood cells (WBC/RBC) (Figure 14C). Treatment with free drugs was associated with a slight decrease in body weight, whereas P(RGD)/Pal,Alp (50%) produced a stable body weight [58]. Reductions in blood count parameters such as WBC and RBC, which are paramount in clinical assessment as they increase the risk of infections and anemia, respectively, were more pronounced with free drugs compared to P(RGD)/Pal,Alp (50%) (Figure 14C). Thus, the results suggest that delivering the combination of drugs by NCs is safer.

There are several studies on the combination of Pal and Alp [59,62], and a clinical study involving this potentially synergistic combination is ongoing [63]. However, to our knowledge, this was the first study investigating this combination using a delivery system that targets the tumor (Pal in combination with other drugs encapsulated in a nanocarrier) [64]. The study provides preclinical in vivo PDX evidence which supports the continued evaluation of P(RGD)Pal,Alp NCs.

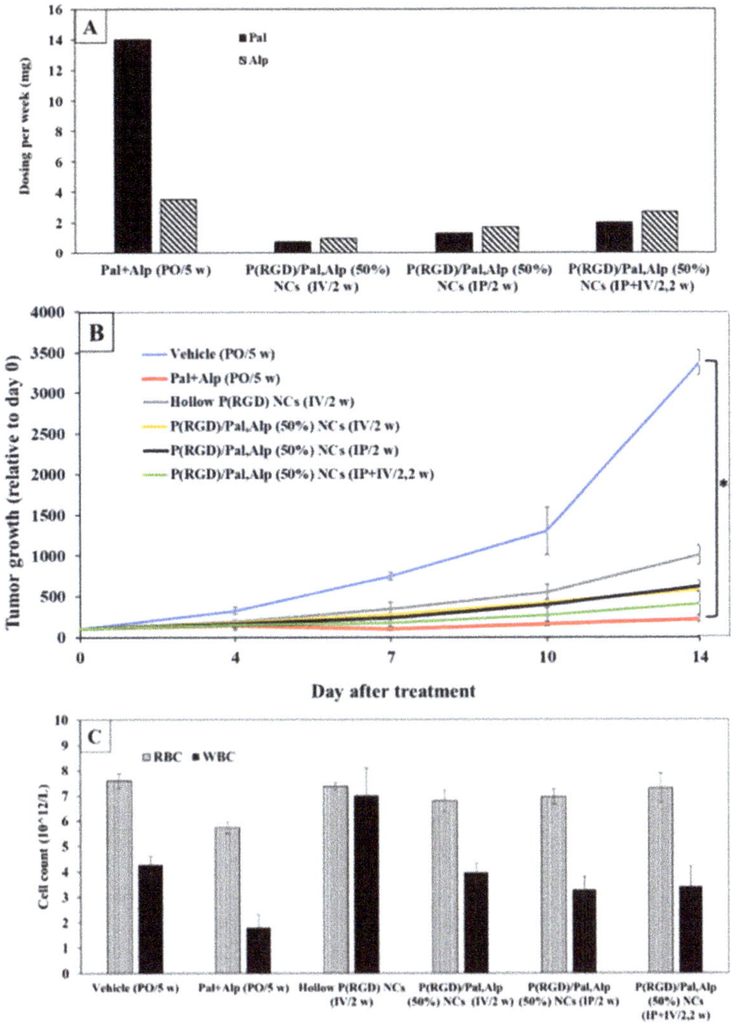

Figure 14. Treatment (14 days) by Pal + Alp and P(RGD) NCs inhibits tumor growth in a PDX colon cancer model (vehicle and drugs administered PO 5 days a week, NCs IP or IV, (**A**)). Tumor growth (**B**) and WBC/RBC at Day 14 (**C**). Data are means with standard errors. * $p < 0.005$ relative to control (5 mice per group) [57].

6. Summary

Thermal step-growth polymerization of suitable amino acids yields a uniform batch of protein-like high molecular weight polymers. Spherical NCs with narrow diameter distribution obtained by self-assembly may encapsulate a variety of molecules including drugs and/or imaging agents.

Proteinoid NCs have great advantages for biomedicine, including non-toxicity, biodegradability, biocompatibility and non-immunogenicity [1–4,10]. Recent in vitro and in vivo studies that are surveyed in this review report low cost and simple preparation of such NCs from proteinoid polymers (P(EF-PLLA), P(KRHF-PLLA) and P(RGD)) for cancer diagnostics, therapy and theranostics with reduced toxicity.

Amino acids and additives such as PLLA can be tailored for various applications, e.g., diagnostics and therapy. Introducing PLLA into the proteinoid backbone resulted in a proteinoid that was stable and augmented the hydrophobic inner region, leading to the formation of smaller, hollow NCs. Encapsulation of NIR ICG within NCs showed promising potential for in vivo diagnosis, due to low background auto-fluorescence and ability to penetrate deep into biomatrices. Another NC used for therapy is P(KRHF-PLLA) NCs which encapsulate Dox and show successful conjugation of PEG to improve their stability and prevent drug leakage. These studies led us to investigate theranostic NCs which co-encapsulate anti-cancer drugs and fluorescence dye as well as specific delivery to tumors by using RGD NCs [12,13].

P(RGD) can serve a dual purpose as a drug carrier through drug encapsulation and a targeting delivery system due to the presence of the RGD sequence on the proteinoid shell. This holds promise for reduced side effects [18]. The delivery system enables targeted transportation to the site of action, reducing the impact on essential tissues and minimizing unwanted side effects [18,57]. In addition, the system shields drugs from rapid degradation or clearance, amplifying their concentration in target tissues, and potentially allowing for lower dose [57].

The terminal amines on the surface of the P(RGD) NCs can be used for conjugation of bioactive compounds. Another potential therapeutic and targeted method was formed by conjugating TRAIL to hollow P(RGD) NCs. The use of Dox/P(RGD) and TRAIL-P(RGD) holds significant potential for targeted cancer therapy, including the possibility of dual therapy that combines the benefits of reduced side effects of Dox and the increased stability of TRAIL. Additionally, RGD proteinoids may have the potential to treat other medical conditions beyond cancer.

The simple way of obtaining NCs which are stable over time enables encapsulating various drugs and/or dyes. In vitro and in vivo experiments indicate that the NCs are capable of penetrating various types of cells, being taken up by different kinds of cells such as breast, colon, glioma and lung cancer cells, and even crossing the blood–brain barrier [9]. Proteinoid-based nanocarriers clearly hold great potential for diagnostic, therapeutic and theranostic applications towards cancer and other indications. Very recent in vivo work in our laboratories with mice illustrated that acidic proteinoid NCs are stable in the stomach, efficiently cross the blood–brain barrier (BBB) and can be administered orally for cancer targeting. Our main future efforts are to illustrate the use of these fluorescent and non-fluorescent NCs for brain cancer theranostics.

Author Contributions: Conceptualization, E.I., N.M., S.M.S. and S.M.; methodology, E.I. and S.M.; software, E.I.; validation, E.I. and N.M.; data curation, E.I. and N.M.; writing—original draft preparation, E.I., S.M. and Y.E.; writing—review and editing, E.I., S.M. and Y.E.; visualization, E.I. and Y.E.; supervision, S.M.S and S.M.; project administration, S.M. and S.M.S. All authors have read and agreed to the published version of the manuscript.

Funding: This research received no external funding.

Data Availability Statement: Not applicable.

Conflicts of Interest: The authors declare no conflict of interest.

References

1. Fox, S.W.; Jungck, J.R.; Nakashima, T. From proteinoid microsphere to contemporary cell: Formation of internucleotide and peptide bonds by proteinoid particles. *Orig. Life* **1974**, *5*, 227–237. [CrossRef]
2. Fox, S.W.; Mccauley, R.J.; Fukushim, T.; Windsor, C.R.; Montgome, P.O. Selective action in boundaries of particles of thermal proteinoid. *Fed. Proc.* **1967**, *26*, 749.
3. Fox, S.W. The proteinoid theory of the origin of life and competing ideas. *Am. Biol. Teach* **1974**, *36*, 161–172. [CrossRef]
4. Harada, K.; Fox, S.W. The thermal condensation of glutamic acid and glycine to linear peptides. *J. Am. Chem. Soc.* **1958**, *80*, 2694–2697. [CrossRef]
5. Matsuno, K. Electrical excitability of proteinoid microspheres composed of basic and acidic proteinoids. *Biosystems* **1984**, *17*, 11–14. [CrossRef]
6. Przybylski, A.T. Excitable cell made of thermal proteinoids. *Biosystems* **1985**, *17*, 281–288. [CrossRef]

7. Kumar, A.B.M.; Jayakumar, R.; Rao, P.K. Synthesis and aggregational behavior of acidic proteinoid. *Sci. Polym. Chem.* **1996**, *34*, 2915–2924. [CrossRef]
8. Kumar, A.B.M.; Rao, P.K. Preparation and characterization of pH-sensitive proteinoid microspheres for the oral delivery of methotrexate. *Biomaterials* **1998**, *19*, 725–732. [CrossRef]
9. Lugasi, L.; Grinberg, I.; Margel, S. Targeted delivery of CBD-loaded poly(RGD) proteinoid nanoparticles for antitumor therapy. *J. Nanomed. Nanotech.* **2020**, *18*, 552.
10. Kiel, S.; Kolitz-Domb, M.; Corem-Salkmon, E.; Margel, S. Engineered doxorubicin delivery system using proteinoid-poly(L-lactic acid) polymeric nanoparticles of narrow size distribution and high molecular weight for cancer treatment. *Int. J. Nanotechnol. Nanomed.* **2017**, *2*, 1–11.
11. Kolitz-Domb, M.; Margel, S. Recent advances of novel proteinoids and proteinoid nanoparticles and their applications in biomedicine and industrial uses. *Isr. J. Chem.* **2018**, *58*, 1277–1285. [CrossRef]
12. Hadad, E.; Rudnick-Glick, S.; Ithaki, E.; Avivi, M.Y.; Grinberg, I.; Elias, Y.; Margel, S. Engineering of doxorubicin-encapsulating and TRAIL-conjugated poly(RGD) proteinoid nanocapsules for drug delivery applications. *Polymers* **2020**, *12*, 2996. [CrossRef] [PubMed]
13. Hadad, E.; Rudnick-Glick, S.; Grinberg, I.; Kolitz-Domb, M.; Chill, J.H.; Margel, S. Synthesis and characterization of Poly(RGD) proteinoid polymers and NIR fluorescent nanoparticles of optimal D, L-configuration for drug-delivery applications— in vitro study. *ACS Omega* **2020**, *5*, 23568–23577. [CrossRef] [PubMed]
14. Kolitz-Domb, M.; Grinberg, I.; Corem-Salkmon, E.; Margel, S. Engineering of near infrared fluorescent proteinoid-poly(L-lactic acid) particles for in vivo colon cancer detection. *J. Nanobiotechnol.* **2014**, *12*, 30. [CrossRef] [PubMed]
15. Kwon, K.; Park, D.; Kim, J.C. Disulfide proteinoid micelles responsive to reduction. *J. Dispers. Sci. Technol.* **2018**, *40*, 1413–1422. [CrossRef]
16. Adamatzky, A. Towards proteinoid computers. hypothesis paper. *Biosystems* **2021**, *208*, 104480. [CrossRef]
17. Sharma, S.; Mougoyannis, P.; Tarabella, G.; Adamatzky, A. A review on the protocols for the synthesis of proteinoids. *arXiv* **2022**, arXiv:2212.02261.
18. Fox, S.W.; Harada, K. Thermal copolymerization of amino acids in the presence of phosphoric acid. *Arch. Biochem. Biophys.* **1960**, *86*, 281–285. [CrossRef]
19. Itzhaki, E.; Hadad, E.; Moskovits, N.; Stemmer, S.M.; Margel, S. Tumor-targeted fluorescent proteinoid nanocapsules encapsulating synergistic drugs for personalized cancer therapy. *Pharmaceuticals* **2021**, *6*, 648. [CrossRef]
20. Das, P.; Ganguly, S.; Margel, S.; Gedanken, A. Tailor made magnetic nanolights: Fabrication to cancer theranostics applications. *Nanoscale Adv.* **2021**, *3*, 6762. [CrossRef]
21. Van den Bos, J.; Wieringa, F.P.; Bouvy, N.D.; Stassen, L.P.S. Optimizing the image of fluorescence cholangiography using ICG: A systematic review and ex vivo experiments. *Surg. Endosc.* **2018**, *32*, 4820–4832. [CrossRef] [PubMed]
22. Bhavane, R.; Starosolski, Z.; Stupin, I.; Ghaghada, K.B.; Annapragada, A. NIR-II fluorescence imaging using indocyanine green nanoparticles. *Sci. Rep.* **2018**, *81*, 14455. [CrossRef] [PubMed]
23. Lwin, T.M.; Murakami, T.; Miyake, K.; Yazaki, P.J.; Shivley, J.E.; Hoffman, R.M.; Bouvet, M. Tumor-specific labeling of pancreatic cancer using a humanized anti-CEA antibody conjugated to a near-infrared fluorophore. *Ann. Surg. Oncol.* **2018**, *25*, 1079–1085. [CrossRef]
24. Makwana, V.; Karanjia, J.; Haselhorst, T.; Anoopkumar-Dukie, S.; Rudrawar, S. Liposomal doxorubicin as targeted delivery platform: Current trends in surface functionalization. *Int. J. Pharm.* **2021**, *593*, 120117. [CrossRef]
25. Neun, B.W.; Barenholz, Y.; Szebeni, J.; Dobrovolskaia, M.A. Understanding the role of anti-PEG antibodies in the complement activation by Doxil in vitro. *Molecules* **2018**, *23*, 1700. [CrossRef] [PubMed]
26. Bavli, Y.; Winkler, I.; Chen, B.M.; Roffler, S.; Cohen, R.; Szebeni, J.; Barenholz, Y. Doxebo (doxorubicin-free Doxil-like liposomes) is safe to use as a pre-treatment to prevent infusion reactions to PEGylated nanodrugs. *J. Control. Release* **2019**, *306*, 138–148. [CrossRef]
27. Pierschbacher, M.D.; Ruoslahti, E. Cell attachment activity of fibronectin can be duplicated by small synthetic fragments of the Molecule. *Nature* **1984**, *309*, 30–33. [CrossRef] [PubMed]
28. Temming, K.; Schiffelers, R.M.; Molema, G.; Kok, R.J. RGD-based strategies for selective delivery of therapeutics and imaging agents to the tumour vasculature. *Drug Resist. Updat.* **2005**, *8*, 381–402. [CrossRef]
29. Ruoslahti, E. RGD and other recognition sequences for integrins. *Annu. Rev. Cell Dev. Biol.* **1996**, *12*, 697–715. [CrossRef]
30. Ruoslahti, E.; Pierschbacher, M. New perspectives in cell adhesion: RGD and integrins. *Science* **1987**, *238*, 491–497. [CrossRef]
31. Asati, S.; Pandey, V.; Soni, V. RGD Peptide as a targeting moiety for theranostic purpose: An update study. *Int. J. Pept. Res. Ther.* **2018**, *25*, 49–65. [CrossRef]
32. Kieffer, B.; Mer, G.; Mann, A.; Lefèvre, J.F. Structural studies of two antiaggregant RGDW peptides by ^{1}H and ^{13}C NMR. *Int. J. Pept. Protein Res.* **1994**, *44*, 70–79. [CrossRef]
33. Zeng, C.; Shang, W.; Wang, K.; Chi, C.; Jia, X.; Feng, C.; Yang, D.; Ye, J.; Fang, C.; Tian, J. Intraoperative identification of liver cancer microfoci using a targeted near-infrared fluorescent probe for imaging-guided surgery. *Sci. Rep.* **2016**, *6*, 21959. [CrossRef]
34. Oddo, L.; Paradossi, G.; Cerroni, B.; Ben-Harush, C.; Ariel, E.; Di Meco, F.; Ram, Z.; Grossman, R. In vivo biodistribution of engineered lipid microbubbles in rodents. *ACS Omega* **2019**, *4*, 13371–13381. [CrossRef]

35. Xie, J.; Yan, C.; Yan, Y.; Chen, L.; Song, L.; Zang, F.; An, Y.; Teng, G.; Gu, N.; Zhang, Y. Multi-modal Mn–Zn ferrite nanocrystals for magnetically-induced cancer targeted hyperthermia: A comparison of passive and active targeting effects. *Nanoscale* **2016**, *8*, 16902–16915. [CrossRef] [PubMed]
36. Hadad, E.; Rudnick-Glick, S.; Grinberg, I.; Yehuda, R.; Margel, S. Engineering of NIR fluorescent PEGylated poly(RGD) proteinoid polymers and nanoparticles for drug delivery applications in chicken embryo and mouse models. *RSC Adv.* **2020**, *10*, 34364–34372. [CrossRef]
37. Bossion, A.; Heifferon, K.V.; Meabe, L.; Zivic, N.; Taton, D.; Hedrick, J.L.; Long, T.E.; Sardon, H. Opportunities for organocatalysis in polymer synthesis via step-growth methods. *Prog. Polym. Sci.* **2019**, *90*, 164–210. [CrossRef]
38. Kim, S.J.; Bae, P.K.; Chung, B.H. Self-assembled levan nanoparticles for targeted breast cancer imaging. *Chem. Commun.* **2015**, *51*, 107–110. [CrossRef] [PubMed]
39. Kim, T.H.; Chen, Y.; Mount, C.W.; Gombotz, W.R.; Li, X.; Pun, S.H. Evaluation of temperature-sensitive, indocyanine green-encapsulating micelles for noninvasive near-infrared tumor imaging. *Pharm. Res.* **2010**, *27*, 1900–1913. [CrossRef]
40. Nowak-Sliwinska, P.; Alitalo, K.; Allen, E.; Anisimov, A.; Aplin, A.C.; Auerbach, R.; Augustin, H.G.; Bates, D.O.; van Beijnum, J.R.; Bender, R.H.F.; et al. Consensus guidelines for the use and interpretation of angiogenesis assays. *Angiogenes* **2018**, *213*, 425–532. [CrossRef]
41. Mezu-Ndubuisi, O.J.; Maheshwari, A. The role of integrins in inflammation and angiogenesis. *Pediatr. Res.* **2020**, *897*, 1619–1626. [CrossRef]
42. Liu, Y.; Ran, R.; Chen, J.; Kuang, Q.; Tang, J.; Mei, L.; Zhang, Q.; Gao, H.; Zhang, Z.; He, Q. Paclitaxel loaded liposomes decorated with a multifunctional tandem peptide for glioma targeting. *Biomaterials* **2014**, *35*, 4835–4847. [CrossRef]
43. Yin, H.Q.; Mai, D.S.; Gan, F.; Chen, X.J. One-step synthesis of linear and cyclic RGD conjugated gold nanoparticles for tumour targeting and imaging. *RSC Adv.* **2014**, *4*, 9078–9085. [CrossRef]
44. Baek, A.; Kim, Y.; Lee, J.W.; Lee, S.C.; Cho, S.R. Effect of polydeoxyribonucleotide on angiogenesis and wound healing in an in vitro model of osteoarthritis. *Cell Transplant.* **2018**, *27*, 1623–1633. [CrossRef]
45. Kazunori, K.; Glenn, S.; Masayuki, Y.; Teruo, O.; Yasuhisa, S. Block copolymer micelles as vehicles for drug delivery. *J. Control. Release* **1993**, *24*, 119–132. [CrossRef]
46. Aggarwal, P.; Hall, J.B.; McLeland, C.B.; Dobrovolskaia, M.A.; McNeil, S.E. Nanoparticle interaction with plasma proteins as it relates to particle biodistribution, biocompatibility and therapeutic efficacy. *Adv. Drug Deliv. Rev.* **2009**, *61*, 428. [CrossRef]
47. Peng, C.; Zheng, L.; Chen, Q.; Shen, M.; Guo, R.; Wang, H.; Cao, X.; Zhang, G.; Shi, X. PEGylated dendrimer-entrapped gold nanoparticles for in vivo blood pool and tumor imaging by computed tomography. *Biomaterials* **2012**, *33*, 1107–1119. [CrossRef]
48. Klymchenko, A.S.; Liu, F.; Collot, M.; Anton, N. Dye-loaded nanoemulsions: Biomimetic fluorescent nanocarriers for bioimaging and nanomedicine. *Adv. Healthc. Mater.* **2021**, *10*, 2001289. [CrossRef]
49. Thi, T.T.H.; Pilkington, E.H.; Nguyen, D.H.; Lee, J.S.; Park, K.D.; Truong, N.P. The importance of poly(ethylene glycol) alternatives for overcoming PEG immunogenicity in drug delivery and bioconjugation. *Polymers* **2020**, *12*, 298.
50. Proulx, S.T.; Luciani, P.; Derzsi, S.; Rinderknecht, M.; Mumprecht, V.; Leroux, J.C.; Detmar, M. Quantitative imaging of lymphatic function with liposomal indocyanine green. *Cancer Res.* **2010**, *70*, 7053–7062. [CrossRef]
51. Kue, C.S.; Tan, K.Y.; Lam, M.L.; Lee, H.B. Chick embryo chorioallantoic membrane (CAM): An alternative predictive model in acute toxicological studies for anti-cancer drugs. *Exp. Anim.* **2015**, *64*, 129. [CrossRef]
52. Ge, Z.; Chen, Q.; Osada, K.; Liu, X.; Tockary, T.A.; Uchida, S.; Dirisala, A.; Ishii, T.; Nomoto, T.; Toh, K.; et al. Targeted gene delivery by polyplex micelles with crowded PEG palisade and cRGD moiety for systemic treatment of pancreatic tumors. *Biomaterials* **2014**, *35*, 3416–3426. [CrossRef]
53. Vu, M.N.; Kelly, H.G.; Wheatley, A.K.; Peng, S.; Pilkington, E.H.; Veldhuis, N.A.; Davis, T.P.; Kent, S.J.; Truong, N.P. Cellular interactions of liposomes and PISA nanoparticles during human blood flow in a microvascular network. *Small* **2020**, *16*, 2002861. [CrossRef] [PubMed]
54. Zhao, J.; Lu, H.; Wong, S.; Lu, M.; Xiao, P.; Stenzel, M.H. Influence of nanoparticle shapes on cellular uptake of paclitaxel loaded nanoparticles in 2D and 3D cancer models. *Polym. Chem.* **2017**, *8*, 3317–3326. [CrossRef]
55. Martin-Banderas, L.; Muñoz-Rubio, I.; Prados, J.; Álvarez-Fuentes, J.; Calderón-Montaño, J.M.; López-Lázaro, M.; Arias, J.L.; Leiva, M.C.; Holgado, M.A.; Fernández-Arévalo, M. In vitro and in vivo evaluation of Δ9-tetrahidrocannabinol/PLGA nanoparticles for cancer chemotherapy. *Int. J. Pharm.* **2015**, *487*, 205–212. [CrossRef]
56. Kruyt, F.A.E. TRAIL and cancer therapy. *Cancer Lett.* **2008**, *263*, 14–25. [CrossRef] [PubMed]
57. Itzhaki, E.; Chausky-Barzakh, E.; Atkins, A.; Bareket-Samish, A.; Stemmer, S.M.; Moskovits, N.; Margel, S. Tumor-targeted poly(ArgGlyAsp) nanocapsules for personalized cancer therapy-In vivo study. *Adv. Ther.* **2023**, 1–34. [CrossRef]
58. Turner, P.V.; Brabb, T.; Pekow, C.; Vasbinder, M.A. Administration of substances to laboratory animals: Routes of administration and factors to consider. *J. Am. Assoc. Lab. Anim. Sci.* **2011**, *50*, 600. [PubMed]
59. Nur Husna, S.M.; Tan, H.-T.T.; Mohamud, R.; Dyhl-Polk, A.; Wong, K.K. Inhibitors targeting CDK4/6, PARP and PI3K in breast cancer: A review. *Ther. Adv. Med. Oncol.* **2018**, *10*, 1–21. [CrossRef] [PubMed]
60. Sarkisian, S.; Markosian, C.; Ali, Z.; Rizvi, M. Palbociclib-induced pneumonitis: A case report and review of the literature. *Cureus* **2020**, *12*, e8929. [CrossRef]
61. Anonymous. *Alpelisib Palbociclib HPLC Report*; Faculty of Life Sciences, Bar-Ilan University: Ramat Gan, Israel, 2019; pp. 6–7, 18–19.

52. Bardia, A.; Chandarlapaty, S.; Linden, H.M.; Ulaner, G.A.; Gosselin, A.; Cartot-Cotton, S.; Cohen, P.; Doroumian, S.; Paux, G.; Celanovic, M.; et al. AMEERA-1 phase 1/2 study of amcenestrant, SAR439859, in postmenopausal women with ER-positive/HER2-negative advanced breast cancer. *Nat. Commun.* **2022**, *13*, 4116. [CrossRef] [PubMed]
53. Ji, Y.; Liu, X.; Li, J.; Xie, X.; Huang, M.; Jiang, J.; Liao, Y.P.; Donahue, T.; Meng, H. Use of ratiometrically designed nanocarrier targeting CDK4/6 and autophagy pathways for effective pancreatic cancer treatment. *Nat. Commun.* **2020**, *11*, 4249. [CrossRef] [PubMed]
54. Moskovits, N.; Peretz, I.; Chausky, E.; Itzhaki, E.; Shmuel, N.; Meerson, R.; Tarasenko, N.; Kaufman, A.; Stemmer, A.; Yaffe, R.; et al. Palbociclib in combination with sunitinib exerts a synergistic anti-cancer effect in patient-derived xenograft models of various human cancers types. *Cancer Lett.* **2022**, *536*, 215665. [CrossRef] [PubMed]

Journal of
Functional Biomaterials

MDPI

Review

The Use of Specialized Pro-Resolving Mediators in Biomaterial-Based Immunomodulation

Ana Beatriz Sousa [1,2,3] and Judite N. Barbosa [1,2,3,*]

1 i3S—Instituto de Inovação e Investigação em Saúde, Universidade do Porto, Rua Alfredo Allen, 208, 4200-125 Porto, Portugal
2 INEB—Instituto de Engenharia Biomédica, Rua Alfredo Allen, 208, 4200-125 Porto, Portugal
3 ICBAS—Instituto de Ciências Biomédicas Abel Salazar, Universidade do Porto, Rua de Jorge Viterbo Ferreira 228, 4050-313 Porto, Portugal
* Correspondence: judite@ineb.up.pt; Tel.: +351-22-0428105

Abstract: The implantation of a biomaterial will lead to the immediate onset of an acute inflammatory response, which is of key importance in shaping the quality of the repair process. However, the return to homeostasis is critical to prevent a chronic inflammatory response that may impair the healing process. The resolution of the inflammatory response is now recognized as an active and highly regulated process, being described as specialized immunoresolvents that have a fundamental role in the termination of the acute inflammatory response. These mediators collectively coined as specialized pro-resolving mediators (SPMs) are a family of endogenous molecules that include lipoxins (Lx), resolvins (Rv), protectins (PD), maresins (Mar), Cysteinyl-SPMs (Cys-SPMs) and n-3 docosapentaenoic acid-derived SPMs (n-3 DPA-derived SPMs). SPMs have important anti-inflammatory and pro-resolutive actions such as decreasing the recruitment of polymorphonuclear leukocytes (PMNs), inducing the recruitment of anti-inflammatory macrophages, and increasing macrophage clearance of apoptotic cells through a process known as efferocytosis. Over the last years, the trend in biomaterials research has shifted towards the engineering of materials that are able to modulate the inflammatory response and thus stimulate appropriate immune responses, the so-called immunomodulatory biomaterials. These materials should be able to modulate the host immune response with the aim of creating a pro-regenerative microenvironment. In this review, we explore the potential of using of SPMs in the development of new immunomodulatory biomaterials and we propose insights for future research in this field.

Keywords: acute inflammation; resolution of inflammation; specialized pro-resolution mediators (SPMs); immunomodulatory biomaterials

Citation: Sousa, A.B.; Barbosa, J.N. The Use of Specialized Pro-Resolving Mediators in Biomaterial-Based Immunomodulation. *J. Funct. Biomater.* **2023**, *14*, 223. https://doi.org/10.3390/jfb14040223

Academic Editor: Christie Ying Kei Lung

Received: 23 February 2023
Revised: 12 April 2023
Accepted: 13 April 2023
Published: 15 April 2023

1. The Acute Inflammatory Response to Biomaterials and Its Resolution

Implanted biomaterials will be recognized by the host immune system as a foreign body, triggering an immediate set of biological responses, leading ultimately to the development of an inflammatory and fibrotic reaction that is known as the Foreign Body Reaction (FBR) that presents a major challenge in the development and application of biomaterials and medical devices [1–3].

The FBR can be subdivided into different and sequential phases, starting with interactions between blood and the biomaterial leading to protein adsorption at the surface of the implant, followed by an acute inflammatory response that if persistent will lead to a chronic inflammation that will ultimately cause the formation of a fibrous capsule. We will herein focus on the onset of the inflammatory response, the acute inflammatory reaction as well as on its resolution.

The acute inflammatory reaction is of short duration and typically lasts from minutes to a few days. There are available in the literature extensive reviews on the different phases of the inflammatory response to biomaterials [1,4–8]. We will address general aspects of the acute response. The acute inflammatory response is mainly characterized by the formation of edema, through the exudation of fluid and proteins from plasma, together with the migration of leukocytes to the implant site [1]. The implantation of a biomaterial causes injury in the cells surrounding the implant site. This will lead to two different effects. Firstly and almost immediately, blood proteins adsorb to the biomaterial surface causing the activation of the coagulation cascade, the triggering of the complement system, and the activation and recruitment of inflammatory cells. Secondly, the injured cells will have an important role in the further recruitment of inflammatory cells through the release of DAMPs (Danger-associated molecular patterns), also known as danger signals or alarmins [8,9]. Polymorphonuclear leukocytes (PMNs) are the predominant inflammatory cell type in an acute response and the first ones to arrive at the implant site. PMNs will secrete ROS (Reactive oxygen species) and proteolytic enzymes that can corrode the surface of the biomaterial. Upon activation, PMNs also secrete interleukin (IL)-8 that will activate and recruit additional PMNs to the implant site. PMNs will also act as chemoattractants for monocytes and macrophages and for immature dendritic cells and lymphocytes through the production of the chemokines MCP-1 (Monocyte chemoattractant protein)-1 and MIP (Macrophage inflammatory protein)-1β. After two days, PMNs will disappear from the implant site, and the macrophage will be the dominant inflammatory cell [6,8,10]. At the implant site, macrophages can either polarize into M1 classically activated macrophages that will produce pro-inflammatory cytokines and release enzymes and ROS in an attempt to degrade the material or can polarize into M2 alternatively activated macrophages that will synthesize anti-inflammatory cytokines and growth factors that will facilitate tissue repair [11,12]. It is important to clarify that the M1/M2 macrophage phenotypic polarization is in fact a continuum of activation between these two phenotypes [12]. Additionally, M2 macrophages can be divided into different sub-groups that express different surface markers and have different associated functions: M2a, M2b, M2c, M2d, and $M2_{eff}$. Among these sub-groups, M2a macrophages are the most associated with tissue repair and regeneration and interestingly $M2_{eff}$ macrophages are induced by efferocytosis, a process that is prompted by SPMs as described in this review [13].

The termination of an inflammatory response was considered for many years to be due to the passive disappearance of the pro-inflammatory stimuli. It is now widely accepted that the resolution of an inflammatory response is an active and endogenous process that has the key role of preventing excessive and /or continued inflammation that would lead to a chronic state [14]. After an acute inflammatory response, the goal is to return to homeostasis in a complex process that involves a large number of molecules and cells of the immune system.

During an acute inflammatory response, cyclooxygenase-2 (COX-2) increases and leads to the production of pro-inflammatory mediators, such as leukotrienes (LTs) and prostaglandins (PGs) from arachidonic acid in phospholipidic membranes. Either these pro-inflammatory mediators can perpetuate the inflammatory response leading to a chronic state, or through the prostaglandin subtypes, PGE2 and PGD2 a change in the production of arachidonic acid-derived molecules is stimulated and anti-inflammatory mediators will be produced [15].

Therefore, the resolution of an inflammatory response can be explained in a simplistic way as a molecular switch of the lipid mediators from the production of pro-inflammatory mediators, such as LTs and PGs, to anti-inflammatory mediators originating from the same precursor. These anti-inflammatory mediators are collectively called specialized pro-resolving mediators (SPMs) [16].

SPMs are a group of lipidic molecules that have the ability to reduce inflammation through the activation of resolution pathways. In 1984 Serhan et al. [17] reported the isolation of the first SPM, lipoxin. Currently, this group of molecules, listed in Table 1, includes lipoxins (Lx), resolvins (Rv), protectins (PD), maresins (Mar), Cysteinyl-SPMs (Cys-SPMs) and n-3 docosapentaenoic acid (DPA)-derived SPMs (n-3 DPA-derived SPMs) [18]. SPMs are synthesized in the local inflammatory microenvironment at the onset of acute inflammation to compensate for the various pro-inflammatory mediators. Inflammation is a natural and important response of the host to injury and therefore SPMs will not block the process. When the immune system has performed its purpose SPMs will down-regulate the inflammatory response and ultimately lead to its resolution [19–21].

Table 1. Specialized pro-resolving mediators (SPMs) family.

SPM(s)	Subtypes	Precursor
Lipoxins	LxA4 and LxB4	Arachidonic acid
Resolvins	E-series Resolvins (RvE1–RvE4) D-series Resolvins (RvD1–RvD6)	Eicosapentaenoic acid Docosahexaenoic acid
Protectins	Protectin D1/Neuroprotectin D1 PDX	Docosahexaenoic acid
Maresins	Mar1 and Mar2	Docosahexaenoic acid
Cysteinyl-SPMs	MCTR1–MCTR3 PCTR1–PCTR3 RCTR1–RCTR3	Docosahexaenoic acid
n-3 DPA-derived SPMs	(RvD1, RvD2, RvD5)$_{n\text{-}3DPA}$ (PD1, PD2)$_{n\text{-}3DPA}$ (Mar1–Mar3)$_{n\text{-}3DPA}$ RvT1–RvT4	n-3 Docosapentaenoic acid

A generalized decrease of pro-inflammatory cytokines together with the phagocytosis of apoptotic PMNs and the removal of inflammatory debris are indicators of the resolution of the inflammatory response. SPMs will also facilitate tissue repair and regeneration through the decrease of pro-inflammatory mediators. When the resolution of an inflammatory response fails fibrosis and chronic inflammation will take place [22,23].

2. Mediators of the Resolution of the Inflammatory Response

Specialized pro-resolving mediators (SPMs) have a dual role since they have the ability to contain an acute inflammatory response and promote its resolution. SPMs are produced from dietary ω-3 polyunsaturated fatty acids (PUFAs) through the action of different enzymes. PUFAs are fundamental components of membrane phospholipids, and among other functions, they act as a precursor pool for lipid mediators since their main components include eicosapentaenoic acid (EPA) and docosahexaenoic acid (DHA) [24]. The different subtypes of SPMs are derived from arachidonic acid, EPA, DHA, and n-3 docosapentaenoic acid (DPA) as presented in Figure 1 [25–27].

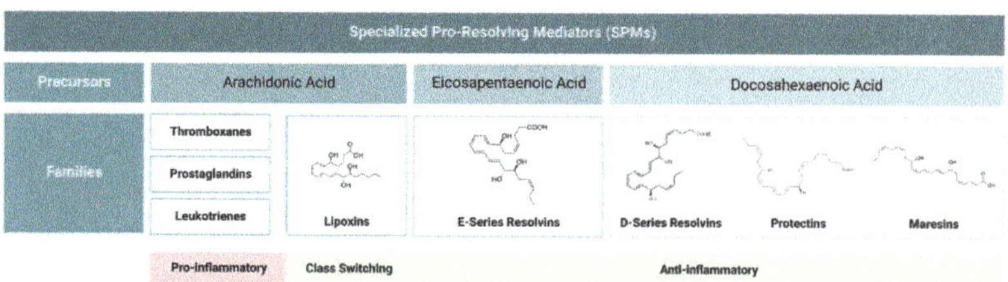

Figure 1. Specialized pro-resolving lipid mediators (SPMs) are produced from essential fatty acids. In the initiation of an inflammatory response, pro-inflammatory mediators such as prostaglandins and leukotrienes will induce the recruitment of inflammatory cells and the production of pro-inflammatory mediators. Afterward, a switch to the production of lipoxins will lead to the onset of the resolution of the inflammatory response.

2.1. Lipoxins (Lx)

The designation lipoxin is an acronym for "lipoxygenase interaction products". There are two different Lx: LxA4 and LxB4. Lx can be synthesized by the lipoxygenation of arachidonic acid in epithelial cells, PMNs, and monocytes or through platelet/leukocyte interactions [19,28]. Lx is able to inhibit PMNs recruitment and activation by reducing vascular permeability; they also cause the non-phlogistic recruitment of monocytes, without degranulation or release of ROS, allowing the phagocytosis of apoptotic inflammatory cells as PMNs [29–33]. Interestingly, Lx also induces an increased release of transforming growth factor β1 (TGF-β1) by macrophages together with a decrease in IL-8 and MCP-1 [34]. Lx also has an effect on macrophage polarization from an M1 pro-inflammatory phenotype toward an M2 anti-inflammatory phenotype [31]. It has also been suggested that Lx can have anti-fibrotic properties [35].

2.2. Resolvins (Rv)

The label resolvins derive from "resolution-phase interaction products". There are D-series resolvins (RvD1–RvD6) and E-series resolvins (RvE1–RvE4). Rv participates in different functions in the process of inflammation resolution as the regulation of the production of ROS and of different cytokines; the decrease of PMNs recruitment; promote the phagocytosis of apoptotic PMNs and promote debris clearance. Taken together, these effects will decrease the intensity of the response and facilitate tissue repair [29,36]. Additionally, it has been described that RvD1 promotes macrophage polarization to M2 anti-inflammatory phenotype [36,37] and that RvD4 decreases the release of neutrophil extracellular traps (NETs) [38].

2.3. Protectins (PD)

The name protectins were proposed for their general anti-inflammatory and protective actions. Protectin D1 (PD1) or neuroprotectin D1 (NPD1) and protectin X (PDX) were identified. In the case of PD1, if it is of neural origin, the designation NPD1 is used whereas the term PD1 is used to indicate its peripheral actions [39]. PD1/NPD1 decreases PMNs influx and strengthen the removal of apoptotic PMNs by macrophages [40]. NPD1 exhibits neuroprotective effects in the retina and in the central nervous system [41]. It has been reported that PD1 promotes M2 macrophage polarization and inhibits macrophage TNF (tumor necrosis factor)-α expression [42]. It is described that PDX has general anti-inflammatory and pro-resolving effects, and regulates inflammatory cell infiltration, namely PMNs [43]. Additionally, PDX causes a decrease in the expression of MIP-2 and MCP-1 in macrophages [44]

2.4. Maresins (Mar)

The term maresins come from "macrophage mediators in resolving inflammation". Mar is produced by macrophages and has homeostatic functions. Mar1 and Mar2 have been identified. Maresins inhibit PMN influx and ROS production [45] and are powerful stimulators of macrophage efferocytosis leading to the clearance of apoptotic PMNs by macrophages [41,46]. Mar promotes the polarization of macrophages to an M2-phenotype [47]. Mar-1 activates leucine-rich repeat-containing G protein-coupled receptor 6 (LGR6), which is present in different tissues and promotes tissue repair and regeneration [48]. Additionally, Mar stimulated tissue regeneration in a planaria model [49].

2.5. Cysteinyl-SPMs (Cys-SPMs)

A new group of peptide-lipid conjugated SPMs was recently identified. This group includes three types of molecules: the resolvin conjugates in tissue regeneration (RCTRs); the protectin conjugates in tissue regeneration (PCTRs) and the maresin conjugates in tissue regeneration (MCTRs). It is reported that this group of SPMs presents anti-inflammatory and pro-resolving effects as well as regenerative properties [50]. It has also been reported its ability to stimulate tissue repair in a planaria model [51]. MCTRs have the ability to induce macrophage efferocytosis and to modulate macrophage polarization towards ar M2 phenotype [25].

2.6. n-3 DPA-Derived SPMS

This new group of SPMs is derived from n-3 docosapentaenoic acid (DPA) and is gen erally designated as n-3 DPA-Derived SPMs. This group includes $RvD_{n\text{-}3DPA}$, $MaR_{n\text{-}3DPA}$ $PD_{n\text{-}3DPA}$, and 13-series resolvins (RvTs) [39]. The n-3 DPA-derived SPMs also have impor tant anti-inflammatory and pro-resolutive actions [27,52].

Taking together the actions of the different SPMs described above, we can conclude that they exert rather interesting effects on the cells of the immune system (Figure 2) SPMs influence inflammatory cell recruitment and activation; regulate inflammatory cell apoptosis and have an effect on immune cell polarization. It is therefore of key importance to explore possible applications of these molecules in a therapeutic context to ensure an effective return to tissue homeostasis and to prevent fibrosis after an inflammatory response [24]. We will follow this review by exploring the potential of SPMs application in different medical conditions and in the field of biomaterial-based immunomodulation.

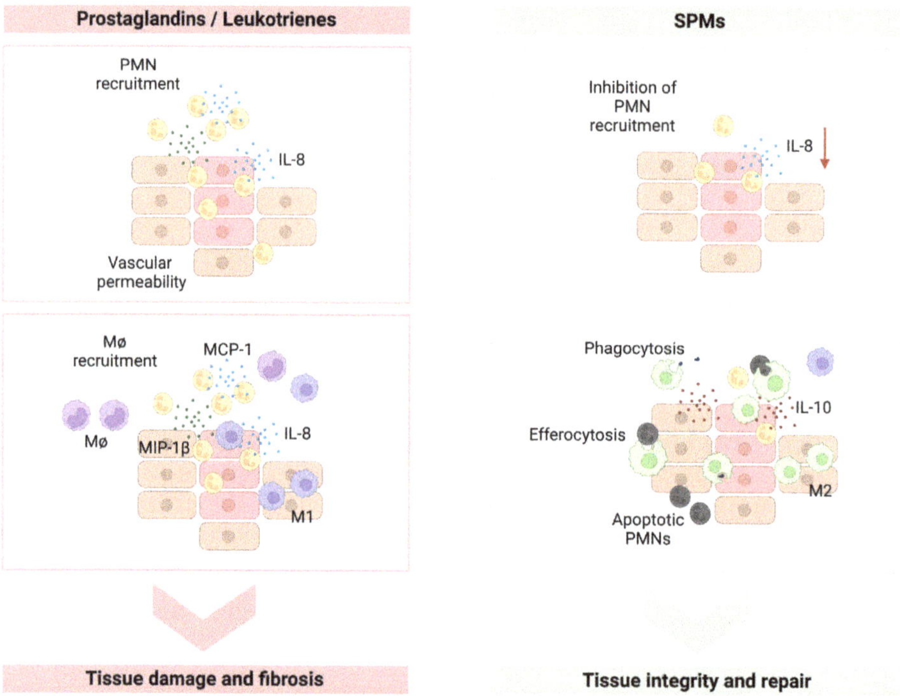

Figure 2. Specialized pro-resolving lipid mediators (SPMs) main actions. In an inflammatory response prostaglandins and leukotrienes will induce the recruitment of polymorphonuclear leukocytes (PMNs) that will produce high levels of IL (interleukin)-8 causing further recruitment of PMNs. PMNs will also produce MCP-1 (monocyte chemoattractant protein-1) and MIP (macrophage inflammatory protein)-1β causing the recruitment of monocytes/macrophages (Mφ), that will polarize into M1 pro-inflammatory macrophages, causing tissue damage. SPMs will cause a decrease in PMN recruitment, a decrease in the production of IL-8, and an increase in IL-10 secretion. SPMs will also lead to macrophage polarization towards an M2 anti-inflammatory phenotype and will induce macrophage efferocytosis. SPMs will cause the return to homeostasis and tissue repair.

3. The Potential of SPMs in Medical Applications

Deficiencies in the resolution of inflammatory responses may lead to the development of chronic immune disorders. Furthermore, SPMs have an important potential in preventing and treating chronic inflammation and immune conditions [53]. The inflammatory response is in essence a protective reaction. Nevertheless, when exacerbated or when perpetuated in time can lead to tissue or organ damage or disease. The paradigm that inflammatory conditions occur due to the excessive production of pro-inflammatory mediators has changed. In fact, there is growing evidence that an inaccurate resolution of the inflammatory reaction, or an imbalance between pro-inflammatory mediators and SPMs are key factors in the emergence and advance of inflammatory conditions and several human diseases [54]. Therefore, the use of SPMs in the treatment of immune disorders appears as a rather promising approach.

Reports in the literature suggest that a deficient activity or production of SPMs is related to the pathogenesis of different inflammatory disorders. For example, an imbalance between pro-inflammatory mediators and SPMs was observed in inflammatory bowel disease [55]. In an evaluation of the lipidome of cerebrospinal fluid of Alzheimer's disease patients, a decrease in the levels of SPMs was observed [56]. In the respiratory tracts of patients with chronic obstructive pulmonary disease or asthma, lower levels of LxA4 and PD1 were detected in comparison with healthy individuals [57]. It is also described that low serum levels of SPMs can be related to the development of hypertension [58]. Interestingly, it has also been suggested that SPMs are interesting candidates in the prevention of the cytokine storm in severe acute respiratory syndrome coronavirus 2 (SARS-CoV-2) patients [59].

Due to the growing evidence of the important role of SPMs, researchers have already started to explore the use of several different pro-resolving agents to treat inflammatory diseases. In an animal model of periodontal disease, the topical application of RvD1 in the tissues leads to an improvement in the condition of the disease causing a decrease in bone loss, and a decrease in the infiltration of PMNs [20,60]. In atherosclerosis, the process of macrophage efferocytosis is important to mitigate plaque progression. Since SPMs are known to induce macrophage efferocytosis, they appear as potential candidates to treat atherosclerotic plaques. In fact, RvD1, RvD2, and Mar1 were used in an animal model with established atherosclerosis and induced the decrease of several markers of plaque progression [61,62]. In an animal model of angiotensin-induced cardiac inflammation, the administration of RvD1 was demonstrated to have cardioprotective effects inhibiting cardiac remodeling and hypertension [63]. In rheumatoid arthritis, characterized by persistent joint inflammation, the administration of RvD1 was investigated using an animal model of this disease. The results obtained suggest that RvD1 led to a reduction of joint inflammation and consequently cartilage protection [64]. In an Alzheimer's disease murine model, the administration of combined LxA4 and RvD1 leads to a decrease in inflammation [65]. Also, in an in vivo rat model of Parkinson's disease, RvD2 caused a recovery of neuronal damage through the decreased expression of pro-inflammatory mediators [66,67]. It is currently accepted that inflammation is an important factor in the pathology of depressive disorders and that anti-inflammatory agents may have anti-depressive effects. It has been recently reviewed that intracranial infusions of different resolvins in rodent models of depression caused anti-depressive effects [68].

The use of SPMs is already being evaluated in clinical trials. The use of RvE1 is being assessed for topical application in ocular lesions [69]. The application of LxA4 analogs is under study for the treatment of infantile eczema and for the treatment of asthma [70,71].

This section clearly illustrates that the potential for SPMs application is vast. We have depicted some interesting examples to elucidate the possible range of applications.

4. SPMs in Biomaterial-Based Immunomodulation

The advent of tissue engineering and regenerative medicine led to the development of new biomaterials with the capability of enhancing tissue repair and regeneration. The recognition of the key importance of the immune system in tissue healing moved researchers to the development of a new class of biomaterials that present the ability to modulate inflammatory responses—the immunomodulatory biomaterials. Ideally, these biomaterials will control the host immune response to provide a favorable microenvironment able to promote tissue repair [72,73]. Immunomodulatory biomaterials have an interesting potential to modulate the FBR and therefore promote implant integration and function [74]. The majority of the immunomodulatory biomaterial-based strategies are focused on macrophages because they are highly plastic cells that can exhibit different phenotypes ranging from pro-inflammatory to anti-inflammatory [75,76]. Additionally, macrophages are important coordinators of tissue repair and regeneration [77].

Different strategies can be used to develop biomaterials with immunomodulatory properties such as (i) Changing the chemical properties of the material; (ii) tuning the physical characteristics of the biomaterial; (iii) incorporating bioactive molecules in the biomaterial either pro- or anti-inflammatory molecules or growth factors; (iv) combining biomaterials with cell therapies and (v) develop biomaterials derived from the extracellular matrix (ECM) of tissues or using ECM components [73,78–81].

Biomaterial-based immunomodulation through the use of SPMs is in our opinion a promising tool for the development of new and powerful immunomodulatory biomaterials The acknowledgment of the active nature of the resolution of an inflammatory response, together with the understanding of the important effects of SPMs provides a new paradigm to address inflammatory conditions, improve implant integration and facilitate the return to tissue homeostasis [82], as represented in Figure 3. There are already some interesting and encouraging studies reported in the literature that combine the application of different types of SPMs with biomaterials, which we describe in the following paragraphs and are summarized in Table 2.

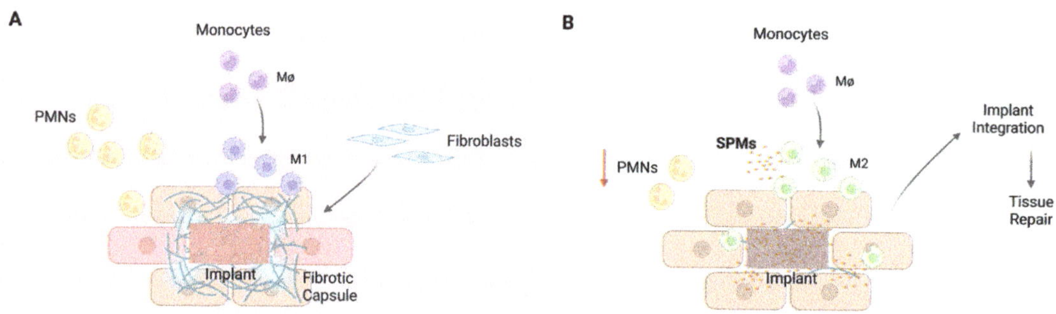

Figure 3. Foreign body response to biomaterials. (**A**) Implanted biomaterials will be considered a foreign body by the host immune system. Briefly, polymorphonuclear leukocytes (PMNs) will be the first immune cells to arrive at the implant site, followed by monocytes/macrophages that will polarize towards a pro-inflammatory M1 phenotype. Fibroblasts will be activated and will secrete collagen leading to the formation of a fibrous capsule around the implanted biomaterial. The implant will be isolated from the surrounding tissues and its function will be impaired. (**B**) An immunomodulatory biomaterial, for example, loaded with specialized pro-resolving lipid mediators (SPMs), will cause a decrease in inflammatory cell recruitment and will lead to macrophage polarization towards an M2 anti-inflammatory phenotype. This will enable implant integration as well as tissue repair.

Regarding the use of lipoxins, Wang et al. [83] have developed a polyisocyanopeptide hydrogel loaded with the antimicrobial agent doxycycline and LxA4 for periodontal treatment. When applied in a dog model, this hydrogel leads to the decrease of sub-gingival bacterial loading and to the decrease in the levels of the pro-inflammatory cytokine IL-8. Reis et al. [84] produced poly-lactic-co-glycolic acid (PLGA) microparticles loaded with LxA4 for the treatment of skin ulcers. When tested in a rat model of skin lesion the microparticles caused a reduction of inflammatory cell numbers, promoted the closure of the wound, reduced the levels of IL-1β and TNF-α, and increased the levels of TGF-β. Sun et al. [85] developed calcium carbonate nanoparticles loaded with a LxA4 agonist for the treatment of endometriosis. They observed an enhancement in macrophage efferocytosis as well as a decrease in the levels of pro-inflammatory cytokines.

We have used LxA4 and RvD1 to dampen the inflammatory response to implanted 3D chitosan (Ch) scaffolds [86]. After implanting the Ch scaffolds in a rodent air-pouch model, we performed local administration of either LxA4 or RvD1 every 12 h for 3 days. We were able to observe a shift in the macrophage polarization profile towards M2, as well as a general decrease in the levels of pro-inflammatory cytokines. Afterward, we incorporated RvD1 in the Ch scaffolds [87]. This biomaterial was evaluated in a rodent air-pouch model and caused a decrease in the recruitment of inflammatory cells together with higher numbers of M2 macrophages and lower numbers of M1 macrophages and a general decrease in the levels of pro-inflammatory cytokines. We have also tested the scaffolds in a rat femoral critical size defect model and concluded that RvD1 contributed to the formation of new bone with improved trabecular thickness [88]. Some other authors have explored the use of resolvins in biomaterial-based immunomodulation. Sok et al. [89] developed a poly(ethylene) glycol (PEG)-based hydrogel to release aspirin-triggered (AT) RvD1 and IL-10 in a murine model of dorsal skinfold window chamber. They observed an enhancement in the recruitment of anti-inflammatory myeloid cells with the combined delivery of AT-RvD1 and IL-10. In another study using the same in vivo model [90], the authors tested PLGA scaffolds loaded with AT-RvD1 and reported a decrease in the recruitment of PMN leukocytes together with an increase of M2 macrophages. Shi et al. [91] developed a polycaprolactone vascular graft loaded with AT-RvD1, and in an in vivo model where they performed the replacement of the rat abdominal aorta by a vascular graft, concluded that the presence of AT-RvD1 in the graft decreased PMN recruitment and activation and caused an increase in the number of M2 macrophages. Yin et al. [92] with the aim of improving bone tissue repair through the control of inflammation during the healing process developed a gold nanocage loaded with RvD1 to induce M2 macrophage polarization. They were able to observe an increase in macrophage polarization towards M2 phenotype both in vitro using a macrophage cell line and in vivo in a mouse femoral defect model. Lu et al. [93] designed a catechol-chitosan hydrogel incorporated with acetalized cyclodextrin nanoparticles loaded with RvE1 for the treatment of chronic wounds. When tested in vitro with a macrophage cell line an increase in the production of the anti-inflammatory cytokine IL-10 was observed. In a rat wound healing model, the hydrogel stimulated the closure of the wound. Jiang et al. [94] investigated the effect of RvD1 in bone regeneration using a rat calvarial defect model. They produced collagen scaffolds impregnated with a pluronic hydrogel with and without the incorporation of RvD1. The results obtained suggest that RvD1 improved bone formation and vascularization. The authors suggest that this effect was due to the ability of RvD1 to control the inflammatory microenvironment although they did not study the effects on immune cells or on immune mediators.

As to the use of maresins, Miranda et al. [95] produced polylactic acid nanoparticles incorporated with Mar2 for intestinal mucosal wound repair. They observed in a biopsy-wounding model that the administration of the nanoparticles promoted mucosal repair and reduced the number of infiltrating immune cells.

We are certain that this is just the first group of studies combining the use of biomaterials and SPMs. The promising and interesting results yielded in these reports will cause increased attention in this area of research. Additionally, there are yet several subtypes of SPMs with powerful anti-inflammatory and pro-resolutive actions that remain to be explored.

Table 2. SPMs are used in biomaterial-based immunomodulation.

Specialized Pro-Resolving Mediators (SPMs)	Biomaterial	Immunomodulatory Outcome	Reference
LxA4	Polyisocyanopeptide hydrogel	Decreased levels of IL-8	[82]
LxA4	Poly-lactic-co-glycolic acid microparticles	Reduced numbers of inflammatory cells Reduced levels of IL-1β and TNF-α Increased levels of TGF-β	[83]
LxA4 agonist	Calcium carbonate nanoparticles	Enhanced macrophages efferocytosis Decreased levels of pro-inflammatory cytokines	[84]
RvD1	Chitosan scaffolds	Higher numbers of M2 macrophages Lower numbers of M1 macrophages Decrease in pro-inflammatory cytokines	[86]
RvD1	Poly(ethylene) glycol-based hydrogel	Decreased recruitment of PMN leukocytes Increased numbers of M2 macrophages	[88]
RvD1	Polycaprolactone vascular graft	Decreased PMN recruitment Increased numbers of M2 macrophages	[89]
RvD1	Gold nanocage	Increased numbers of M2 macrophages	[91]
RvE1	Catechol-chitosan hydrogel	Increased production of IL-10	[92]
Mar2	Polylactic acid nanoparticles	Reduced numbers of infiltrating immune cells	[94]

SPMs are challenging to work with because they are unstable molecules and have a diminutive biological half-life that was observed both in in vitro and in in vivo studies. One possibility to overcome this limitation is the development of more stable analogs. It is reported in the literature for example the development of an oxidation-resistant metabolic stable analog of the n-3 DPA-derived protectin D1 [96]; the production of lipoxin A4 stable analogs [97]; and also of analogs for resolvin D1 [98]. Alternatively, in a previous work of ours with chitosan scaffolds incorporated with RvD1 [87] we used lyophilization which resulted in higher RvD1 activity since this procedure is a lipid stabilizer and is reported to increase the stability of different pharmaceutical formulations. In a different study conducted by de Prinse et al. [99], another solution was explored, SPMs-loaded micelles were developed to address the challenge of the long-term use of these molecules. These micelles were revealed to be non-cytotoxic and their potential use in vivo must now be explored.

5. Conclusions and Future Perspectives

SPMs have an essential role in the resolution of an inflammatory acute reaction and in the return to homeostasis. SPMs have important modulatory effects in different cells of the immune system, and although their actions are broad, it is possible to summarize in three key functions the main effects of this family of pro-resolutive molecules:

(i). SPMs can inhibit the recruitment and activation of immune cells. These molecules have an important role for example in the recruitment of PMNs to the inflammatory site and can decrease the production of pro-inflammatory cytokines by inflammatory cells.

(ii). SPMs have an important effect on the modulation of the polarization of immune cells. For example, these molecules induce the polarization of macrophages towards an M2 anti-inflammatory phenotype.

(iii). SPMs regulate macrophage phagocytic capacity, leading to increased elimination of apoptotic immune cells by macrophages in a process called efferocytosis.

Taking into consideration these important effects, we believe that SPMs represent a rather promising strategy to stop an inflammatory response and mitigate the potential undesired side effects such as tissue damage or fibrosis. Moreover, SPMs seem to be advantageous in comparison with other anti-inflammatory or immunosuppressive molecules because they will promote the resolution of an inflammatory reaction without the unwanted immunosuppressive side effect. Additionally, SPMs will not suppress the positive effects of an inflammatory reaction in terms of host response against pathogens and in the promotion of tissue repair and return to homeostasis.

The potential action of SPMs must be thoroughly evaluated. In our opinion, research efforts should be made to clarify several different aspects. In the following part of the manuscript, we list some important research lines that are in our opinion worth exploring:

1. Extend the studies already carried out to all SPMs and address possible new applications: SPMs represent a large family of lipidic molecules, and many of them were not tested in biomaterial-based applications. In Section 4, where we performed a review of the SPMs used for biomaterial-based immunomodulation we were only able to find research works using LxA4, RvD1, RvE1, and Mar2. Many other SPMs still need to be explored.
2. Explore the potential synergic effect of SPMs: The combination of more than one SPM within a biomaterial could have interesting synergic effects. In Section 3, we discuss the use of combined LxA4 and RvD1 in an animal model of Alzheimer's disease that caused a significant decrease in the inflammatory response. Therefore, the combination of different SPMs should be addressed.
3. Investigate the potential of protectins: The application of protectins in the development of new functional biomaterials is clearly underexplored. More specifically, neuroprotectins that have been identified as having neuroprotective effects could represent an advance in the development of biomaterials for neuroscience applications.
4. Understand the regenerative capacity of the cysteinyl-SPMs: The recently identified subclass of SPMs, the cysteinyl-SPMs, includes the resolvin, protectin, and maresin conjugates in tissue regeneration. It is reported that this group of molecules besides having pro-resolutive actions have regenerative properties. Since we are seeking ideal biomaterials for tissue engineering and regenerative medicine, these molecules appear as ideal candidates for the development of new biomaterials with immunomodulatory properties.

A comprehensive understanding of the biological actions of SPMs will provide key information for the regulation of inflammation, its resolution, and the return to homeostasis. We are certain that the application of SPMs in the development of new biomaterials will be a milestone in the field of immuno-engineering.

Author Contributions: A.B.S.: Writing—Original draft, image creation; J.N.B.: Writing—Original draft, Review & Editing. All authors have read and agreed to the published version of the manuscript.

Funding: This research was funded by Norte Regional Operational Program-Structured R&D&I Projects—Unorte.pt. HEALTH-UNORTE; NORTE-01-0145-FEDER-000039.

Data Availability Statement: Not applicable.

Conflicts of Interest: The authors have no conflict of interest to declare.

Abbreviations

FBR: Foreign Body Reaction; PMNs: Polymorphonuclear leukocytes; DAMPs: Danger-associated molecular patterns; ROS: Reactive oxygen species; IL: interleukin; MCP: Monocyte chemoattractant protein; MIP: Macrophage inflammatory protein; COX-2: cyclooxygenase-2; LTs: Leukotrienes; PGs: Prostaglandins; SPMs: Specialized pro-resolving mediators; Lx: Lipoxins; Rv: Resolvins; PD: Protectins; Mar: Maresins; Cys-SPMs: Cysteinyl- Specialized pro-resolving mediators; DPA: Docosapentaenoic acid; PUFAs: Polyunsaturated fatty acids; EPA: Eicosapentaenoic acid; DHA: Docosahexaenoic acid; TGF: transforming growth factor; NETs: Neutrophil extracellular traps; NPD:

Neuroprotectin; TNF: tumor necrosis factor; RCTRs: Resolvin conjugates in tissue regeneration; PCTRs: Protectin conjugates in tissue regeneration; MCTRs: Maresin conjugates in tissue regeneration SARS-CoV 2: Severe acute respiratory syndrome coronavirus 2; PLGA: Poly-lactic-co-glycolic acid Ch: Chitosan; PEG: Poly(ethylene) glycol; AT: Aspirin-triggered.

References

1. Anderson, J.M. Biological responses to materials. *Annu. Rev. Mater. Res.* **2001**, *31*, 81–110. [CrossRef]
2. Carnicer-Lombarte, A.; Chen, S.T.; Malliaras, G.G.; Barone, D.G. Foreign Body Reaction to Implanted Biomaterials and Its Impact in Nerve Neuroprosthetics. *Front. Bioeng. Biotechnol.* **2021**, *9*, 622524. [CrossRef]
3. Barone, D.G.; Carnicer-Lombarte, A.; Tourlomousis, P.; Hamilton, R.S.; Prater, M.; Rutz, A.L.; Dimov, I.B.; Malliaras, G.G.; Lacour S.P.; Robertson, A.A.B.; et al. Prevention of the foreign body response to implantable medical devices by inflammasome inhibition *Proc. Natl. Acad. Sci. USA* **2022**, *119*, e2115857119. [CrossRef]
4. Anderson, J.M. Inflammatory response to implants. *ASAIO Trans.* **1988**, *34*, 101–107. [CrossRef]
5. Anderson, J.M.; Rodriguez, A.; Chang, D.T. Foreign body reaction to biomaterials. *Semin. Immunol.* **2008**, *20*, 86–100. [CrossRef]
6. Christo, S.N.; Diener, K.R.; Bachhuka, A.; Vasilev, K.; Hayball, J.D. Innate Immunity and Biomaterials at the Nexus: Friends or Foes. *Biomed. Res. Int.* **2015**, *2015*, 342304. [CrossRef]
7. Badylak, S.E.; Gilbert, T.W. Immune response to biologic scaffold materials. *Semin. Immunol.* **2008**, *20*, 109–116. [CrossRef]
8. Franz, S.; Rammelt, S.; Scharnweber, D.; Simon, J.C. Immune responses to implants—A review of the implications for the design of immunomodulatory biomaterials. *Biomaterials* **2011**, *32*, 6692–6709. [CrossRef]
9. Bianchi, M.E. DAMPs, PAMPs and alarmins: All we need to know about danger. *J. Leukoc. Biol.* **2007**, *81*, 1–5. [CrossRef]
10. Anderson, J.M.; McNally, A.K. Biocompatibility of implants: Lymphocyte/macrophage interactions. *Semin. Immunopathol.* **2011**, *33*, 221–233. [CrossRef]
11. Brown, B.N.; Badylak, S.F. Expanded applications, shifting paradigms and an improved understanding of host-biomaterial interactions. *Acta Biomater.* **2013**, *9*, 4948–4955. [CrossRef]
12. Brown, B.N.; Ratner, B.D.; Goodman, S.B.; Amar, S.; Badylak, S.F. Macrophage polarization: An opportunity for improved outcomes in and regenerative medicine. *Biomaterials* **2012**, *33*, 3792–3802. [CrossRef] [PubMed]
13. Martin, K.E.; Garcia, A.J. Macrophage phenotypes in tissue repair and the foreign body response: Implications for biomaterial-based regenerative medicine strategies. *Acta Biomater.* **2021**, *133*, 4–16. [CrossRef] [PubMed]
14. Panigrahy, D.; Gilligan, M.M.; Serhan, C.N.; Kashfi, K. Resolution of inflammation: An organizing principle in biology and medicine. *Pharmacol. Ther.* **2021**, *227*, 107879. [CrossRef] [PubMed]
15. Widgerow, A.D. Cellular resolution of inflammation—Catabasis. *Wound Repair. Regen.* **2012**, *20*, 2–7. [CrossRef]
16. Serhan, C.N.; Gupta, S.K.; Perretti, M.; Godson, C.; Brennan, E.; Li, Y.; Soehnlein, O.; Shimizu, T.; Werz, O.; Chiurchiu, V.; et al. The Atlas of Inflammation Resolution (AIR). *Mol. Asp. Med.* **2020**, *74*, 100894. [CrossRef]
17. Serhan, C.N.; Hamberg, M.; Samuelsson, B. Lipoxins: Novel series of biologically active compounds formed from arachidonic acid in human leukocytes. *Proc. Natl. Acad. Sci. USA* **1984**, *81*, 5335–5339. [CrossRef]
18. Feehan, K.T.; Gilroy, D.W. Is Resolution the End of Inflammation? *Trends Mol. Med.* **2019**, *25*, 198–214. [CrossRef]
19. Maderna, P.; Godson, C. Lipoxins: Resolutionary road. *Br. J. Pharmacol.* **2009**, *158*, 947–959. [CrossRef]
20. Freire, M.O.; Van Dyke, T.E. Natural resolution of inflammation. *Periodontology 2000* **2013**, *63*, 149–164. [CrossRef]
21. Serhan, C.N.; Brain, S.D.; Buckley, C.D.; Gilroy, D.W.; Haslett, C.; O'Neill, L.A.; Perretti, M.; Rossi, A.G.; Wallace, J.L. Resolution of inflammation: State of the art, definitions and terms. *FASEB J.* **2007**, *21*, 325–332. [CrossRef]
22. Serhan, C.N. Pro-resolving lipid mediators are leads for resolution physiology. *Nature* **2014**, *510*, 92–101. [CrossRef]
23. Serhan, C.N.; Chiang, N.; Dalli, J. The resolution code of acute inflammation: Novel pro-resolving lipid mediators in resolution. *Semin. Immunol.* **2015**, *27*, 200–215. [CrossRef]
24. Pan, G.; Zhang, P.; Yang, J.; Wu, Y. The regulatory effect of specialized pro-resolving mediators on immune cells. *Biomed. Pharm.* **2022**, *156*, 113980. [CrossRef]
25. Julliard, W.A.; Myo, Y.P.A.; Perelas, A.; Jackson, P.D.; Thatcher, T.H.; Sime, P.J. Specialized pro-resolving mediators as modulators of immune responses. *Semin. Immunol.* **2022**, *59*, 101605. [CrossRef]
26. Serhan, C.N.; Savill, J. Resolution of inflammation: The beginning programs the end. *Nat. Immunol.* **2005**, *6*, 1191–1197. [CrossRef]
27. Vik, A.; Dalli, J.; Hansen, T.V. Recent advances in the chemistry and biology of anti-inflammatory and specialized pro-resolving mediators biosynthesized from n-3 docosapentaenoic acid. *Bioorg. Med. Chem. Lett.* **2017**, *27*, 2259–2266. [CrossRef]
28. Serhan, C.N.; Clish, C.B.; Brannon, J.; Colgan, S.P.; Chiang, N.; Gronert, K. Novel functional sets of lipid-derived mediators with antiinflammatory actions generated from omega-3 fatty acids via cyclooxygenase 2-nonsteroidal antiinflammatory drugs and transcellular processing. *J. Exp. Med.* **2000**, *192*, 1197–1204. [CrossRef]
29. Serhan, C.N. Resolution phase of inflammation: Novel endogenous anti-inflammatory and proresolving lipid mediators and pathways. *Annu. Rev. Immunol.* **2007**, *25*, 101–137. [CrossRef]
30. Romano, M. Lipoxin and Aspirin-Triggered Lipoxins. *Sci. World J.* **2010**, *10*, 1048–1064. [CrossRef]

31. Mitchell, S.; Thomas, G.; Harvey, K.; Cottell, D.; Reville, K.; Berlasconi, G.; Petasis, N.A.; Erwig, L.; Rees, A.J.; Savill, J.; et al. Lipoxins, aspirin-triggered epi-lipoxins, lipoxin stable analogues, and the resolution of inflammation: Stimulation of macrophage phagocytosis of apoptotic neutrophils in vivo. *J. Am. Soc. Nephrol.* **2002**, *13*, 2497–2507. [CrossRef]
32. Lee, T.H.; Horton, C.E.; Kyan-Aung, U.; Haskard, D.; Crea, A.E.; Spur, B.W. Lipoxin A4 and lipoxin B4 inhibit chemotactic responses of human neutrophils stimulated by leukotriene B4 and N-formyl-L-methionyl-L-leucyl-L-phenylalanine. *Clin. Sci.* **1989**, *77*, 195–203. [CrossRef]
33. Maddox, J.F.; Hachicha, M.; Takano, T.; Petasis, N.A.; Fokin, V.V.; Serhan, C.N. Lipoxin A4 stable analogs are potent mimetics that stimulate human monocytes and THP-1 cells via a G-protein-linked lipoxin A4 receptor. *J. Biol. Chem.* **1997**, *272*, 6972–6978. [CrossRef]
34. Godson, C.; Mitchell, S.; Harvey, K.; Petasis, N.A.; Hogg, N.; Brady, H.R. Cutting edge: Lipoxins rapidly stimulate nonphlogistic phagocytosis of apoptotic neutrophils by monocyte-derived macrophages. *J. Immunol.* **2000**, *164*, 1663–1667. [CrossRef]
35. Rodgers, K.; McMahon, B.; Mitchell, D.; Sadlier, D.; Godson, C. Lipoxin A4 modifies platelet-derived growth factor-induced pro-fibrotic gene expression in human renal mesangial cells. *Am. J. Pathol.* **2005**, *167*, 683–694. [CrossRef]
36. Titos, E.; Rius, B.; Gonzalez-Periz, A.; Lopez-Vicario, C.; Moran-Salvador, E.; Martinez-Clemente, M.; Arroyo, V.; Claria, J. Resolvin D1 and its precursor docosahexaenoic acid promote resolution of adipose tissue inflammation by eliciting macrophage polarization toward an M2-like phenotype. *J. Immunol.* **2011**, *187*, 5408–5418. [CrossRef]
37. Hsiao, H.M.; Sapinoro, R.E.; Thatcher, T.H.; Croasdell, A.; Levy, E.P.; Fulton, R.A.; Olsen, K.C.; Pollock, S.J.; Serhan, C.N.; Phipps, R.P.; et al. A novel anti-inflammatory and pro-resolving role for resolvin D1 in acute cigarette smoke-induced lung inflammation. *PLoS ONE* **2013**, *8*, e58258. [CrossRef]
38. Cherpokova, D.; Jouvene, C.C.; Libreros, S.; DeRoo, E.P.; Chu, L.; de la Rosa, X.; Norris, P.C.; Wagner, D.D.; Serhan, C.N. Resolvin D4 attenuates the severity of pathological thrombosis in mice. *Blood* **2019**, *134*, 1458–1468. [CrossRef]
39. Chiang, N.; Serhan, C.N. Specialized pro-resolving mediator network: An update on production and actions. *Essays Biochem.* **2020**, *64*, 443–462.
40. Ariel, A.; Serhan, C.N. Resolvins and protectins in the termination program of acute inflammation. *Trends Immunol.* **2007**, *28*, 176–183. [CrossRef]
41. Serhan, C.N.; Dalli, J.; Colas, R.A.; Winkler, J.W.; Chiang, N. Protectins and maresins: New pro-resolving families of mediators in acute inflammation and resolution bioactive metabolome. *Biochim. Biophys. Acta* **2015**, *1851*, 397–413. [CrossRef]
42. Xia, H.; Chen, L.; Liu, H.; Sun, Z.; Yang, W.; Yang, Y.; Cui, S.; Li, S.; Wang, Y.; Song, L.; et al. Protectin DX increases survival in a mouse model of sepsis by ameliorating inflammation and modulating macrophage phenotype. *Sci. Rep.* **2017**, *7*, 99. [CrossRef]
43. Hu, X.; Zhang, Y.A.; Chen, B.; Jin, Z.; Lin, M.L.; Li, M.; Mei, H.X.; Lu, J.C.; Gong, Y.Q.; Jin, S.W.; et al. Protectin DX promotes the inflammatory resolution via activating COX-2/L-PGDS-PGD(2) and DP(1) receptor in acute respiratory distress syndrome. *Int. Immunopharmacol.* **2022**, *102*, 108348. [CrossRef]
44. Ye, Y.; Zhang, H.W.; Mei, H.X.; Xu, H.R.; Xiang, S.Y.; Yang, Q.; Zheng, S.X.; Smith, F.G.; Jin, S.W.; Wang, Q. PDX regulates inflammatory cell infiltration via resident macrophage in LPS-induced lung injury. *J. Cell. Mol. Med.* **2020**, *24*, 10604–10614. [CrossRef]
45. Marcon, R.; Bento, A.F.; Dutra, R.C.; Bicca, M.A.; Leite, D.F.; Calixto, J.B. Maresin 1, a proresolving lipid mediator derived from omega-3 polyunsaturated fatty acids, exerts protective actions in murine models of colitis. *J. Immunol.* **2013**, *191*, 4288–4298. [CrossRef]
46. Serhan, C.N.; Yang, R.; Martinod, K.; Kasuga, K.; Pillai, P.S.; Porter, T.F.; Oh, S.F.; Spite, M. Maresins: Novel macrophage mediators with potent antiinflammatory and proresolving actions. *J. Exp. Med.* **2009**, *206*, 15–23. [CrossRef]
47. Daniel, B.; Nagy, G.; Czimmerer, Z.; Horvath, A.; Hammers, D.W.; Cuaranta-Monroy, I.; Poliska, S.; Tzerpos, P.; Kolostyak, Z.; Hays, T.T.; et al. The Nuclear Receptor PPARgamma Controls Progressive Macrophage Polarization as a Ligand-Insensitive Epigenomic Ratchet of Transcriptional Memory. *Immunity* **2018**, *49*, 615–626.e6. [CrossRef]
48. Chiang, N.; Libreros, S.; Norris, P.C.; de la Rosa, X.; Serhan, C.N. Maresin 1 activates LGR6 receptor promoting phagocyte immunoresolvent functions. *J. Clin. Investig.* **2019**, *129*, 5294–5311. [CrossRef]
49. Serhan, C.N.; Dalli, J.; Karamnov, S.; Choi, A.; Park, C.K.; Xu, Z.Z.; Ji, R.R.; Zhu, M.; Petasis, N.A. Macrophage proresolving mediator maresin 1 stimulates tissue regeneration and controls pain. *FASEB J.* **2012**, *26*, 1755–1765. [CrossRef]
50. Dalli, J.; Colas, R.A.; Serhan, C.N. Novel n-3 immunoresolvents: Structures and actions. *Sci. Rep.* **2013**, *3*, 1940. [CrossRef]
51. Dyall, S.C.; Balas, L.; Bazan, N.G.; Brenna, J.T.; Chiang, N.; da Costa Souza, F.; Dalli, J.; Durand, T.; Galano, J.M.; Lein, P.J.; et al. Polyunsaturated fatty acids and fatty acid-derived lipid mediators: Recent advances in the understanding of their biosynthesis, structures, and functions. *Prog. Lipid Res.* **2022**, *86*, 101165. [CrossRef]
52. Hansen, T.V.; Dalli, J.; Serhan, C.N. The novel lipid mediator PD1(n-3 DPA): An overview of the structural elucidation, synthesis, biosynthesis and bioactions. *Prostaglandins Other Lipid Mediat.* **2017**, *133*, 103–110. [CrossRef]
53. Perretti, M.; Dalli, J. Resolution Pharmacology: Focus on Pro-Resolving Annexin A1 and Lipid Mediators for Therapeutic Innovation in Inflammation. *Annu. Rev. Pharmacol. Toxicol.* **2023**, *63*, 449–469. [CrossRef]
54. Bento, A.F.; Claudino, R.F.; Dutra, R.C.; Marcon, R.; Calixto, J.B. Omega-3 fatty acid-derived mediators 17(R)-hydroxy docosahexaenoic acid, aspirin-triggered resolvin D1 and resolvin D2 prevent experimental colitis in mice. *J. Immunol.* **2011**, *187*, 1957–1969. [CrossRef]

55. Do, K.V.; Hjorth, E.; Wang, Y.; Jun, B.; Kautzmann, M.I.; Ohshima, M.; Eriksdotter, M.; Schultzberg, M.; Bazan, N.G. Cerebrospinal Fluid Profile of Lipid Mediators in Alzheimer's Disease. *Cell. Mol. Neurobiol.* **2023**, *43*, 797–811. [CrossRef]
56. Levy, B.D.; Serhan, C.N. Resolution of acute inflammation in the lung. *Annu. Rev. Physiol.* **2014**, *76*, 467–492. [CrossRef]
57. Yucel, H.; Ozdemir, A.T. Low LXA4, RvD1 and RvE1 levels may be an indicator of the development of hypertension. *Prostaglandins Leukot. Essent. Fat. Acids* **2021**, *174*, 102365. [CrossRef]
58. Lee, C.H. Role of specialized pro-resolving lipid mediators and their receptors in virus infection: A promising therapeutic strategy for SARS-CoV-2 cytokine storm. *Arch. Pharm. Res.* **2021**, *44*, 84–98. [CrossRef]
59. Mustafa, M.; Zarrough, A.; Bolstad, A.I.; Lygre, H.; Mustafa, K.; Hasturk, H.; Serhan, C.; Kantarci, A.; Van Dyke, T.E. Resolvin D1 protects periodontal ligament. *Am. J. Physiol. Cell. Physiol.* **2013**, *305*, C673–C679. [CrossRef]
60. Fredman, G.; Tabas, I. Boosting Inflammation Resolution in Atherosclerosis: The Next Frontier for Therapy. *Am. J. Pathol.* **2017**, *187*, 1211–1221. [CrossRef]
61. Viola, J.R.; Lemnitzer, P.; Jansen, Y.; Csaba, G.; Winter, C.; Neideck, C.; Silvestre-Roig, C.; Dittmar, G.; Doring, Y.; Drechsler, M.; et al. Resolving Lipid Mediators Maresin 1 and Resolvin D2 Prevent Atheroprogression in Mice. *Circ. Res.* **2016**, *119*, 1030–1038. [CrossRef] [PubMed]
62. Olivares-Silva, F.; De Gregorio, N.; Espitia-Corredor, J.; Espinoza, C.; Vivar, R.; Silva, D.; Osorio, J.M.; Lavandero, S.; Peiro, C.; Sanchez-Ferrer, C.; et al. Resolvin-D1 attenuation of angiotensin II-induced cardiac inflammation in mice is associated with prevention of cardiac remodeling and hypertension. *Biochim. Biophys. Acta Mol. Basis Dis.* **2021**, *1867*, 166241. [CrossRef] [PubMed]
63. Norling, L.V.; Headland, S.E.; Dalli, J.; Arnardottir, H.H.; Haworth, O.; Jones, H.R.; Irimia, D.; Serhan, C.N.; Perretti, M. Proresolving and cartilage-protective actions of resolvin D1 in inflammatory arthritis. *JCI Insight* **2016**, *1*, e85922. [CrossRef] [PubMed]
64. Kantarci, A.; Aytan, N.; Palaska, I.; Stephens, D.; Crabtree, L.; Benincasa, C.; Jenkins, B.G.; Carreras, I.; Dedeoglu, A. Combined administration of resolvin E1 and lipoxin A4 resolves inflammation in a murine model of Alzheimer's disease. *Exp. Neurol.* **2018**, *300*, 111–120. [CrossRef] [PubMed]
65. Tian, Y.; Zhang, Y.; Zhang, R.; Qiao, S.; Fan, J. Resolvin D2 recovers neural injury by suppressing inflammatory mediators expression in lipopolysaccharide-induced Parkinson's disease rat model. *Biochem. Biophys. Res. Commun.* **2015**, *460*, 799–805. [CrossRef] [PubMed]
66. Valente, M.; Dentoni, M.; Bellizzi, F.; Kuris, F.; Gigli, G.L. Specialized Pro-Resolving Mediators in Neuroinflammation: Overview of Studies and Perspectives of Clinical Applications. *Molecules* **2022**, *27*, 4836. [CrossRef]
67. Deyama, S.; Kaneda, K.; Minami, M. Resolution of depression: Antidepressant actions of resolvins. *Neurosci. Res.* **2022**. [CrossRef]
68. Basil, M.C.; Levy, B.D. Specialized pro-resolving mediators: Endogenous regulators of infection and inflammation. *Nat. Rev. Immunol.* **2016**, *16*, 51–67. [CrossRef]
69. Wu, S.H.; Chen, X.Q.; Liu, B.; Wu, H.J.; Dong, L. Efficacy and safety of 15(R/S)-methyl-lipoxin A(4) in topical treatment of infantile eczema. *Br. J. Dermatol.* **2013**, *168*, 172–178. [CrossRef]
70. Levy, B.D.; Kohli, P.; Gotlinger, K.; Haworth, O.; Hong, S.; Kazani, S.; Israel, E.; Haley, K.J.; Serhan, C.N. Protectin D1 is generated in asthma and dampens airway inflammation and hyperresponsiveness. *J. Immunol.* **2007**, *178*, 496–502. [CrossRef]
71. Sadtler, K.; Estrellas, K.; Allen, B.W.; Wolf, M.T.; Fan, H.N.; Tam, A.J.; Patel, C.H.; Luber, B.S.; Wang, H.; Wagner, K.R.; et al. Developing a pro-regenerative biomaterial scaffold microenvironment requires T helper 2 cells. *Science* **2016**, *352*, 366–370. [CrossRef] [PubMed]
72. Vishwakarma, A.; Bhise, N.S.; Evangelista, M.B.; Rouwkema, J.; Dokmeci, M.R.; Ghaemmaghami, A.M.; Vrana, N.E.; Khademhosseini, A. Engineering Immunomodulatory Biomaterials To Tune the Inflammatory Response. *Trends Biotechnol.* **2016**, *34*, 470–482. [CrossRef] [PubMed]
73. Kharbikar, B.N.; Chendke, G.S.; Desai, T.A. Modulating the foreign body response of implants for diabetes treatment. *Adv. Drug Deliv. Rev.* **2021**, *174*, 87–113. [CrossRef] [PubMed]
74. Mantovani, A.; Biswas, S.K.; Galdiero, M.R.; Sica, A.; Locati, M. Macrophage plasticity and polarization in tissue repair and remodelling. *J. Pathol.* **2013**, *229*, 176–185. [CrossRef]
75. Sica, A.; Mantovani, A. Macrophage plasticity and polarization: In vivo veritas. *J. Clin. Investig.* **2012**, *122*, 787–795. [CrossRef]
76. Wynn, T.A.; Vannella, K.M. Macrophages in Tissue Repair, Regeneration, and Fibrosis. *Immunity* **2016**, *44*, 450–462. [CrossRef] [PubMed]
77. Salthouse, D.; Novakovic, K.; Hilkens, C.M.U.; Ferreira, A.M. Interplay between biomaterials and the immune system: Challenges and opportunities in regenerative medicine. *Acta Biomater.* **2023**, *155*, 1–18. [CrossRef]
78. Rowley, A.T.; Nagalla, R.R.; Wang, S.W.; Liu, W.F. Extracellular Matrix-Based Strategies for Immunomodulatory Biomaterials Engineering. *Adv. Healthc. Mater.* **2019**, *8*, e1801578. [CrossRef]
79. Chen, Y.; Sun, W.; Tang, H.; Li, Y.; Li, C.; Wang, L.; Chen, J.; Lin, W.; Li, S.; Fan, Z.; et al. Interactions Between Immunomodulatory Biomaterials and Immune Microenvironment: Cues for Immunomodulation Strategies in Tissue Repair. *Front. Bioeng. Biotechnol.* **2022**, *10*, 820940. [CrossRef]
80. Lynch, R.I.; Lavelle, E.C. Immuno-modulatory biomaterials as anti-inflammatory therapeutics. *Biochem. Pharmacol.* **2022**, *197*, 114890. [CrossRef]

81. Serhan, C.N.; Petasis, N.A. Resolvins and protectins in inflammation resolution. *Chem. Rev.* **2011**, *111*, 5922–5943. [CrossRef] [PubMed]
82. Wang, B.; Booij-Vrieling, H.E.; Bronkhorst, E.M.; Shao, J.; Kouwer, P.H.J.; Jansen, J.A.; Walboomers, X.F.; Yang, F. Antimicrobial and anti-inflammatory thermo-reversible hydrogel for periodontal delivery. *Acta Biomater.* **2020**, *116*, 259–267. [CrossRef] [PubMed]
83. Reis, M.B.; Pereira, P.A.T.; Caetano, G.F.; Leite, M.N.; Galvao, A.F.; Paula-Silva, F.W.G.; Frade, M.A.C.; Faccioli, L.H. Lipoxin A4 encapsulated in PLGA microparticles accelerates wound healing of skin ulcers. *PLoS ONE* **2017**, *12*, e0182381. [CrossRef] [PubMed]
84. Sun, Q.; Lei, Y.; Zhang, H.; Ding, X.; Yang, M.; Zhang, T.; Chen, J.; Huang, Z.; Wang, L.; Lan, J.; et al. A multifunctional nanoparticle for efferocytosis and pro-resolving-mediated endometriosis therapy. *Colloids Surf. B Biointerfaces* **2022**, *220*, 112893. [CrossRef]
85. Vasconcelos, D.P.; Costa, M.; Amaral, I.F.; Barbosa, M.A.; Aguas, A.P.; Barbosa, J.N. Modulation of the inflammatory response to chitosan through M2 macrophage polarization using pro-resolution mediators. *Biomaterials* **2015**, *37*, 116–123. [CrossRef]
86. Vasconcelos, D.P.; Costa, M.; Amaral, I.F.; Barbosa, M.A.; Aguas, A.P.; Barbosa, J.N. Development of an immunomodulatory biomaterial: Using resolvin D1 to modulate inflammation. *Biomaterials* **2015**, *53*, 566–573. [CrossRef]
87. Vasconcelos, D.P.; Costa, M.; Neves, N.; Teixeira, J.H.; Vasconcelos, D.M.; Santos, S.G.; Aguas, A.P.; Barbosa, M.A.; Barbosa, J.N. Chitosan porous 3D scaffolds embedded with resolvin D1 to improve in vivo bone healing. *J. Biomed. Mater. Res. Part A* **2018**, *106*, 1626–1633. [CrossRef]
88. Sok, M.C.P.; Baker, N.; McClain, C.; Lim, H.S.; Turner, T.; Hymel, L.; Ogle, M.; Olingy, C.; Palacios, J.I.; Garcia, J.R.; et al. Dual delivery of IL-10 and AT-RvD1 from PEG hydrogels polarize immune cells towards pro-regenerative phenotypes. *Biomaterials* **2021**, *268*, 120475. [CrossRef]
89. Sok, M.C.P.; Tria, M.C.; Olingy, C.E.; Emeterio, C.L.S.; Botchwey, E.A. Aspirin-Triggered Resolvin D1-modified materials promote the accumulation of pro-regenerative immune cell subsets and enhance vascular remodeling. *Acta Biomater.* **2017**, *53*, 109–122. [CrossRef]
90. Shi, J.; Zhang, X.; Jiang, L.; Zhang, L.; Dong, Y.; Midgley, A.C.; Kong, D.; Wang, S. Regulation of the inflammatory response by vascular grafts modified with Aspirin-Triggered Resolvin D1 promotes blood vessel regeneration. *Acta Biomater.* **2019**, *97*, 360–373. [CrossRef]
91. Yin, C.; Zhao, Q.; Li, W.; Zhao, Z.; Wang, J.; Deng, T.; Zhang, P.; Shen, K.; Li, Z.; Zhang, Y. Biomimetic anti-inflammatory nano-capsule serves as a cytokine blocker and M2 polarization inducer for bone tissue repair. *Acta Biomater.* **2020**, *102*, 416–426. [CrossRef]
92. Lu, B.; Han, X.; Zou, D.; Luo, X.; Liu, L.; Wang, J.; Maitz, M.F.; Yang, P.; Huang, N.; Zhao, A. Catechol-chitosan/polyacrylamide hydrogel wound dressing for regulating local inflammation. *Mater. Today Bio* **2022**, *16*, 100392. [CrossRef]
93. Jiang, X.; Liu, J.; Li, S.; Qiu, Y.; Wang, X.; He, X.; Pedersen, T.O.; Mustafa, K.; Xue, Y.; Mustafa, M.; et al. The effect of resolvin D1 on bone regeneration in a rat calvarial defect model. *J. Tissue Eng. Regen. Med.* **2022**, *16*, 987–997. [CrossRef]
94. Miranda, J.; Brazil, J.C.; Morris, A.H.; Parkos, C.A.; Quiros, M.; Nusrat, A. Maresin-2 promotes mucosal repair and has therapeutic potential when encapsulated in thermostable nanoparticles. *Proc. Natl. Acad. Sci. USA* **2023**, *120*, e2218162120. [CrossRef]
95. Furutani, K.; Chen, O.; McGinnis, A.; Wang, Y.; Serhan, C.N.; Hansen, T.V.; Ji, R.R. Novel proresolving lipid mediator mimetic 3-oxa-PD1n-3 docosapentaenoic acid reduces acute and chronic itch by modulating excitatory and inhibitory synaptic transmission and astroglial secretion of lipocalin-2 in mice. *Pain* **2022**. [CrossRef]
96. Hodges, R.R.; Li, D.; Shatos, M.A.; Bair, J.A.; Lippestad, M.; Serhan, C.N.; Dartt, D.A. Lipoxin A(4) activates ALX/FPR2 receptor to regulate conjunctival goblet cell secretion. *Mucosal Immunol.* **2017**, *10*, 46–57. [CrossRef]
97. Tang, H.; Liu, Y.; Yan, C.; Petasis, N.A.; Serhan, C.N.; Gao, H. Protective actions of aspirin-triggered (17R) resolvin D1 and its analogue, 17R-hydroxy-19-para-fluorophenoxy-resolvin D1 methyl ester, in C5a-dependent IgG immune complex-induced inflammation and lung injury. *J. Immunol.* **2014**, *193*, 3769–3778. [CrossRef]
98. de Prinse, M.; Qi, R.; Amsden, B.G. Polymer micelles for the protection and delivery of specialized pro-resolving mediators. *Eur. J. Pharm. Biopharm.* **2023**, *184*, 159–169. [CrossRef]
99. Chiang, N.; Serhan, C.N. Structural elucidation and physiologic functions of specialized pro-resolving mediators and their receptors. *Mol. Asp. Med.* **2017**, *58*, 114–129. [CrossRef]

Journal of *Functional Biomaterials*

MDPI

Article

Solid Lipid–Polymer Hybrid Nanoplatform for Topical Delivery of siRNA: In Vitro Biological Activity and Permeation Studies

Margarete Moreno de Araujo, Livia Neves Borgheti-Cardoso, Fabíola Garcia Praça, Priscyla Daniely Marcato and Maria Vitória Lopes Badra Bentley *

School of Pharmaceutical Sciences of Ribeirao Preto, University of São Paulo, Av. do Café, s/n, Ribeirão Preto 14040-903, SP, Brazil; margaretemoreno@gmail.com (M.M.d.A.); pmarcato@fcfrp.usp.br (P.D.M.)
* Correspondence: vbentley@usp.br; Tel./Fax: +55-16-33154301

Abstract: Small interfering RNA (siRNA) molecules have limited transfection efficiency and stability necessitating the use of delivery systems to be effective in gene knockdown therapies. In this regard lipid–polymeric nanocarriers have emerged as a promising class of nanoparticles for siRNA delivery, particularly for topical applications. We proposed the use of solid lipid–polymer hybrid nanoparticles (SLPHNs) as topical delivery systems for siRNA. This approach was evaluated by assessing the ability of SLPHNs–siRNA complexes to internalize siRNA molecules and both to penetrate skin layers in vitro and induce gene knocking down in a skin cell line. The SLPHNs were formed by a specific composition of solid lipids, a surfactant polymer as a dispersive agent, and a cationic polymer as a complexing agent for siRNA. The optimized nanocarriers exhibited a spherical shape with a smooth surface. The average diameter of the nanoparticles was found to be 200 nm, and the zeta potential was measured to be +20 mV. Furthermore, these nanocarriers demonstrated excellent stability when stored at 4 °C over a period of 90 days. In vitro and in vivo permeation studies showed that SLPHNs increased the cutaneous penetration of fluorescent-labeled siRNA, which reached deeper skin layers. Efficacy studies were conducted on keratinocytes and fibroblasts, showing that SLPHNs maintained cell viability and high cellular uptake. Furthermore, SLPHNs complexed with siRNA against Firefly luciferase (siLuc) reduced luciferase expression, proving the efficacy of this nanocarrier in providing adequate intracellular release of siRNA for silencing specific genes. Based on these results, the developed carriers are promising siRNA delivery systems for skin disease therapy.

Keywords: solid lipid–polymer hybrid nanoparticles; siRNA; nucleic acid; gene silencing; topical delivery; skin penetration

Citation: de Araujo, M.M.; Borgheti-Cardoso, L.N.; Praça, F.G.; Marcato, P.D.; Bentley, M.V.L.B. Solid Lipid–Polymer Hybrid Nanoplatform for Topical Delivery of siRNA: In Vitro Biological Activity and Permeation Studies. *J. Funct. Biomater.* **2023**, *14*, 374. https://doi.org/10.3390/jfb14070374

Academic Editor: Kai Yang

Received: 1 June 2023
Revised: 30 June 2023
Accepted: 7 July 2023
Published: 17 July 2023

1. Introduction

The use of RNA interference (RNAi) is a promising therapeutic approach to modulate or inhibit endogenous gene expression [1–5] through synthetic small interfering nucleic acid (siRNA) therapeutics [6,7]. RNAi is mediated by double-stranded siRNA molecules consisting of 21–27 base pairs and has demonstrated a remarkable ability to effectively suppress the expression of specific genes [8]. However, one of the major challenges in siRNA therapy is the effective transfer of (naked) nucleic acid across cell membranes due to the anionic charge of this molecule as well as its poor stability in biological environments [9–11].

The administration route is another issue that must be considered for the success of the siRNA clinical approach. Topical administration of siRNA targeting genes involved in several skin disorders offers a fresh and innovative therapeutic strategy for addressing inherited skin conditions, viral infections, skin cancer, and atopic dermatitis [10,12,13]. However, the clinical application of siRNA into the skin is limited by the efficient barrier properties of the skin provided by the stratum corneum (SC) [14]. To surmount the inherent obstacles associated with the delivery of siRNA to the skin, nanoparticles have been devised [2,12–17]. Through these nanotechnologies, high penetration rates of siRNA into the skin can be achieved by multiple pathways, including intercellular, transcellular, and appendage pathways. For instance, lipid nanoparticles can interact with intercellular lipids in the SC, which is the outermost layer of the skin, and disrupt the organization of the SC lipids, creating temporary gaps that allow siRNA penetration. This pathway involves siRNA release from the nanoparticles and its diffusion among corneocytes (flat, dead skin cells) through the lipid-rich intercellular spaces. siRNA can also penetrate corneocytes via a transcellular route. Cellular uptake of siRNA-loaded nanoparticles occurs through endocytosis or other internalization mechanisms. Once inside the cells, nanoparticles can release their cargo (siRNA) into the cytoplasm, where silencing of target genes takes place. Moreover, skin appendages such as hair follicles and sweat glands provide an alternate route for penetration. siRNA carried by nanoparticles can enter hair follicles or sweat ducts and then travel through the surrounding skin layers to reach the target cells [2].

Several nanotechnology platforms have been studied for topical delivery of siRNA, including nanostructured delivery systems that can be produced using different compounds. Among them, polyethylenimine (PEI) has shown remarkable efficacy in delivering siRNA due to its amino groups that can be protonated at physiological pH and its strong buffering capacity [18,19]. Nevertheless, the formation of PEI/siRNA complexes frequently leads to elevated cytotoxicity [12,20]. Thus, the clinical application of siRNA requires the use of effective and biodegradable delivery systems that simultaneously exhibit low cytotoxicity. In the present work, we developed a solid lipid–polymer hybrid nanoparticle (SLPHN) for the effective delivery of siRNA. The SLPHNs combine appropriate siRNA delivery characteristics of PEI and solid lipid nanoparticles (SLN), which have high biocompatibility and bioavailability, controlled drug release, physical stability, and protection of incorporated labile drugs from degradation, while their production process is easily scalable [21–24].

Our optimized SLPHNs consist of Compritol® 888 ATO as a structural lipid, PEI as a cationic agent, and poloxamer 188 as a surfactant. Several combinations of these components were investigated in order to select the best composition that provided small particles of homogeneous sizes (polydispersity index) and positive charge surface (zeta potential). Then, the optimized SLPHN samples were characterized regarding their physicochemical properties by DLS, NTA, and AFM. The efficiency of the siRNA–SLPHN complex, siRNA release ability, cytotoxicity, and cellular internalization in a human immortalized keratinocyte cell line (HaCaT) and a mouse embryo fibroblast cell line (NIH/3T3) were also evaluated. Furthermore, in vitro skin penetration and silencing efficiency studies were carried out.

2. Materials and Methods

2.1. Materials

Compritol® 888 ATO (glyceryl behenate) was kindly supplied by Gattefossé (Paramount, NJ, USA) and Cremer Oleo GmbH & Co.KG (Hamburg, Germany). Tris(hydroxymethyl)aminomethane (Tris–HCl) was purchased from Merck KGaA (Darmstadt, Germany), and branched polyethylenimine (PEI) (25 kDa) and poloxamer 188 were purchased from Sigma Aldrich Co. (St. Louis, MO, USA). UltraPure™ Agarose was purchased from Invitrogen (Carlsbad, CA, USA). The heparin used in this study was obtained from Blausiegel (Cotia, SP, Brazil). Loading buffer (6× DNA loading dye #R0611) was acquired from Thermo Scientific (Rockford, IL, USA). Fetal bovine serum (FBS) and Opti-MEM™ media were obtained from Gibco (Grand Island, NY, USA), and Dulbecco's modified Eagle's medium (DMEM) and antibiotic-antimycotic cell culture solutions were obtained from Sigma Aldrich Co (St. Louis, MO, USA). Silencer™ Negative Control siRNA (siRNA), Silencer™ FAM (carboxyfluorescein)-labeled, and Alexa-fluor® 647-labeled siRNA were purchased from Ambion® (Austin, TX, USA). Mouse Fibroblast cells (BALB/NIH/3T3 strain clone A31 and HaCaT cells) were obtained from the Rio de Janeiro cell bank (Rio de Janeiro, RJ, Brazil). Lipofectamine™ 2000 and Prolong™ Gold antifade reagent with DAPI were obtained from Life Technologies (Paisley, U.K.). The pCMV-Luc vector, which expresses Firefly luciferase (FL), was obtained from PlasmidFactory (Bielefeld, Germany). The pGL4.74 (hRluc/TK) vector, which expresses Renilla luciferase (RL), and Dual Luciferase were obtained from Sigma Aldrich (Darmstadt, Germany). The Reporter Assay System kit was acquired from Promega (Fitchburg, WI, USA).

2.2. SLPHNs Preparation

SLPHNs were prepared by hot emulsion high shear homogenization followed by sonication methods. Briefly, the lipid matrix containing Compritol® 888 ATO and Tris–HCl buffer (100 mM, pH 6.5) as aqueous phase, PEI, and poloxamer 188 were heated separately to 75 °C. The mixture of aqueous phase and the oily phase was dispersed under high shear homogenization at 30,000 rpm for 2.5 min by Ultra-turrax (IKA T10 basic, Staufen, Germany) and then by ultrasonic processing for 5 min (Sonics VCX 750, Newtown, PA, USA, 13 mm probe and 40% amplitude). SLPHNs were formed after cooling the dispersion to 25 °C. In order to optimize the average size and polydispersity index (PdI), SLPHN samples were prepared using varying concentrations (%) of Compritol® 888 ATO, PEI, and poloxamer.

2.3. Characterization of SLPHNs

2.3.1. Particle Size, Zeta Potential, Nanoparticle Concentration, and Morphology

Nanoparticles were analyzed by dynamic light scattering (DLS) in NanoSize ZS (Malvern Instruments, Malvern, U.K.) using the DTS Nano software v5.03. The colloidal solutions were loaded into disposable plastic cuvettes and examined using backscattering light at a 90° angle. Size and distribution were measured in triplicate and expressed as Z-average diameter and PdI. The zeta potential of nanoparticles was also determined by observing their electrophoretic mobility in an electric field using the same equipment.

Nanoparticle tracking analysis (NTA) was determined using a NanoSight LM20 (NanoSight, Amesbury, U.K.). The samples were appropriately diluted and introduced into the sample chamber using sterile syringes. All measurements were carried out at room temperature.

The size and morphology of SLPHNs were determined by atomic force microscopy (AFM). Nanocarrier samples were prepared by depositing dilute particle dispersions on a freshly cleaved mica plate and drying with argon. Images were obtained using a Shimadzu Scanning Probe Microscope SPM-9600 model (Kyoto, Japan) equipped with a 100 µm tripod scanner and pyramidal cantilevers with silicon probes (force constant: 10–130 N/m) at a resonance frequency of 204–497 kHz. All measurements were performed in intermittent-contact mode at a scan speed of approximately 1 Hz to avoid damage to the sample surface

2.3.2. Physical Stability of SLPHNs

The physical stability of SLPHNs (n = 3) was assessed for 90 days under various conditions according to the Drug Stability FDA Guidelines [25]. These conditions included a low temperature of 4 °C and a high temperature of 30 °C with a relative humidity of approximately 75%. The evaluation of physical stability was based on monitoring the mean Z-average diameter and zeta potential of the formulations using DLS analysis.

2.4. Electrophoretic Mobility Shift Assay

2.4.1. Evaluation of siRNA Binding and Polyanion Competition

The ability of siRNA to bind to SLPHNs was studied using an agarose gel assay [15,26] For this, free siRNA and SLPHNs (1:4 v/v) with and without siRNA (10 µM) were prepared and incubated for 30 min. Electrophoresis was carried out in 2% agarose gel at 100 V for 20 min in Tris–acetate–EDTA (TAE) buffer (pH 8.0) after the addition of loading buffer (Bromophenol blue (0.25% w/v), xylene cyanol FF (0.25% w/v), Orange G (0.25% w/v), Tris–HCl pH 7.5 (10 mmol/L), ethylenediamine tetra-acetic acid (10 mmol/L), and saccharose (0.65% w/v)). The siRNA mobility shift was observed and photographed under UV light after addition of ethidium bromide (5 mg/mL) to the gel. Image acquisition was performed using the Quantity One software v.4.6.6.

To assess the stability of siRNA after decomplexation from SLPHNs, a polyanion competition assay was performed using heparin as a competitor [26,27]. Briefly, siRNAs–SLPHNs were prepared as previously described, and 10 μL of heparin (5000 IU/mL) was added. The samples were subjected to electrophoresis on agarose gels after incubation at 37 °C for 1 h using the same conditions described above.

2.4.2. Serum Stability Study

The protective effect of SLPHNs against siRNA degradation was evaluated by incubating siRNA–SLPHN samples with 25% fetal bovine serum (FBS) for 24 h at 37 °C. Simultaneously, free siRNA and Lipofectamine™–siRNA samples were also incubated with 25% FBS at 37 °C for 24 h, serving as negative and positive controls, respectively. Following incubation, all samples were further treated with 10 μL of heparin (5000 IU/mL) for 1 h to ensure complete release of siRNA from the formulations [28]. The remaining percentage of intact siRNA was then analyzed using agarose gel electrophoresis as described in Section 2.4.1.

2.5. In Vitro Skin Penetration Study

In vitro permeation studies were conducted following the protocol outlined in OECD Guideline 428 [29]. Porcine ear skin was used as the biological membrane, and FAM-labeled siRNA [30] was employed in the study. Vertical Franz diffusion cells with a diffusion area of 0.68 cm^2 were utilized. Each cell was filled with 3 mL of 100 mM phosphate buffer (pH 7.4 $\pm$ 0.2) at a temperature of 32 °C. The cells were maintained under continuous stirring at 400 rpm throughout the experiment. The dermatomized porcine ear skins (500 μm) were mounted in a Franz diffusion cell with SC facing the donor compartment, where 100 μL of the samples (naked siRNA–FAM and siRNA–FAM–SLPHN–0.25%) were applied. After 24 h of permeation, the skins were carefully removed from the Franz diffusion cell, and the surface was thoroughly washed with distilled water to eliminate any excess formulation. Subsequently, the skins were frozen in Tissue-Tek® (Pelco International, Redding, CA, USA) embedding compound with acetone at −30 °C. A cryostat microtome (Leica, Wetzlar, Germany) was used to section the skin into 20 μm thick slices at a temperature of −20 °C, and slices were mounted on glass slides. The fluorescence intensity and depth of FAM-labeled siRNA within the skin were examined using fluorescence microscopy (Axioskop 2 plus, Carl Zeiss, Gottingen, Germany) along 640 nm and 730 nm band-pass excitation and emission filters. Images analysis was performed using the AxioVision software v.4. Due to skin auto-fluorescence, skin sections treated with PBS were used as a control. Consistent sensitivity and exposure settings were applied to all samples during the fluorescence microscopy analysis.

2.6. Cellular Studies

2.6.1. Cell Culture Conditions

Non-tumorigenic keratinocytes (HaCaT) and murine fibroblasts (NIH/3T3) were individually cultured in Dulbelco's modified Eagle's medium (DMEM) rich in glucose supplemented with 10% heat-inactivated fetal bovine serum (FBS) and 1% (v/v) of an antibiotic solution containing 10,000 IU penicillin, 10 mg streptomycin, and 25 μg amphotericin B/mL. Cells were incubated at 37 °C in a humidified incubator (95% relative humidity) with 5% CO_2 atmosphere.

2.6.2. Cellular Viability

For the relative viability assay of HaCaT and NIH/3T3 cells, the resazurin assay was employed after 24 h of incubation. When the cells reached approximately 80% confluence, they were harvested from T-flasks using 0.05% (v/v) trypsin. Subsequently, 5×10^4 cells/well for HaCaT cells and 1×10^4 for NIH/3T3 cells were seeded onto 96-well plates and incubated at 37 °C and 5% CO_2 atmosphere for 24 h to allow for cell attachment. Once the cells were attached, the culture medium was replaced with formulations containing SLPHN–0.15 PEI and SLPHN–0.25% PEI, along with various nanoparticle concentrations ranging from 1.59×10^{10} to 52.4×10^{10} particles/mL.

After an additional 24 h of incubation, the cells were washed with PBS, and resazurin solution (0.025 mg/mL) was added to each well. Following 4 h of incubation, fluorescence was measured using a Biotek plate reader (Synergy model) with an excitation wavelength set at 540 and emission at 590 nm. Untreated cells served as the negative control, and three repeated measurements were performed for each sample. The results were expressed as a percentage of cell viability.

2.6.3. Cellular Internalization by Flow Cytometry

The cellular uptake of Alexa Fluor® 647-labeled siRNA (siRNA Alexa Fluor® 647) mediated by SLPHN was quantified by flow cytometry. HaCaT and NIH/3T3 cells were seeded onto 12-well tissue culture plates at a density of 3×10^5 cells/well and 1.25×10^5 cells/well respectively. After 24 h, 100 μL of naked siRNA Alexa Fluor® 647 (40 pmol/well) (free siRNA) or siRNA complexed with SLPHNs was added to the cells and incubated for 24 h at 37 °C with 5% CO_2. After this period, cells were washed twice with PBS buffer and trypsinized, and cell uptake was assessed by flow cytometry using a Facscanto BD FACSCalibur™ Cytometer equipped with 3 lasers (488, 633, and 405 nm) and a 660/10 nm emission filter for Alexa Fluor® 647. Propidium iodide (50 μg/mL) was used as an indicator of cell viability. Untreated cells (siRNA control) and cells treated with Lipofectamine™–siRNA were used as controls; 10,000 events were collected per analyzed sample.

2.6.4. Intracellular Localization of SLPHNs by Confocal Microscopy

In vitro cellular localization of siRNA–SLPHNs was investigated by confocal microscopy using a LEICA—TCS SP2 microscope (Leica, Heidelberg, Germany). For this, a siRNA-labeled Alexa Fluor® 647 was used in the formulation samples (siRNA–SLPHNs), and naked siRNA (free siRNA) was used as a negative control. HaCaT and NIH/3T3 cells were seeded onto coverslips (previously treated with glutaraldehyde) at 3×10^5 cells/well and 1.25×10^5 cells/well, respectively, in 12-well plates to guarantee 60–80% confluence. After 24 h, 100 μL of the siRNA Alexa Fluor® 647 (40 pmol/well) naked or complexed with SLPHNs was added to the cells and incubated for 24 h. Cells were washed with PBS twice and fixed overnight with a 1% formaldehyde solution in PBS. After fixation, cells were washed twice with PBS and mounted on glass cover slides using Prolong™ Gold containing the blue fluorescent stain 4′,6-diamidino-2-phenylindole (DAPI) for nucleus labeling. Cells treated with Lipofectamine™–siRNA were used as a positive control. Finally, confocal images of the cells were taken using a 63× oil-immersion objective under emission filters of 350/470 and 660/10 nm, suitable for DAPI and siRNA Alexa Fluor® 647, respectively.

2.6.5. In Vitro Silencing Efficiency of siRNA–SLPHNs

In the luciferase reporter assay, HaCaT cells (1×10^6 cells) were transfected with reporter vectors containing Firefly luciferase (pCMV-Luc) and Renilla luciferase (pGL4.74 hRluc/TK) using Lipofectamine™ 2000 in antibiotic-free medium. After 24 h, transfected cells were seeded onto 96-well plates at a density of 5×10^4 cells/well. Once cells reached approximately 70% confluence, they were treated with SLPHN (containing 0.15 and 0.25% PEI) or Lipofectamine™ 2000. Both complexed with 50 pmol of siRNA per well. The siRNA used targeted Firefly luciferase (siLuc) or nonspecific siRNA (ns-siRNA). Cells were then incubated for an additional 24 h, washed with PBS, and maintained in fresh medium for another 24 h at 37 °C in a 5% CO_2 atmosphere.

Following incubation, cells were lysed with Passive Lysis Buffer (Promega) using 200 μL per well. To measure luciferase activity, 10 μL of cell lysates and 50 μL of Luciferase Assay Reagent were added to a white 96-well plate. The luminescence was immediately measured using a plate reader, with an initial mixing period of 2 s, followed by luminescence measurements at 0.5 s intervals for 5 s. Subsequently, 50 μL of Stop & Glo Reagent was added to each well, and luminescence measurements were repeated.

To obtain normalized data, the Firefly luciferase luminescence was divided by the Renilla luciferase luminescence using the formula (Firefly luminescence AUC 5 s/Renilla luminescence AUC 5 s) [28,31].

2.7. Statistics

Results are presented as mean ± standard error of the mean (SD). One-way analysis of variance (ANOVA) followed by Tukey's multiple comparison test was used as statistical analysis to compare all groups studied. Significance levels were reported as * $p < 0.05$, ** $p < 0.01$, and *** $p < 0.001$. All statistical analyses were conducted using GraphPad Prism software v.8.4.

3. Results and Discussion

SLPHNs have been shown to be novel nanocarriers for topical and transdermal drug delivery. They have a dual advantage over other lipid nanoparticles due to their vesicular and particulate nature, the high biocompatibility of lipids [32,33], and the structural contribution of PEI molecules. This study demonstrated the successful preparation of SLPHNs and their effective utilization as nanocarriers for the delivery of siRNA to the skin.

3.1. Preparation and DLS Characterization of SLPHNs

3.1.1. SLPHNs Preparation

A hot emulsion high shear homogenization followed by the sonication method was used to prepare the SLPHNs developed in this study. We selected Compritol® 888 ATO, poloxamer 188, and PEI as SLPHN components due to their favorable characteristics for topical application and ease of formation of nanoscale delivery systems. PEI was incorporated into the systems to form complexes with nucleic acid by non-covalent bonds [34]. To assess the influence of each component of the SLPHNs on particle sizes and polydispersity index, 15 different samples were prepared with different concentrations of Compritol® 888, PEI, and poloxamer 188 (Table 1).

Table 1. Physicochemical characteristics of SLPHNs in different compositions (n = 3).

Samples	PEI %	Compritol® 888 ATO %	Poloxamer 188%	Aqueous Phase %	Size ± SD (nm)	PdI [1] ± SD [3]	ZP [2] ± SD [3] (mV)
F1	0.00	2.00	1.50	96.50	143.20 ± 20.79	0.34 ± 0.03	−8.33 ± 7.88
F2	1.00	1.00	1.50	96.50	182.80 ± 13.45	0.37 ± 0.03	38.75 ± 2.14
F3	1.00	2.00	1.50	95.50	219.35 ± 8.84	0.42 ± 0.00	36.60 ± 0.28
F4	1.00	3.00	1.50	94.50	236.35 ± 14.24	0.43 ± 0.09	36.83 ± 1.56
F5	1.00	4.00	1.50	93.50	250.95 ± 59.25	0.45 ± 0.07	37.18 ± 1.74
F6	1.00	5.00	1.50	92.50	267.55 ± 43.18	0.44 ± 0.02	38.63 ± 3.21
F7	1.00	2.00	0.50	96.50	977.05 ± 67.81	0.39 ± 0.08	36.70 ± 0.85
F8	1.00	2.00	3.00	94.00	201.90 ± 0.42	0.36 ± 0.03	30.85 ± 0.78
F9	1.00	2.00	5.00	92.00	111.30 ± 15.41	0.37 ± 0.01	30.13 ± 0.66
F10	0.10	2.00	1.50	96.40	174.20 ± 1.68	0.23 ± 0.02	4.46 ± 0.75
F11	0.15	2.00	1.50	96.35	164.03 ± 4.74	0.27 ± 0.05	11.50 ± 1.15
F12	0.20	2.00	1.50	96.30	162.70 ± 11.01	0.38 ± 0.04	20.90 ± 2.87
F13	0.25	2.00	1.50	96.25	175.15 ± 17.71	0.29 ± 0.12	27.18 ± 1.55
F14	0.50	2.00	1.50	96.00	180.93 ± 5.85	0.48 ± 0.05	34.10 ± 3.55
F15	0.75	2.00	1.50	95.75	212.68 ± 15.85	0.46 ± 0.02	32.38 ± 3.53

[1] PdI: polydispersity index; [2] ZP: zeta potential; [3] SD: standard deviation.

3.1.2. Effect of Compritol and Poloxamer on Nanoparticle Composition

First, the influence of different amounts of Compritol® 888 (1 to 5%) on particle size and surface charge of the SLPHNs was evaluated. For that purpose, the amount of poloxamer 188 and PEI concentration were fixed at 1.5 and 1% (w/w) in the samples (F2 to F6). Particle size and polydispersity index (PdI) increased proportionally with the addition of Compritol® 888, perhaps due to the macroscopically observed high viscosity of the lipid phase, which contributed to the increase in superficial tension, leading to the formation of larger nanoparticles. Ribeiro and coworkers reached a similar result when they used a factorial design to investigate the role of each excipient in the desired responses (size, zeta potential, and PdI) in natural lipid-based nanocarriers [35,36]. Our findings are also consistent with those of Yalcin et al. when the authors optimized gemcitabine hydrochloride-loaded LPHNs based on an experimental approach. The study employed PLGA, soya phosphatidylcholine, and DSE-PEG as independent variables (components). It was observed that as the lipid/polymer ratio increased from low to high levels, there was a noticeable upward trend in particle size values [37].

Compritol® 888 is composed of different esters of behenic acid with glycerol (15–23% monoglycerides, 40–60% diglycerides, 21–35% triglycerides) [38]. This lipid composition is regarded as a safe and biocompatible excipient that is widely used for the preparation of lipid nanocarriers because of various advantages, such as sustained release, better lymphatic targeting, and chemical stability, resulting in a suitable ability to load lipophilic and/or hydrophilic drugs [38,39]. It is noteworthy that lipid carriers containing Compritol® exhibit a platelet-like shape, which allows them to easily slide between corneocytes after penetrating the intercellular lipids of the skin. This unique shape provides these carriers with adhesiveness and promotes skin hydration properties [40]. In terms of nanocarrier characteristics, Wang et al. prepared LPHNs using six types of solid lipids and observed that glycerides showed superior colloidal properties in terms of smaller particle size and narrower size distribution than fatty acids [41].

Based on our findings (Table 1), 2% Compritol® 888 was selected and then combined with poloxamer 188 at four concentrations (0.5, 1.5, 3, and 5% *w/w*, samples F7 to F9). Thus, particle size decreased with increasing surfactant concentration, possibly due to a reduction in interfacial tension [24,42]. Furthermore, it has been reported that the presence of a surfactant such as poloxamer increases particle stability by covering the particle surface and preventing particle agglomeration [43].

3.1.3. Influence of Branched PEI-Coated Nanoparticles

The F3 formulation containing Compritol® 888 (2% *w/w*) and Poloxamer 188 (1.5% *w/w*) was selected for evaluation of PEI influence on particle size and PdI. Particle size and PdI of samples F10 to F15 increased with increasing PEI concentration. Similar results were previously reported when PEI was incorporated on both PLGA (poly (lactic-acid coglycol)) and PEG (polyethylene glycol)-based nanoparticles [44,45].

Furthermore, the zeta potential of the nanoparticles changed from negative in the particles without PEI (F1) to positive in the SLPHNs (F2 to F15). This positive surface charge of the particles is a result of the alignment of the PEI on the nanoparticle surface. The positively charged particle surface can facilitate particle adherence to negatively charged cellular membranes. This interaction promotes and enhances intracellular uptake of the particles [18,46]. Taking this into account, the SLPHNs consisting of Compritol® 888 (2% *w/w*), poloxamer 188 (1.5% *w/w*), and PEI concentrations of 0.15% (F11) and 0.25% *w/w* (F13) were selected for further physicochemical characterization and in vitro studies.

3.2. *Nanoparticle Tracking Analysis (NTA) and Morphology of SLPHN*

Increasing the PEI concentration from 0.15% to 0.25% had no influence on particle concentration, which was similar for SLPHNs–0.15% PEI and SLPHNs–0.25% PEI (2.54×10^{13} to 2.62×10^{13} particles/mL, respectively). Also, a non-significant difference was observed between particle size determined by NTA and that determined by DLS. These were relevant parameters and important findings for the next in vitro studies since the recommended metric is the number of particles/volume.

As expected, the AFM images showed spherical particles with smooth surfaces (Figure 1). The chemically homogeneous lipid allows for the formation of perfect crystals that have a well-defined platelet-like pattern of the β-modification. Spherical lipid nanoparticles can be obtained with a combination of heterogeneous surfactants and heterogeneous lipids [47,48]. The morphology of nanoparticles is typically influenced by the composition of the lipid matrix and can manifest as a spheroidal, anisometric, or flat shape [46]. The specific shape that the nanoparticle assume is determined by both their size and the polymorphic shape of the lipid used in the formulation [43].

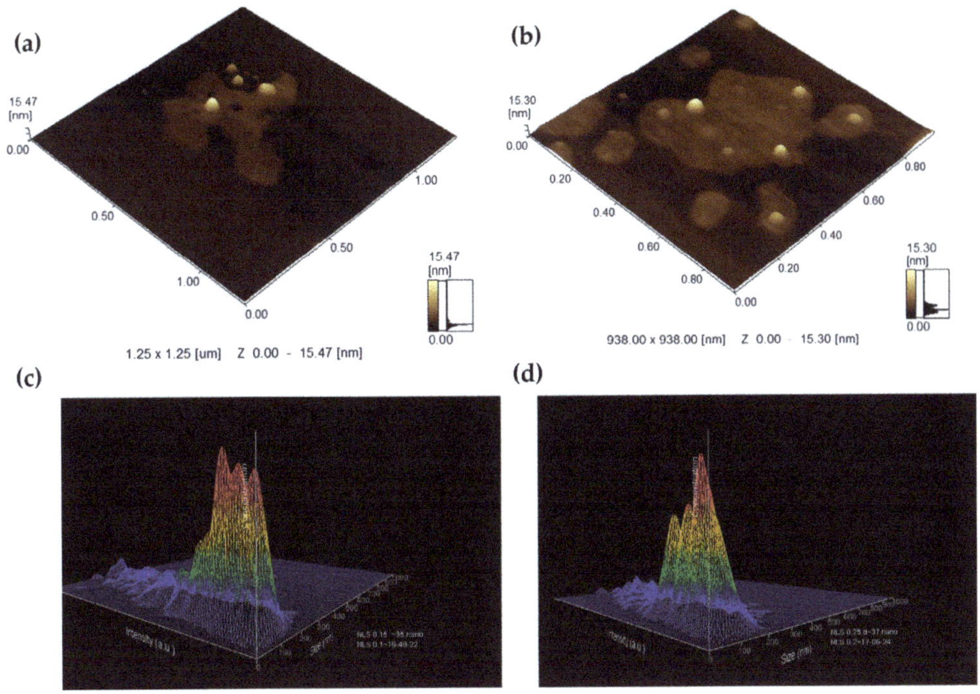

Figure 1. 3D size distribution profile obtained by AFM (**a**,**b**) and NTA 3D (**c**,**d**) topographic image of SLPHN 0.15% PEI (**a**,**c**) and SLPHN 0.25% PEI (**b**,**d**).

The average diameter of 58 nm and 67.51 nm determined by AFM for SLPHNs–0.15% PEI and SLPHNs–0.25% PEI, respectively, is smaller than the diameter values determined by DLS and NTA, which were in the order of 170 nm approximately. In general, this difference is related to the principles of each technique. On the one hand, the size values obtained by the DLS and NTA techniques are calculated using the Stokes–Einstein equation, which considers the hydrodynamic diameter of the particles composed by the particle diameter and its stabilization layer formed by the presence of the surfactant used [49]. Indeed, NTA is a technique that provides information about the concentration of nanoparticles and measures their size based on the analysis of light scattering and Brownian motion of individual nanoparticles [47,50]. With AFM, on the other hand, the microscope allows for the measurement of the diameter of nanoparticles without the stabilization layer [51].

3.3. Physical Stability Studies

Monitoring changes in zeta potential, particle size, appearance, and viscosity over an extended storage period allows assessment of the physical properties of lipid nanoparticles. External parameters such as temperature appear to be of primary importance for long-term stability [23]. Here, the physical stability properties of SLPHN–0.0%, –0.15%, and –0.25% PEI were assessed over a period of 90 days. When particle sizes were compared, SLPHN–0.25% PEI samples exhibited physical stability at both low temperatures (4 °C) and accelerated conditions (30 °C), as shown in Figure 2a,c. When lower amounts of PEI (0.15%) were inserted into the SLPHN, a progressive increase in particle size was observed throughout the experimental period, demonstrating the instability of the samples in these experimental conditions. These results could be related to the amount of PEI molecules on the nanoparticles, which presumably changes the structural conformation of the nanoparticles over the days. We believe that a PEI of 0.25% favors the stability of the formulation by a combination of steric hindrance and electrostatic repulsion and is, therefore, more stable over long periods than SLPHNs with less PEI. Also, the changes in the zeta potential of the SLPHNs–0.25% PEI were not significant over 90 days under experimental conditions (Figure 2b,d). Previously, our research group has demonstrated that polymers such as PEI can also act as a stabilizing agent in lipid-based nanoparticle systems and modify the surface charge [12,14,15,52].

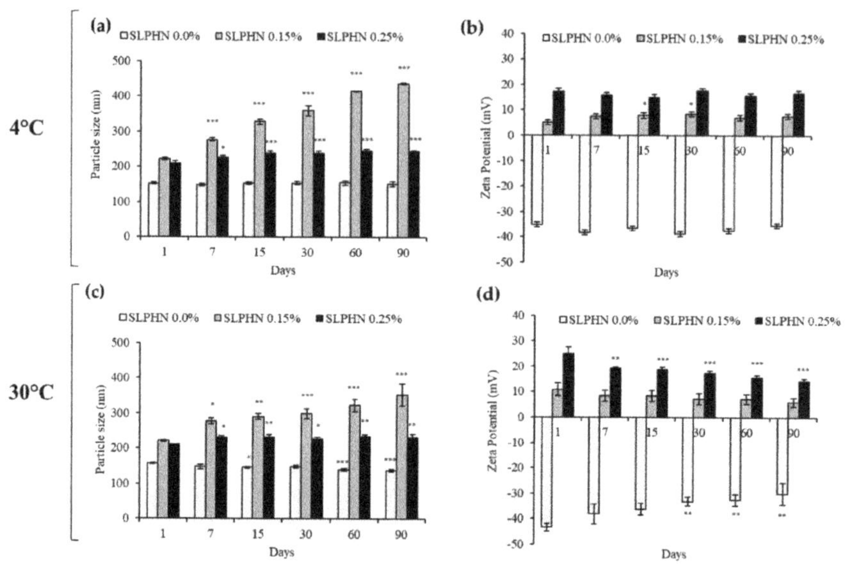

Figure 2. Physical stability data based on particle size (nm) (**a**,**c**) and zeta potential (mV) (**b**,**d**) of SLPHNs in the absence (SLPHN 0.0) and presence of PEI (SLPHN 0.15 and SLPHN 0.25) over 90 days at 4 °C (top, **a**,**b**) and 30 °C (bottom, **c**,**d**). * $p < 0.05$; ** $p < 0.01$; *** $p < 0.001$ compared to day 1 of each formulation. One-way ANOVA (95% confidence interval) and Tukey's test.

Another contribution to the long-term stability of solid lipid nanoparticles is the presence of poloxamer 188, which, because of its nonionic nature, stabilizes the nanoparticles by forming a coat on their surfaces [53]. Therefore, we attributed the physical stability of the SLPHN observed in this work to the SLPHN composition (lipid, polymer, and surfactant) and their relative ratios.

3.4. siRNA Binding and Stability of the SLPHNs–siRNA Complex

Figure 3 summarizes the ability of SLPHNs to complex siRNA and the consequent inhibition of its electrophoretic migration. In contrast, the siRNA Control (free siRNA), which has a negative charge, migrated through the gel, whereas positively charged complexes remained at the top of the gel (SLPHN 0.15% PEI and SLPHN 0.25% PEI). The presence of PEI in SLPHNs was crucial to complex siRNA, as SLPHNs lacking PEI did not form complexes with siRNA (Figure 3a). Complex formation leads to the electroneutralization of the negative charge of siRNA, which is interesting for stronger cellular internalization and endosomal escape. The complexation of siRNA by PEI has been previously demonstrated in studies using liquid crystalline systems [12,14,15,26,52], providing a positive surface charge to nanoparticles and thus promoting electrostatic interactions between nanoparticles and siRNA.

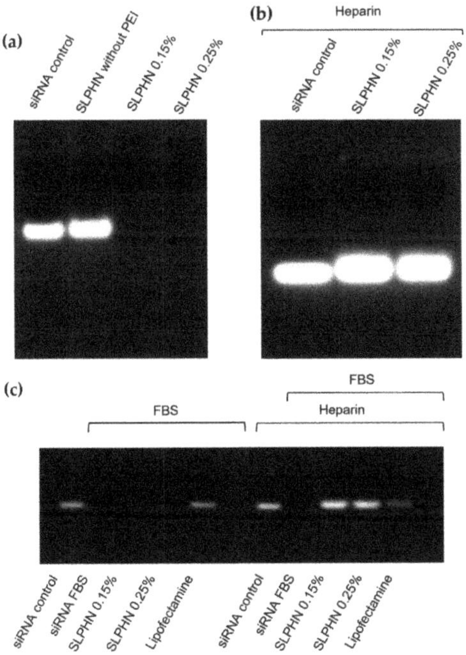

Figure 3. Electrophoretic mobility of siRNA after complexation with SLPHN without PEI, SLPHN with 0.15% PEI, and SLPHN with 0.25% PEI in the absence of heparin (**a**) and in the presence of heparin (**b**) and evaluation of the stability of siRNA under serum conditions (**c**). The siRNA control refers to free siRNA without FBS treatment, and siRNA-FBS refers to free siRNA after FBS treatment.

After determining that SLPHNs formulated with PEI successfully complexed siRNA, heparin (5000 IU/mL) was added to the complexes, and a polyanion competition between siRNA and heparin for PEI amine groups was observed, suggesting dissociation of the siRNA–SLPHNs complex (Figure 3b). Generally, for effective delivery of intact siRNA, it is important that the complexes remain stable during the delivery process. However, it is also necessary for siRNA to be released into the cytosol to achieve gene silencing [54]. Polyanions, such as heparin, can disassemble gene complexes, facilitating the release of siRNA from the complexes [55]. In this study, the siRNA released from SLPHNs demonstrated a similar migration distance as the siRNA control, indicating that SLPHNs can readily release the siRNA without causing its degradation. This finding confirms the successful preservation of siRNA integrity during the release process from SLPHNs.

3.5. Protection of siRNA in SLPHN from Serum Degradation

The ability to resist nuclease degradation is crucial for successful siRNA delivery, both in vitro and especially in vivo [1,56,57]. In addition to us, several authors have successfully submitted their developed nanocarriers to 50% FBS to mimic serum conditions [58]. Furthermore, evaluation of the interplay between SLPHNs and serum proteins can provide important guidelines for predicting their performance in biological systems [59].

After a 24 h exposure to serum (Figure 3c), free siRNA (10 μM) was not stable in the presence of 25% FBS and underwent complete degradation. This finding highlights the susceptibility of free siRNA to degradation in the presence of serum, emphasizing the need for protective delivery systems to ensure its stability and effectiveness. In contrast, the siRNA complexed with SLPHNs or Lipofectamine™ was released intact after the heparin polyanion assay, proving that SLPHNs protect siRNA from serum degradation.

3.6. In Vitro Skin Penetration Study

Next, we evaluated the delivery of siRNA into the skin by the SLPHNs. The penetration of siRNA into the skin is challenging because of its negative charge and high molecular weight. Additionally, the SC, which serves as a highly efficient barrier, controls the penetration of molecules and microorganisms through the skin [60]. To overcome this barrier and facilitate the delivery of siRNA into the deeper layers of the skin, nanoparticles have been extensively investigated as topical or transdermal formulations [2,61].

The in vitro siRNA penetration of SLPHN–0.25% PEI was evaluated using vertical diffusion cells and porcine ear skin as a biological membrane, which is a suitable alternative when barrier integrity is assured [30]. Figure 4 shows that naked siRNA displayed weak or no fluorescence on the skin surface. However, when the skin was treated with siRNA-SLPHN 0.25% PEI, a prominent green fluorescence signal was observed throughout the tissue, including the viable epidermis and dermis, indicating the presence of fluorescein-labeled siRNA. This observation suggests that the siRNA–SLPHN formulation allowed effective transport of siRNA across the stratum corneum, enabling the genetic material to be retained throughout the skin.

Figure 4. Distribution of labeled siRNA under light microscopy of in vitro penetration studies 24 h after application of naked FAM-siRNA (free siRNA) and FAM-siRNA–SLPHN–0.25% PEI (complexed siRNA) to untreated skin. The microscopic sections were visualized (Axioskop 2 plus microscope) at 640 nm and 730 nm band-pass excitation and emission filters, respectively, through a 10× objective The formulation baths were tested in triplicate, and representative images are shown. EP: epidermis including stratum corneum, DE: dermis.

According to this finding, the enhanced penetration of siRNA into the deeper skin layers facilitated by SLPHNs is due to the combination of two factors. First, the positive charge of SLPHN increases the interaction of this nanocarrier with the skin surface, and second, the occlusion caused by solid lipid nanoparticles on the stratum corneum surface increases its permeability [62]. The interaction of these two factors may have further contributed to weakening the skin barrier and thus enhanced the skin penetration of siRNA

3.7. Cellular Studies

3.7.1. Cell Viability Study

The cell viability of SLPHNs was evaluated by the resazurin reduction assay to verify their action on representative skin cells (keratinocytes (HaCaT) and fibroblasts cells (NIH/3T3)) to determine the appropriate nanoparticle concentration for the next in vitro studies. The concentration of nanoparticles used in this experiment was determined from the NTA results (described in Section 3.1.1), assuming that the exposure dose of the particles, expressed in number per volume, is the best dose measure [63].

Figure 5 shows the potential cytotoxicity of SLPHNs–0.15 PEI and SLPHNs–0.25% PEI after 24 h of treatment in HaCaT and NIH/3T3 cells. Overall, total lipid content or PEI concentration in SLPHNs applied to NIH/3T3 cells promoted a crucial toxicity factor causing cell toxicity. However, when SLPHNs were applied in a series of dilutions, this effect was less prominent, suggesting a dose-dependent cytotoxicity of SLPHNs.

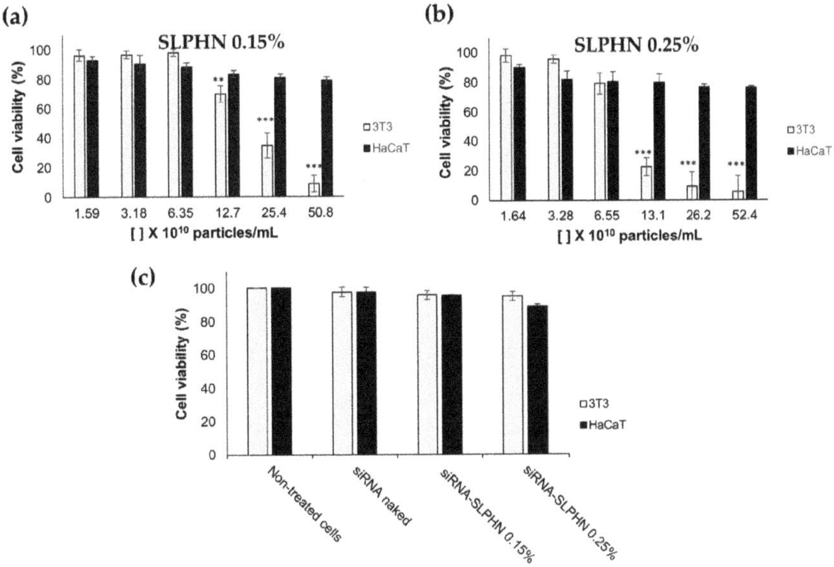

Figure 5. Cell viability of NIH/3T3 and HaCaT cells treated with different concentrations of SLPHN 0.15% PEI (**a**) and SLPHN 0.25% PEI (**b**) and cell viability of naked siRNA, siRNA–SLPHN–0.15% PEI and siRNA–SLPHN–0.25% PEI in NIH/3T3 and HaCaT cells (**c**). One-way ANOVA followed by Tukey's post-test was used as statistical analysis. ** $p < 0.01$; *** $p < 0.001$ compared to the lowest particle concentration ($\times 10^{10}$ particles/mL) in each group.

In fact, all SLPHNs tested on HaCaT did not present a cytotoxic effect, but a cytotoxic effect on NIH/3T3 cells was observed with 25.4×10^{10} particles/mL for SLPHN–0.15% PEI and 13.1×10^{10} particles/mL for SLPHN–0.25% PEI (Figure 5a,b). These results suggest that NIH/3T3 cells were more susceptible to both the physical effect of sedimentation of low-density nanoparticles on the cell membrane and the chemical effect of their composition [64] compared with HaCaT cells, mainly because PEI can be cytotoxic at certain concentrations [26]. These findings are consistent with the findings of Lee and coworkers (2002) regarding phototoxicity [65]. Finally, cell viability above 80% was achieved in HaCaT and NIH/3T3 cells when using SLPHNs at approximately 6×10^{10} and 3×10^{10} particles/mL, respectively (Figure 5c). Under these conditions, both SLPHN formulations did not present cytotoxicity when complexed with siRNA.

3.7.2. Cellular Uptake

The ability of SLPHNs to promote cellular uptake of siRNA was assessed by quantitative flow cytometry, which quantifies siRNA Alexa Fluor® 647 within the cell after incubation of HaCaT and NIH/3T3 cells with different preparations. Figure 6 shows that siRNA cell internalization is above 85% for all tested SLPHNs, regardless of the cell line. For instance, the percentage of cells containing siRNA Alexa Fluor® 647 was 92.93 ± 2.71% for SLPHN–0.15% and 85.37 ± 3.94% for SLPHN–0.25% in HaCaT cells (Figure 6b) and 99.70 ± 0.20% for SLPHN–0.15% and 99.93 ± 0.06% for SLPHN–0.25% in NIH/3T3 cells (Figure 6d). Notably, naked siRNA did not have a significant internalization ($p < 0.001$).

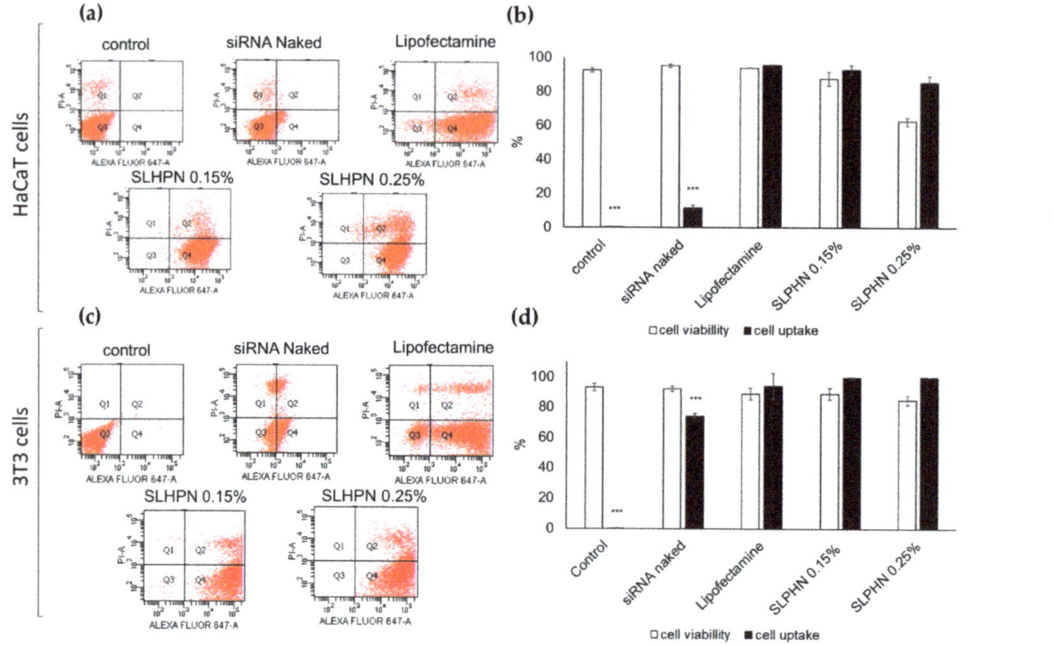

Figure 6. Representative plot obtained by flow cytometry (**a**,**c**) and cellular uptake/viability (**b**,**d**) in total cells using HaCaT and NIH/3T3 cells and siRNA Alexa Fluor® 647 at a concentration of 40 pmol/well. Results represent the mean ± SD (n = 3). One-way ANOVA followed by Tukey's post-test was used as statistical analysis. *** $p < 0.001$ compared with the Lipofectamine group.

In fact, the strategy of releasing siRNA from SLPHN–0.15% PEI, SLPHNs–0.25% PEI, and Lipofectamine™ resulted in 5-fold higher fluorescence ($p < 0.001$) compared with naked siRNA. These findings suggest that the developed SLPHNs associated with PEI promote cell internalization through endocytosis due to their opposite charge to the cell membrane [55], allowing transfection of siRNA, a molecule that has limited cell penetration due to its hydrophilicity, high molecular weight, and negative charge [66,67].

3.7.3. Intracellular Localization by Confocal Microscopy

The intracellular localization of siRNA–SLPHN in HaCaT and NIH/3T3 cells was performed by confocal microscopy. siRNA–SLPHNs were efficiently internalized by the cells and showed cytoplasmic localization (Figure 7). In this study, we used the siRNA–Lipofectamine™ 2000® as a positive control of transfection [14].

The presence and distribution of siRNA Alexa® 647 red fluorescence around the DAPI-stained nucleus (blue) can be visualized in the merged images (Figure 7), evidencing the cell-wide distribution of siRNA–SLPHNs and siRNA–Lipofectamine™ in the cytoplasm of both NIH/3T3 and HaCaT cells. Stronger red fluorescence intensity was observed for siRNA Alexa® 647 in keratinocyte cells (HaCaT) compared to fibroblast cells (NIH/3T3) in all tested samples. On the other hand, naked Alexa® 647 siRNA exhibited minimal red fluorescence, indicating that there was no significant cellular uptake of free siRNA.

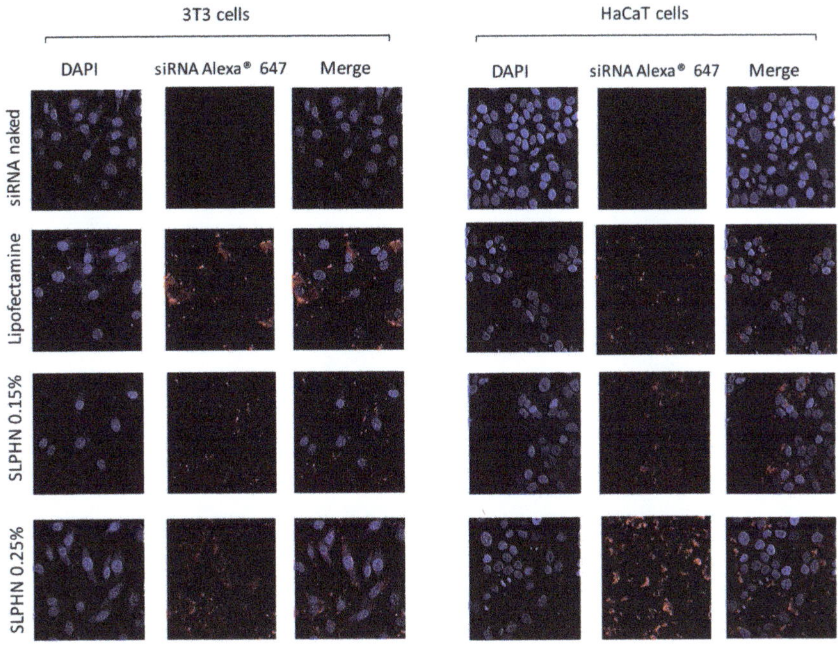

Figure 7. Cellular transfection in NIH/3T3 and HaCaT cells treated with free siRNA labeled with Alexa Fluor® 647, Lipofectamine™, and SLPHNs. The cell nucleus was stained with DAPI (blue), and siRNA Alexa® complexed with the internalized nanostructures was stained red.

3.7.4. In Vitro Silencing Efficiency of siRNA–SLPHN

The in vitro silencing study was performed in HaCaT cells stably expressing Firefly luciferase. The Renilla luciferase assay is commonly used to monitor transfection and silencing efficiency [28,31,68]. The silencing of genes expressing luciferase is measured by the ability of luciferase to convert chemical energy to light energy by catalyzing biological reactions using luciferin as a substrate. Thus, firefly luciferase activity in cells treated with Firefly luciferase-specific siRNA (siLuc) and nonspecific siRNA (ns-siRNA) was normalized by Renilla luciferase activity.

As demonstrated in the previous section, naked siRNA is unable to penetrate the cell membrane and reach the cytoplasm of HaCaT cells, so this sample was not used in this study. Figure 8 summarizes the silencing effect of anti-luciferase siRNA complexed with SLPHN and Lipofectamine™. All tested SLPHN samples demonstrated effective delivery of siLuc to HaCaT cells. Moreover, these formulations exhibited a significant reduction in Firefly luciferase expression in comparison to cells that were treated with ns-siRNA ($p < 0.05$).

Cells treated with siLuc–SLPHN and 0.25% PEI showed a 1.67-fold and 2.06-fold reduction in Firefly luciferase expression than Lipofectamine™ and SLPHN–0.15% PEI, respectively. SLPHN–0.25% PEI has a greater cationic charge, as shown by the zeta potential (Table 1), which allows a greater amount of siRNA to be complexed and increases cellular uptake (Figure 7). It is importante also to consider the contribution of the PEI interaction with particle components, mainly due to its interaction with poloxamer, being part of its constitution. The interaction of PEI with the poloxamer molecule was previously described [69], suggesting an interaction between protonated PEI amines and free electrons of the oxygen atoms of poloxamer.

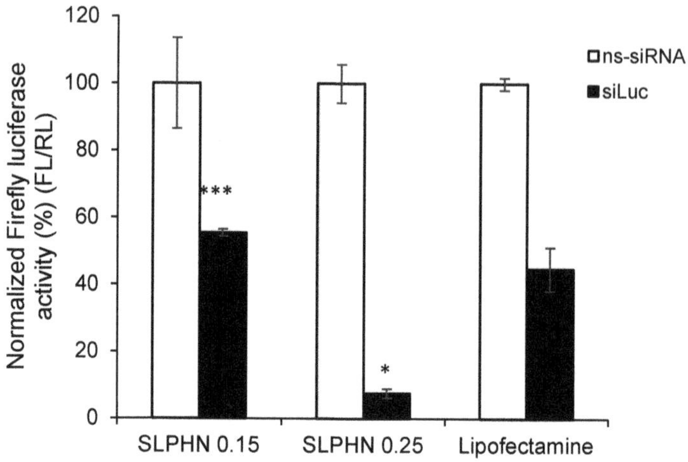

Figure 8. Percentage of inhibition of Firefly luciferase activity normalized by Renilla luciferase after treatment of HaCaT cells with siLuc (siRNA luciferase) and ns-siRNA (nonspecific siRNA (100 pmol/1 × 10^5 cells) complexed with SLPHNs (0.15% and 0.25% PEI) and Lipofectamine™ *** $p < 0.001$ and * $p < 0.05$ compared to cells treated with siLuc–Lipofectamine™ complex (100 pmol/1 × 10^5 cells), ($n = 3$).

All these characteristics resulted in greater silencing efficiency. By measuring luciferase activity, it is possible to determine whether the therapeutic genes carried by the nanoparticles are successfully delivered and expressed in the target cells, providing quantitative information about treatment efficacy. Therefore, we believe that before selecting a suitable inflammatory gene target, an appropriate luciferase assay must be conducted as a proof-of-concept silencing experiment.

Considering the data obtained in this work, we believe that the SLPHNs we developed stand out compared to several other lipid-based nanoparticles carrying siRNA for topical delivery [2] because of their favorable features, such as the biocompatibility of the components. The simple and reproducible preparation method favors the transfer of the production of these nanoparticles from the laboratory scale to the industrial scale. To our knowledge, this is the first time SLPHNs composed of Compritol®/PEI/poloxamer for topical delivery of nucleic acids have been reported, and they were found to be highly effective in complexing with siRNA and facilitating its release into the cytoplasm, leading to a therapeutic effect in vitro.

4. Conclusions

Novel and non-cytotoxic SLPHNs for siRNA delivery and topical application were successfully designed. The optimized SLPHN formulation exhibited a particle size of less than 200 nm and positive zeta potential. These particles remained stable when stored at 4 °C for a period of 90 days. This delivery system efficiently bound siRNA and showed high cellular transfection and cytoplasmic localization. Furthermore, the siRNA–SLPHNs released the specific siRNAs, reduced luciferase expression in HaCaT cells, and demonstrated significant knockdown of Firefly luciferase reporter expression. Moreover, the in vitro penetration assay demonstrated that SLPHN favored the penetration of siRNA into the epidermis, where several pathological processes of skin diseases occur.

Our findings demonstrate that the developed SLPHN–0.25% PEI is a promising nanoplatform for the cutaneous delivery of siRNA as it favored skin penetration of siRNA through the SC barrier and kept it in the epidermis, which is the desirable skin layer for topical application. The system also prevents transdermal delivery of siRNA, ensuring the safety of SLPHN as a carrier system of siRNA for topical delivery. In addition, SLPHNs–

0.25% PEI formed stable siRNA complexes that efficiently transfected cells and promoted gene knockdown in vitro.

Author Contributions: Conceptualization, M.M.d.A., P.D.M. and M.V.L.B.B.; methodology, M.M.d.A., L.N.B.-C., F.G.P.; software, M.M.d.A., L.N.B.-C., F.G.P.; writing—original draft preparation, M.M.d.A., L.N.B.-C., F.G.P., P.D.M. and M.V.L.B.B.; writing—review and editing, M.M.d.A., F.G.P. and M.V.L.B.B.; supervision, M.V.L.B.B.; project administration, M.M.d.A. and M.V.L.B.B.; funding acquisition, M.V.L.B.B. All authors have read and agreed to the published version of the manuscript.

Funding: This work was conducted as part of the National Institute of Science and Technology of Pharmaceutical Nanotechnology (INCT-Nanofarma) and received support from the São Paulo Research Foundation (FAPESP, Brazil, grant #2014/50928-2) and Conselho Nacional de Desenvolvimento Científico e Tecnológico (CNPq, Brazil, grant #465687/2014-8). Margarete Moreno de Araujo was the recipient of a CAPES fellowship.

Data Availability Statement: Not applicable.

Acknowledgments: The authors would like to thank José Orestes Del Ciampo and Henrique Diniz for technical assistance and Kamilla Swiech Antonietto, Laboratory of Flow Cytometry of FCFRP-USP, Brazil, for assistance with the flow cytometry.

Conflicts of Interest: The authors declare no conflict of interest.

References

1. Sajid, M.I.; Moazzam, M.; Kato, S.; Cho, K.Y.; Tiwari, R.K. Overcoming Barriers for siRNA Therapeutics: From Bench to Bedside. *Pharmaceuticals* **2020**, *13*, 294. [CrossRef] [PubMed]
2. Rosa, J.; Suzuki, I.; Kravicz, M.; Caron, A.; Pupo, A.V.; Praça, F.G.; Bentley, M.V. Current Non-viral siRNA Delivery Systems as a Promising Treatment of Skin Diseases. *Curr. Pharm. Des.* **2018**, *24*, 2644–2663. [CrossRef] [PubMed]
3. Baran-Rachwalska, P.; Torabi-Pour, N.; Sutera, F.M.; Ahmed, M.; Thomas, K.; Nesbit, M.A.; Welsh, M.; Moore, C.T.; Saffie-Siebert, S.R. Topical siRNA delivery to the cornea and anterior eye by hybrid silicon-lipid nanoparticles. *J. Control. Release* **2020**, *326*, 192–202. [CrossRef] [PubMed]
4. Yu, C.; Li, L.; Hu, P.; Yang, Y.; Wei, W.; Deng, X.; Wang, L.; Tay, F.R.; Ma, J. Recent Advances in Stimulus-Responsive Nanocarriers for Gene Therapy. *Adv. Sci.* **2021**, *8*, 2100540. [CrossRef]
5. Mehmood, Y.; Shahid, H.; Rashid, A.; Alhamhoom, Y.; Kazi, M. Developing of SiO_2 Nanoshells Loaded with Fluticasone Propionate for Potential Nasal Drug Delivery: Determination of Pro-Inflammatory Cytokines through mRNA Expression. *J. Funct. Biomater.* **2022**, *13*, 229. [CrossRef]
6. Islam, R.A.; Al-Busaidi, H.; Zaman, R.; Abidin, S.A.Z.; Othman, I.; Chowdhury, E.H. Carbonate Apatite and Hydroxyapatite Formulated with Minimal Ingredients to Deliver SiRNA into Breast Cancer Cells In Vitro and In Vivo. *J. Funct. Biomater.* **2020**, *11*, 63. [CrossRef]
7. Lin, Y.-L.; Jiang, G.; Zhang, Z.; Nör, J.E.; ElSayed, M.E.H. Silencing Bcl-2 Expression in Epithelial Cancer Cells Using "Smart" Particles. *J. Funct. Biomater.* **2014**, *5*, 167–182. [CrossRef]
8. Zu, H.; Gao, D. Non-viral Vectors in Gene Therapy: Recent Development, Challenges, and Prospects. *AAPSJ.* **2021**, *23*, 78. [CrossRef]
9. Zhang, M.M.; Bahal, R.; Rasmussen, T.P.; Manautou, J.E.; Zhong, X.B. The growth of siRNA-based therapeutics: Updated clinical studies. *Biochem. Pharmacol.* **2021**, *189*, 114432. [CrossRef]
10. Ain, Q.U.; Campos, E.V.R.; Huynh, A.; Witzigmann, D.; Hedtrich, S. Gene Delivery to the Skin—How Far Have We Come? *Trends Biotechnol.* **2021**, *39*, 474–487. [CrossRef]
11. Aljabbari, A.; Lokras, A.G.; Kirkensgaard, J.J.K.; Rades, T.; Franzyk, H.; Thakur, A.; Zhang, Y.; Foged, C. Elucidating the nanostructure of small interfering RNA-loaded lipidoid-polymer hybrid nanoparticles. *J. Colloid Interface Sci.* **2023**, *633*, 907–922. [CrossRef] [PubMed]
12. Depieri, L.V.; Borgheti-Cardoso, L.N.; Campos, P.M.; Otaguiri, K.K.; Vicentini, F.T.M.D.C.; Lopes, L.B.; Fonseca, M.J.V.; Bentley, M.V.L.B. RNAi mediated IL-6 in vitro knockdown in psoriasis skin model with topical siRNA delivery system based on liquid crystalline phase. *Eur. J. Pharm. Biopharm.* **2016**, *105*, 50–58. [CrossRef] [PubMed]
13. Suzuki, I.L.; de Araujo, M.M.; Bagnato, V.S.; Bentley, M.V.L.B. TNFα siRNA delivery by nanoparticles and photochemical internalization for psoriasis topical therapy. *J. Control. Release* **2021**, *338*, 316–329. [CrossRef] [PubMed]
14. Viegas, J.S.R.; Praça, F.G.; Caron, A.L.; Suzuki, I.; Silvestrini, A.V.P.; Medina, W.S.G.; Del Ciampo, J.O.; Kravicz, M.; Bentley, M.V.L.B. Nanostructured lipid carrier co-delivering tacrolimus and TNF-α siRNA as an innovate approach to psoriasis. *Drug Deliv. Transl. Res.* **2020**, *10*, 646–660. [CrossRef]
15. Petrilli, R.; Eloy, J.O.; Praça, F.S.G.; Del Ciampo, J.O.; Fantini, M.A.C.; Fonseca, M.J.V.; Bentley, M.V.L.B. Liquid Crystalline Nanodispersions Functionalized with Cell-Penetrating Peptides for Topical Delivery of Short-Interfering RNAs: A Proposal for Silencing a Pro-Inflammatory Cytokine in Cutaneous Diseases. *J. Biomed. Nanotechnol.* **2016**, *12*, 1063–1075. [CrossRef]

16. Silvestrini, A.V.P.; Praça, F.G.; Leite, M.N.; Fantini, M.C.d.A.; Frade, M.A.C.; Bentley, M.V.L.B. Liquid crystalline nanoparticles enable a multifunctional approach for topical psoriasis therapy by co-delivering triptolide and siRNAs. *Int. J. Pharm.* **2023**, *640*, 123019. [CrossRef]
17. Ashrafizadeh, M.; Delfi, M.; Hashemi, F.; Zabolian, A.; Saleki, H.; Bagherian, M.; Azami, N.; Farahani, M.V.; Sharifzadeh, S.O. Hamzehlou, S.; et al. Biomedical application of chitosan-based nanoscale delivery systems: Potential usefulness in siRNA delivery for cancer therapy. *Carbohydr. Polym.* **2021**, *260*, 117809. [CrossRef]
18. Zakeri, A.; Kouhbanani, M.A.J.; Beheshtkhoo, N.; Beigi, V.; Mousavi, S.M.; Hashemi, S.A.R.; Zade, A.K.; Amani, A.M.; Savardashtaki, A.; Mirzaei, E.; et al. Polyethylenimine-based nanocarriers in co-delivery of drug and gene: A developing horizon. *Nano Rev. Exp.* **2018**, *9*, 1488497. [CrossRef]
19. Li, R.; Wei, F.; Wu, X.; Zhou, P.; Chen, Q.; Cen, Y.; Xu, G.; Cheng, X.; Zhang, A.; Hu, Q. PEI modified orange emissive carbon dots with excitation-independent fluorescence emission for cellular imaging and siRNA delivery. *Carbon* **2021**, *177*, 403–411 [CrossRef]
20. Djemaa SBen Munnier, E.; Chourpa, I.; David, S. Versatile electrostatically assembled polymeric siRNA nanovectors: Can they overcome the limits of siRNA tumor delivery ? *Int. J. Pharm.* **2019**, *567*, 118432. [CrossRef]
21. Alsaad, A.A.A.; Hussien, A.A.; Gareeb, M.M. Solid Lipid Nanoparticles (SLN) as a Novel Drug Delivery System: A Theoretical Review. *Syst. Rev. Pharm.* **2020**, *11*, 259–273.
22. Dave, V.; Tak, K.; Sohgaura, A.; Gupta, A.; Sadhu, V.; Reddy, K.R. Lipid-polymer hybrid nanoparticles: Synthesis strategies and biomedical applications. *J. Microbiol. Methods* **2019**, *160*, 130–142. [CrossRef] [PubMed]
23. Desfrançois, C.; Auzély, R.; Texier, I. Lipid Nanoparticles and Their Hydrogel Composites for Drug Delivery: A Review. *Pharmaceuticals* **2018**, *11*, 118. [CrossRef] [PubMed]
24. Rapalli, V.K.; Sharma, S.; Roy, A.; Alexander, A.; Singhvi, G. Solid lipid nanocarriers embedded hydrogel for topical delivery of apremilast: In-vitro, ex-vivo, dermatopharmacokinetic and anti-psoriatic evaluation. *J. Drug Deliv. Sci. Technol.* **2021**, *63*, 102442. [CrossRef]
25. EMEA. *ICH Topic Q 1 A (R2). Stability Testing of New Drug Substances and Products Step*; European Medicines Agency: Amsterdam, The Netherlands, 2003; pp. 1–20.
26. Borgheti-Cardoso, L.N.; Depieri, L.V.; Kooijmans, S.A.; Diniz, H.; Calzzani, R.A.; Vicentini, F.T.; van der Meel, R.; Fantini, M.C.; Iyomasa, M.M.; Schiffelers, R.M.; et al. In situ gelling liquid crystalline system based on monoglycerides and polyethylenimine for local delivery of siRNAs. *Eur. J. Pharm. Sci.* **2015**, *74*, 103–117. [CrossRef]
27. Borgheti-Cardoso, L.N.; Depieri, L.V.; Diniz, H.; Calzzani, R.A.; de Abreu Fantini, M.C.; Iyomasa, M.M.; Moura de Carvalho Vicentini, F.T.; Bentley, M.V. Self-assembling gelling formulation based on a crystalline-phase liquid as a non-viral vector for siRNA delivery. *Eur. J. Pharm. Sci.* **2014**, *58*, 72–82. [CrossRef]
28. Borgheti-Cardoso, L.N.; Kooijmans, S.A.A.; Fens, M.H.A.M.; van der Meel, R.; Vicentini, F.T.M.C.; Fantini, M.C.A.; Bentley, M.V.L.B.; Schiffelers, R.M. In Situ Gelling Liquid Crystalline System as Local siRNA Delivery System. *Mol. Pharm.* **2017**, *14*, 1681–1690. [CrossRef]
29. OECD. *OECD Guideline for the Testing of Chemicals. Skin Absorption: In Vitro Method. Test*; OECD: Paris, France, 2004; Volume 428, pp. 1–8.
30. Praça, F.S.G.; Medina, W.S.G.; Eloy, J.O.; Petrilli, R.; Campos, P.M.; Ascenso, A.; Bentley, M.V.L. Evaluation of critical parameters for in vitro skin permeation and penetration studies using animal skin models. *Eur. J. Pharm. Sci.* **2017**, *111*, 121–132. [CrossRef]
31. Leitner, S.; Grijalvo, S.; Solans, C.; Eritja, R.; García-Celma, M.; Calderó, G. Ethylcellulose nanoparticles as a new "in vitro" transfection tool for antisense oligonucleotide delivery. *Carbohydr. Polym.* **2020**, *229*, 115451. [CrossRef]
32. Rouco, H.; García-García, P.; Évora, C.; Díaz-Rodríguez, P.; Delgado, A. Screening strategies for surface modification of lipid-polymer hybrid nanoparticles. *Int. J. Pharm.* **2022**, *624*, 121973. [CrossRef]
33. Arunkumar, R.; Baskaran, V. Lutein Encapsulated in PLGA–Phospholipid Nano-Carrier Effectively Mitigates Cytokines by Inhibiting Tumor Necrosis Factor TNF-α and Nuclear Factor NF-κB in Mice Retina. *J. Funct. Biomater.* **2023**, *14*, 197. [CrossRef] [PubMed]
34. Sawant, R.R.; Sriraman, S.K.; Navarro, G.; Biswas, S.; Dalvi, R.A.; Torchilin, V.P. Biomaterials Polyethyleneimine-lipid conjugate-based pH-sensitive micellar carrier for gene delivery. *Biomaterials* **2012**, *33*, 3942–3951. [CrossRef] [PubMed]
35. Ribeiro, L.N.; Franz-Montan, M.; Breitkreitz, M.C.; Alcântara, A.C.; DE Castro, S.; Guilherme, V.A.; Barbosa, R.; de Paula, E. Nanostructured lipid carriers as robust systems for topical lidocaine-prilocaine release in dentistry. *Eur. J. Pharm. Sci.* **2016**, *93*, 192–202. [CrossRef] [PubMed]
36. Ribeiro, L.N.; Breitkreitz, M.C.; Guilherme, V.A.; da Silva, G.H.; Couto, V.M.; Castro, S.R.; de Paula, B.O.; Machado, D.; de Paula, E. Natural lipids-based NLC containing lidocaine: From pre-formulation to in vivo studies. *Eur. J. Pharm. Sci.* **2017**, *106*, 102–112. [CrossRef]
37. Yalcin, T.E.; Ilbasmis-Tamer, S.; Takka, S. Development and characterization of gemcitabine hydrochloride loaded lipid polymer hybrid nanoparticles (LPHNs) using central composite design. *Int. J. Pharm.* **2018**, *548*, 255–262. [CrossRef]
38. Farid, R.M.; El-Salamouni, N.S.; El-Kamel, A.H.; El-Gamal, S.S. Chapter 16—Lipid-based nanocarriers for ocular drug delivery. In *Nanostructures for Drug Delivery*; Elsevier Inc.: Amsterdam, The Netherlands, 2017; pp. 495–522.

39. Shukla, T.; Upmanyu, N.; Pandey, S.P.; Gosh, D. Chapter 1. Lipid nanocarriers. In *Lipid Nanocarriers for Drug Targeting*; Elsevier Inc.: Amsterdam, The Netherlands, 2018; pp. 1–48.
40. Javadzadeh, Y.; Bahari, L.A. Therapeutic Nanostructures for Dermal and Transdermal Drug Delivery. In *Nano- and Microscale Drug Delivery Systems*; Elsevier BV: Amsterdam, The Netherlands, 2017; pp. 131–146.
41. Wang, T.; Bae, M.; Lee, J.-Y.; Luo, Y. Solid lipid-polymer hybrid nanoparticles prepared with natural biomaterials: A new platform for oral delivery of lipophilic bioactives. *Food Hydrocoll.* **2018**, *84*, 581–592. [CrossRef]
42. Jangde, R.; Elhassan, G.O.; Khute, S.; Singh, D.; Singh, M.; Sahu, R.K.; Khan, J. Hesperidin-Loaded Lipid Polymer Hybrid Nanoparticles for Topical Delivery of Bioactive Drugs. *Pharmaceuticals* **2022**, *15*, 211. [CrossRef]
43. Carvalho, I.; Miranda, M.; Silva, L.; Chrysostomo-Massaro, T.; Paschoal, J.; Bastos, J.; Marcato, P. IN VITRO Anticancer Activity and Physicochemical Properties of SOLANUM LYCOCARPUM Alkaloidic Extract Loaded in Natural Lipid-Based Nanoparticles. *Colloid Interface Sci. Commun.* **2019**, *28*, 5–14. [CrossRef]
44. Bivas-Benita, M.; Romeijn, S.; Junginger, H.E.; Borchard, G. PLGA–PEI nanoparticles for gene delivery to pulmonary epithelium. *Eur. J. Pharm. Biopharm.* **2004**, *58*, 1–6. [CrossRef]
45. Ulasov, A.V.; Khramtsov, Y.V.; Trusov, G.A.; Rosenkranz, A.A.; Sverdlov, E.D.; Sobolev, A.S. Properties of PEI-based Polyplex Nanoparticles That Correlate With Their Transfection Efficacy. *Mol. Ther.* **2009**, *19*, 103–112. [CrossRef]
46. Ahmed, M.M.; Anwer, K.; Fatima, F.; Aldawsari, M.F.; Alalaiwe, A.; Alali, A.S.; Alharthi, A.I.; Kalam, M.A. Boosting the Anticancer Activity of Sunitinib Malate in Breast Cancer through Lipid Polymer Hybrid Nanoparticles Approach. *Polymers* **2022**, *14*, 2459. [CrossRef]
47. Filipe, V.; Hawe, A.; Jiskoot, W. Critical Evaluation of Nanoparticle Tracking Analysis (NTA) by NanoSight for the Measurement of Nanoparticles and Protein Aggregates. *Pharm. Res.* **2010**, *27*, 796–810. [CrossRef]
48. Mehnert, W.; Mäder, K. Solid lipid nanoparticles: Production, characterization and applications. *Adv. Drug. Deliv. Rev.* **2012**, *64*, 83–101. [CrossRef]
49. Panalytical, M. NTA and DLS—Examples of Complementary Nanomaterial Characterization Techniques. *AZO Mater.* **2017**, *1*, 1–11.
50. *ASTM E2834-12R22*; Standard Guide for Measurement of Particle Size Distribution of Nanomaterials in Suspension by Nanoparticle Tracking Analysis (NTA). American Society for Testing and Materials: West Conshohocken, PA, USA, 2012; Volume 9, pp. 1–11.
51. Sitterberg, J.; Özcetin, A.; Ehrhardt, C.; Bakowsky, U. Utilising atomic force microscopy for the characterisation of nanoscale drug delivery systems. *Eur. J. Pharm. Biopharm.* **2010**, *74*, 2–13. [CrossRef] [PubMed]
52. Tofani, L.B.; Depieri, L.V.; Campos, P.M.; Riul, T.B.; Antonietto, K.S.; Fantini, M.C.D.A.; Bentley, M.V.L.B. In Vitro TyRP-1 Knockdown Based on siRNA Carried by Liquid Crystalline Nanodispersions: An Alternative Approach for Topical Treatment of Vitiligo. *Pharm. Res.* **2018**, *35*, 104. [CrossRef] [PubMed]
53. Kushwaha, A.K.; Vuddanda, P.R.; Karunanidhi, P.; Singh, S.K. Development and Evaluation of Solid Lipid Nanoparticles of Raloxifene Hydrochloride for Enhanced Bioavailability. *BioMed Res. Int.* **2013**, *2013*, 584549. [CrossRef] [PubMed]
54. Zhou, L.; Chen, Z.; Chi, W.; Yang, X.; Wang, W.; Zhang, B. Mono-methoxy-poly(3-hydroxybutyrate-co-4-hydroxybutyrate)-graft-hyper-branched polyethylenimine copolymers for siRNA delivery. *Biomaterials* **2012**, *33*, 2334–2344. [CrossRef] [PubMed]
55. Qi, R.; Liu, S.; Chen, J.; Xiao, H.; Yan, L.; Huang, Y.; Jing, X. Biodegradable copolymers with identical cationic segments and their performance in siRNA delivery. *J. Control. Release* **2012**, *159*, 251–260. [CrossRef]
56. Xiong, X.-B.; Uludağ, H.; Lavasanifar, A. Biodegradable amphiphilic poly(ethylene oxide)-block-polyesters with grafted polyamines as supramolecular nanocarriers for efficient siRNA delivery. *Biomaterials* **2009**, *30*, 242–253. [CrossRef]
57. Pauley, K.M.; Cha, S. RNAi Therapeutics in Autoimmune Disease. *Pharmaceuticals* **2013**, *6*, 287–294. [CrossRef]
58. Kim, Y.-M.; Park, M.-R.; Song, S.-C. An injectable cell penetrable nano-polyplex hydrogel for localized siRNA delivery. *Biomaterials* **2013**, *34*, 4493–4500. [CrossRef]
59. Abstiens, K.; Figueroa, S.M.; Gregoritza, M.; Goepferich, A.M. Interaction of functionalized nanoparticles with serum proteins and its impact on colloidal stability and cargo leaching. *Soft Matter* **2019**, *15*, 709–720. [CrossRef] [PubMed]
60. Holmes, T.R.; Paller, A.S. Gene Regulation Using Spherical Nucleic Acids to Treat Skin Disorders. *Pharmaceuticals* **2020**, *13*, 360. [CrossRef] [PubMed]
61. Dharamdasani, V.; Mandal, A.; Qi, Q.M.; Suzuki, I.; Bentley, M.V.L.B.; Mitragotri, S. Topical delivery of siRNA into skin using ionic liquids. *J. Control. Release* **2020**, *323*, 475–482. [CrossRef]
62. Tiwari, N.; Osorio-Blanco, E.R.; Sonzogni, A.; Esporrín-Ubieto, D.; Wang, H.; Calderón, M. Nanocarriers for Skin Applications: Where Do We Stand? *Angew. Chem. Int. Ed. Engl.* **2022**, *61*, e202107960. [CrossRef] [PubMed]
63. Joris, F.; Manshian, B.B.; Peynshaert, K.; De Smedt, S.C.; Braeckmans, K.; Soenen, S.J. Assessing nanoparticle toxicity in cell-based assays: Influence of cell culture parameters and optimized models for bridging the in vitro–in vivo gap. *Chem. Soc. Rev.* **2013**, *42*, 8339–8359. [CrossRef]
64. Khanbeigi, R.A.; Kumar, A.; Sadouki, F.; Lorenz, C.; Forbes, B.; Dailey, L.A.; Collins, H. The delivered dose: Applying particokinetics to in vitro investigations of nanoparticle internalization by macrophages. *J. Control. Release* **2012**, *162*, 259–266. [CrossRef]
65. Lee, J.K.; Lee, E.; Lee, S. Comparison of Sensitivity Between Balb/c 3T3 Cell and HaCaT Cell by NRU Assay to Predict Skin Phototoxicity Potential. *Toxicol. Public Health* **2002**, *18*, 227–232.

66. Gürcan, S.; Tsapis, N.; Reynaud, F.; Denis, S.; Vergnaud, J.; Özer, Ö.; Fattal, E. Combining dexamethasone and TNF-α siRNA within the same nanoparticles to enhance anti-inflammatory effect. *Int. J. Pharm.* **2021**, *598*, 120381. [CrossRef]
67. Karim, M.E.; Chowdhury, E.H. PEGylated Strontium Sulfite Nanoparticles with Spontaneously Formed Surface-Embedded Protein Corona Restrict Off-Target Distribution and Accelerate Breast Tumour-Selective Delivery of siRNA. *J. Funct. Biomater.* **2022**, *13*, 211. [CrossRef]
68. Suñé-Pou, M.; Prieto-Sánchez, S.; El Yousfi, Y.; Boyero-Corral, S.; Nardi-Ricart, A.; Nofrerias-Roig, I.; Pérez-Lozano, P.; García-Montoya, E.; Miñarro-Carmona, M.; Ticó, J.R.; et al. Cholesteryl oleate-loaded cationic solid lipid nanoparticles as carriers for efficient gene-silencing therapy. *Int. J. Nanomed.* **2018**, *13*, 3223–3233. [CrossRef]
69. Liang, W.; Gong, H.; Yin, D.; Lu, S.; Fu, Q. High-Molecular-Weight Polyethyleneimine Conjuncted Pluronic for Gene Transfer Agents. *Chem. Pharm. Bull.* **2011**, *59*, 1094–1101. [CrossRef]

Journal of *Functional Biomaterials*

MDPI

Article

Influence of the Loading with Newly Green Silver Nanoparticles Synthesized Using *Equisetum sylvaticum* on the Antibacterial Activity and Surface Hardness of a Composite Resin

Ionuț Tărăboanță [1,†], Ana Flavia Burlec [2,†], Simona Stoleriu [1,*], Andreia Corciovă [2,*], Adrian Fifere [3], Denisa Batir-Marin [4], Monica Hăncianu [2], Cornelia Mircea [2], Irina Nica [1], Andra Claudia Tărăboanță-Gamen [1] and Sorin Andrian [1]

1 Faculty of Dental Medicine, Grigore T. Popa University of Medicine and Pharmacy, 16 Universitatii Str., 700115 Iasi, Romania; ionut-taraboanta@umfiasi.ro (I.T.); irina.nica@umfiasi.ro (I.N.); andra-claudia.gamen@umfiasi.ro (A.C.T.-G.); sorin.andrian@umfiasi.ro (S.A.)

2 Faculty of Pharmacy, Grigore T. Popa University of Medicine and Pharmacy, 16 University Str., 700115 Iasi, Romania; ana-flavia.l.burlec@umfiasi.ro (A.F.B.); mhancianu@yahoo.com (M.H.); corneliamircea@yahoo.com (C.M.)

3 Centre of Advanced Research in Bionanoconjugates and Biopolymers Department, "Petru Poni" Institute of Macromolecular Chemistry, 41A Grigore Ghica Voda Alley, 700487 Iasi, Romania; fifere@icmpp.ro

4 Faculty of Medicine and Pharmacy, Dunărea de Jos University, 800010 Galati, Romania; denisa.batir@ugal.ro

* Correspondence: simona.stoleriu@umfiasi.ro (S.S.); acorciova@yahoo.com (A.C.)

† These authors contributed equally to this work.

Citation: Tărăboanță, I.; Burlec, A.F.; Stoleriu, S.; Corciovă, A.; Fifere, A.; Batir-Marin, D.; Hăncianu, M.; Mircea, C.; Nica, I.; Tărăboanță-Gamen, A.C.; et al. Influence of the Loading with Newly Green Silver Nanoparticles Synthesized Using *Equisetum sylvaticum* on the Antibacterial Activity and Surface Hardness of a Composite Resin. *J. Funct. Biomater.* **2023**, *14*, 402. https://doi.org/10.3390/jfb14080402

Academic Editor: Christie Ying Kei Lung

Received: 21 June 2023
Revised: 25 July 2023
Accepted: 26 July 2023
Published: 28 July 2023

Abstract: The aim of the study was to evaluate the antibacterial activity and surface hardness of a light-activated microhybrid composite resin modified with green silver nanoparticles (AgNPs). AgNPs were synthesized using an *Equisetum sylvaticum* extract and characterized through different methods such as UV-Vis, EDX, and FTIR. The obtained AgNPs were mixed with a microhybrid composite resin (Herculite XRV, Kerr Corp., Orange, CA, USA) in different concentrations: 0% (group A-control); 0.5% (group B); 1% (group C); and 1.5% (group D). A total of 120 composite resin disk-shaped samples were obtained and divided into 4 groups (n = 30) according to AgNP concentration. Each group was then divided into 2 subgroups: subgroup 1—samples were not soaked in 0.01 M NaOH solution; and subgroup 2—samples were soaked in 0.01 M NaOH solution. The antibacterial activity against *Streptococcus mutans* was determined using a direct contact test. A digital electronic hardness tester was used to determine the composite resin's Vickers surface hardness (VH). Statistical analysis was performed using the Mann–Whitney U and Kruskal–Wallis nonparametric tests with a confidence level of 95%. Groups C and D showed higher antibacterial activity against *S. mutans* when compared to the control group ($p < 0.05$). No significant differences were recorded between VH values ($p > 0.05$). The use of AgNPs synthesized from *Equisetum sylvaticum* as a composite resin filler in 1% wt. and 1.5% wt. reduced the activity of *Streptococcus mutans*. Soaking of the experimental composite resin decreased the antibacterial efficacy. The loading of a microhybrid composite resin with AgNPs in concentrations of 0.5% wt., 1% wt., and 1.5% wt. did not influence the surface hardness.

Keywords: AgNPs; *Equisetum sylvaticum*; *Streptococcus mutans*; microhybrid composite resin; Vickers hardness

1. Introduction

The oral cavity is a conducive environment for development and multiplication of pathogenic microorganisms [1]. Some oral pathogens are associated with the development of carious or periodontal disease [2]. One of the common causes of direct restorations failure is the occurrence of secondary caries due to the adhesion of the bacterial biofilm to the tooth surfaces or to restorative materials [1]. In the absence of an adequate treatment dental caries and periodontal diseases can evolve into systemic diseases such as infectious endocarditis or respiratory diseases [3,4]. Pathogenic microorganisms form the bacterial biofilm that attaches to dental surfaces, to the surrounding tissues, or to the surface of restorative materials [5]. Then it is important for dental materials used for direct restorations to have antimicrobial properties [1]. Some of the mostly used antibacterial components are: monomer MDPB (12-methacryloyloxydodecyl pyridinium bromide), antibiotics, silver nanoparticles (AgNPs), chlorhexidine digluconate (CHX), ursolic acid, or chitosan [1,4].

Silver, initially used in the form of silver nitrate ($AgNO_3$) and later in combination with fluoride (AgF), has been used in oral care for a long time, becoming widespread in the 19th century as a component of dental amalgams. In the middle of the 20th century, when polymer resin appeared on the market, its popularity decreased [6]. In recent years, as a result of developments in the nanotechnology field, the use of silver in the form of nanoparticles smaller than 100 nm has been highlighted in various industries such as pharmaceuticals, healthcare, information technology, and cosmetics [7–9]. The estimates show continuous growth of the silver nanoparticle market in the following years [9]. The use of AgNPs as antimicrobial agents incorporated into the inorganic component of direct dental restorative materials has become very popular due to the prolonged antimicrobial effect, the reduction of biofilm colonization potential [10], and the reduced volatility [1]. The antimicrobial effect of AgNPs relies on the silver content (ranging between 10,000 and 15,000 silver atoms) and on increased surface/volume ratio [11]. A perfect explanation for AgNPs's mechanism of action is not known, but due to their small size, they interact with the peptidoglycan cell wall, increasing membrane permeability, resulting in cell death [10]. According to Gudkov et al. [12], the antimicrobial efficacy of AgNPs is manifested by at least 5 mechanisms, such as the destruction of the cell wall and cytoplasmic components as a result of binding to the cell envelope, inactivation of respiratory chain dehydrogenases, inhibition of ATP synthesis and production of oxidative stress [13,14], exerting a genotoxic activity, blocking the potential of cell multiplication, and a photocatalytic effect [12]. At the same time, such nanoparticles block the multiplication potential of the bacterial DNA by condensing it [13,14] and disrupting the enzymatic activity by attacking thiol groups [15].

Previous studies have demonstrated that the use of AgNPs in direct dental restorative materials offers bacteriostatic or bactericidal effects on oral bacteria such as *Streptococcus mutans* or *Lactobacillus acidophilus* [16,17], increased biocompatibility [10], and reduced cytotoxicity and immunological response [12]. Synthesis of AgNPs can be achieved by chemical, physical, and biological methods [18]. At this time, the synthesis of AgNPs using plant extracts is preferred due to increased accessibility and improved biological properties [19]. The *Equisetum* genus from the *Equisetaceae* family, also known as "horsetail", is found in Europe, North, Central, and South America and is used in the treatment of lung, kidney, or gastrointestinal ailments [20]. Species from this genus contain several bioactive compounds such as flavonoids, quercetin glycosides, phenolic acids, campesterol, phytosterols, alkaloids, and isofucosterol [21,22]. Nonetheless, a small number of studies that focus on species of the *Equisetum* genus can be found in the existing literature [19].

Consequently, the aim of this study was to assess the surface hardness and antibacterial efficacy against *S. mutans* of a microhybrid composite resin loaded with AgNPs obtained from an ethanolic extract of *Equisetum sylvaticum*. The null hypothesis was that there are no differences in hardness and antibacterial effect against *S. mutans* of the new experimental microhybrid composite resin when comparing to an unmodified composite material.

2. Materials and Methods

This study was performed in accordance with the Declaration of Helsinki and complied with all the rules imposed by the Ethics Commission of "Grigore T. Popa" University of Medicine and Pharmacy Iași (no. 291/10.04.2023).

A schematic representation of the study protocol is shown in Figure 1.

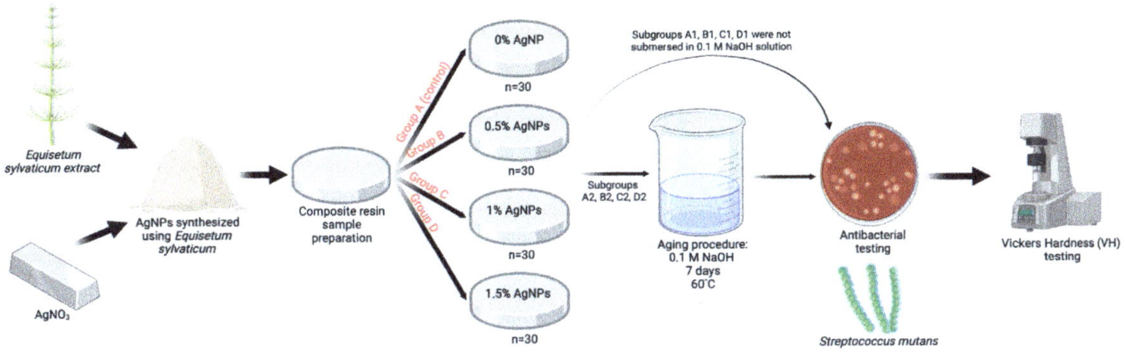

Figure 1. Schematic representation of the study protocol.

2.1. *Synthesis of AgNPs and Optimization and Characterization*

Initially, the *Equisetum sylvaticum* extract was obtained. Ten grams of dried and finely crushed plant material were mixed with 100 mL of water in an ultrasonic bath at 30 °C for 30 min. Afterwards, the mixture was filtered through Whatman no. 1 filter paper. The extract was stored at 4 °C in a dark place until further use.

The extract was used as a source of reducing agent and $AgNO_3$ as a precursor, the reaction being conducted at room temperature by magnetic stirring. For the optimization of synthesis, four parameters were taken into consideration: $AgNO_3$ concentration, pH, extract:$AgNO_3$ volume ratio, and reaction time. After establishing the reaction conditions, the colloidal solution containing AgNPs was centrifuged at 10,000 rpm for 30 min. To remove unwanted impurities, the obtained AgNPs were redispersed in water, centrifuged, and separated, with the operation being repeated twice. The purified AgNPs were dried and stored for further experiments.

To confirm the formation of AgNPs, the reaction mixture's color change (extract: $AgNO_3$) was monitored, and then the solution was examined using a UV-Vis spectrophotometer (Jasco V530, Jasco Inc., Tokyo, Japan) in the 400–500 nm range at various intervals.

Fourier transform infrared spectroscopy (FTIR) spectra were obtained by analyzing the pellet containing extract and AgNPs, respectively, in potassium bromide over the 4000–310 cm^{-1} spectral range using a Bruker Vertex 70 spectrophotometer (Bruker Corp. Berlin, Germany). Qualitative analysis of AgNPs was performed using a Quanta 200 environmental scanning electron microscope (ESEM) with energy dispersive X-ray spectroscopy (EDX). A Delsa Nano submicron particle size analyzer (Beckman Coulter, Brea, CA, USA) was used to measure the average diameter and the zeta potential value.

Transmission electron microscopy (TEM) studies were carried out with a Hitachi High-Tech HT7700 transmission electron microscope (Hitachi, Tokyo, Japan) that operated at a 100 kV accelerating voltage in high-contrast mode. The sample preparation methodology was as follows: a drop (10 µL) was placed on carbon-coated copper grids with 300-mesh size, and then the solvent was allowed to evaporate at room temperature.

2.2. *Composite Resin Sample Preparation*

To determine the sample size, G* Power software (Heinrich-Heine Universitat Dusseldorf, Dusseldorf, Germany) was used with an effect size set to 0.35, considered a medium effect according to Cohen's classification, an alpha value of 0.05, and a power of 0.80. The results obtained indicate the use of a total of 96 samples.

For sample preparation, Herculite XRV microhybrid composite resin (Kerr Corp., Orange, CA, USA) was used. The used composite material is composed of a resin matrix based on BisGMA, TEGDMA, and UDMA monomers and barium/silicium fillers in 79% wt., with an average size of 0.6 µm. A Kern PCB high-precision scale (Kern&Sohn GmbH, Balingen, Germany) was used to weigh the materials. The synthesized AgNPs were incorporated into the composite material in different concentrations of 0% wt., 0.5% wt., 1% wt., and 1.5% wt. The incorporation of AgNPs was performed in sterile containers devoid of light, and for the homogenization process, a magnetic stirrer set at 200 rpm was used for 20 min.

A total of 120 disk-shaped samples of 2 mm height and 5 mm width were divided into 4 groups according to loading level of composite resin with the synthetized AgNPs: group A (n = 30)—0% AgNP load (control); group B (n = 30)—0.5% AgNP load; group C (n = 30)—1% AgNP load; and group D (n = 30)—1.5% AgNP load. Each group was further divided into 2 subgroups: subgroup 1—samples were not soaked in 0.01 M NaOH solution; subgroup 2—samples were soaked in 0.01 M NaOH solution. The samples were prepared at an ambient temperature of 23–25 °C, humidity conditions of 65–75%, and low lighting. To obtain the samples, the microhybrid composite resin was inserted into a cylindrical acrylic mold and covered on both sides with 2 celluloid strips and 2 glass plates. A constant pressure was applied with a 500 g weight for 30 s to remove air voids and obtain a smooth surface. The photoactivation of the material was performed for 40 s using a Bluephase 20i light-curing lamp (Ivoclar, Vivadent, Schaan, Lichtenstein) with a wavelength between 285 and 515 nm and a light intensity of 1200 mW/cm^2. A Bluephase Meter II radiometer (Ivoclar, Vivadent, Schaan, Lichtenstein) was used to measure the light intensity for each light activation. The samples were finished and polished using a SofLex finishing and polishing system (Batch No. NC11346, 3M ESPE, St. Paul, MN, USA). The system consists of two spiral wheels (beige and white) made of thermoplastic elastomer impregnated with aluminum oxide particles. Finishing was performed for 1 min (30 s for each wheel) under a continuous water jet at a conventional speed of 20,000 rpm. Afterwards, the samples were sterilized in an autoclave (Dr. Mayer GmbH, Kempten, Germany) and stored in physiological serum at a constant temperature of 37 °C in an incubator (Biobase BJPXH30II, Biodusty, Shandong, China). Then, the samples were soaked for 7 days in a 0.1 M NaOH solution at a constant temperature of 60 °C, according to a protocol described by Prakki et al. [22]. The solution pH was 13, and it was checked every 24 h using a portable pH meter (Thermo Scientific Eutech pH 5+, Vernon Hills, IL, USA).

2.3. Antibacterial Analysis

The analysis of antimicrobial efficacy against *S. mutans* (ATCC®25175™) was performed using a direct contact test [23]. The bacterial suspension, with a content of 1.5×10^8 bacteria/1 mL, was cultured in a brain–heart infusion (BHI) broth with a concentration of 0.5 McFarland. The McFarland suspension was subsequently diluted 10^3 times to a final concentration of 1.5×10^5 to reduce the number of bacterial colonies and facilitate their visual quantification. A 0.01 mL volume of bacterial suspension was applied to the surfaces of sterile composite resin discs, after which the samples were placed in test tubes containing 0.5 mL of BHI and incubated for 12 h in an incubator (Biobase BJPXH30II, Biodusty, Jinan, China) at 37 °C in a 5% CO_2 atmosphere. Afterwards, 0.01 mL of liquid was collected from each culture medium and applied to a 5% sheep blood agar plate. After the incubation period, the colony-forming units (CFUs) were calculated using the following formula:

$$\text{CFU} = \frac{\text{Number of colonies} \times \text{Dilution factor}}{\text{Volume of culture plate}}$$

This assay was performed in triplicate.

2.4. Vickers Hardness Evaluation

To determine the Vickers surface hardness on both surfaces of the samples, a digital electronic hardness tester (Micro-Vickers Hardness System CV-400DMTM, CV Instruments Namicon, Otopeni, Romania) equipped with special indentation tips and a software program for measurement and evaluation were used. For each sample, 3 successive indentations were made on both sides of the composite resin discs, respecting a distance of 1 mm between them.

A dwell time of 30 s and a load of 50 g was applied to the determination tip. The criteria upon which the indentations were accepted consisted of the existence of a uniform appearance, no irregularities, and sharp, diagonal edges. To determine the hardness, the diagonal of the indentation was measured, and the results were expressed in Vickers hardness numbers (VHN), calculated based on the formula $D = 1854.4 \times F/d^2$ (N/mm^2), where D is the Vickers hardness value, F is the test force, and d is the length of the diagonal. The VHN values of each sample resulted from the mean of the 3 determinations.

2.5. Statistical Analysis

Statistical analysis was performed using SPSS 29.0 software (IBM SPSS Inc., Chicago, IL, USA). The Shapiro–Wilk test was used to test the normality of distribution and Levene's test was used to assess the homogeneity of variances. The nonparametric statistical tests Mann-Whitney U and Kruskal–Wallis were used due to the heterogeneity of the variances and large dispersion of the data, with a significance level of 0.05.

3. Results

3.1. Optimization of AgNP Synthesis and Physico-Chemical Characterization

Firstly, the synthesis was demonstrated by the color modification of the reaction mixture (extract:$AgNO_3$) from yellow to dark brown in 300 min and confirmed by comparing the UV-Vis absorbance spectra of the extract, $AgNO_3$ solution, and mixture (Figure 2). No absorbance peak was observed for the extract or $AgNO_3$ in the 400–500 nm range, but a peak appeared at 440 nm for the mixture.

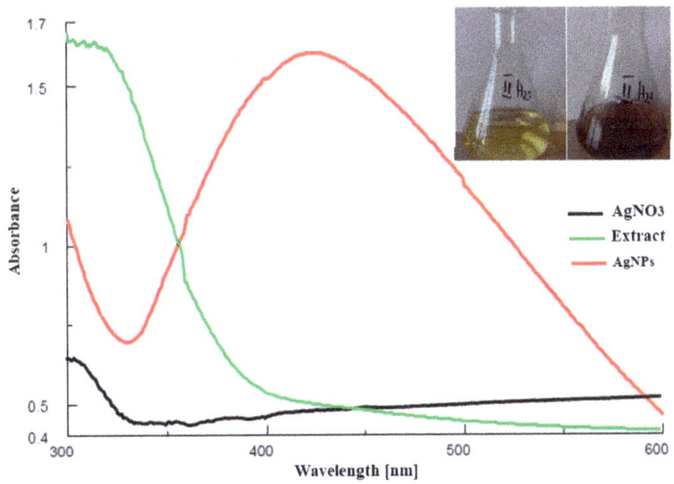

Figure 2. UV-Vis comparative spectra of the extract, $AgNO_3$ solution, final mixture. Inset: color modification: initial vs. final.

In the process of optimization, the extract concentration was maintained constant (10 g%), but different $AgNO_3$ concentrations were used (1 mM, 3 mM, and 5 mM) (Figure 3A). It was observed that, in all three cases, the AgNPs peak appeared, but its aspect was different: for 1 mM, the peak was large, while for 3 mM and 5 mM, the peak appeared sharper and without a significant increase in the absorbance value. Therefore, 3 mM $AgNO_3$ was used for further studies.

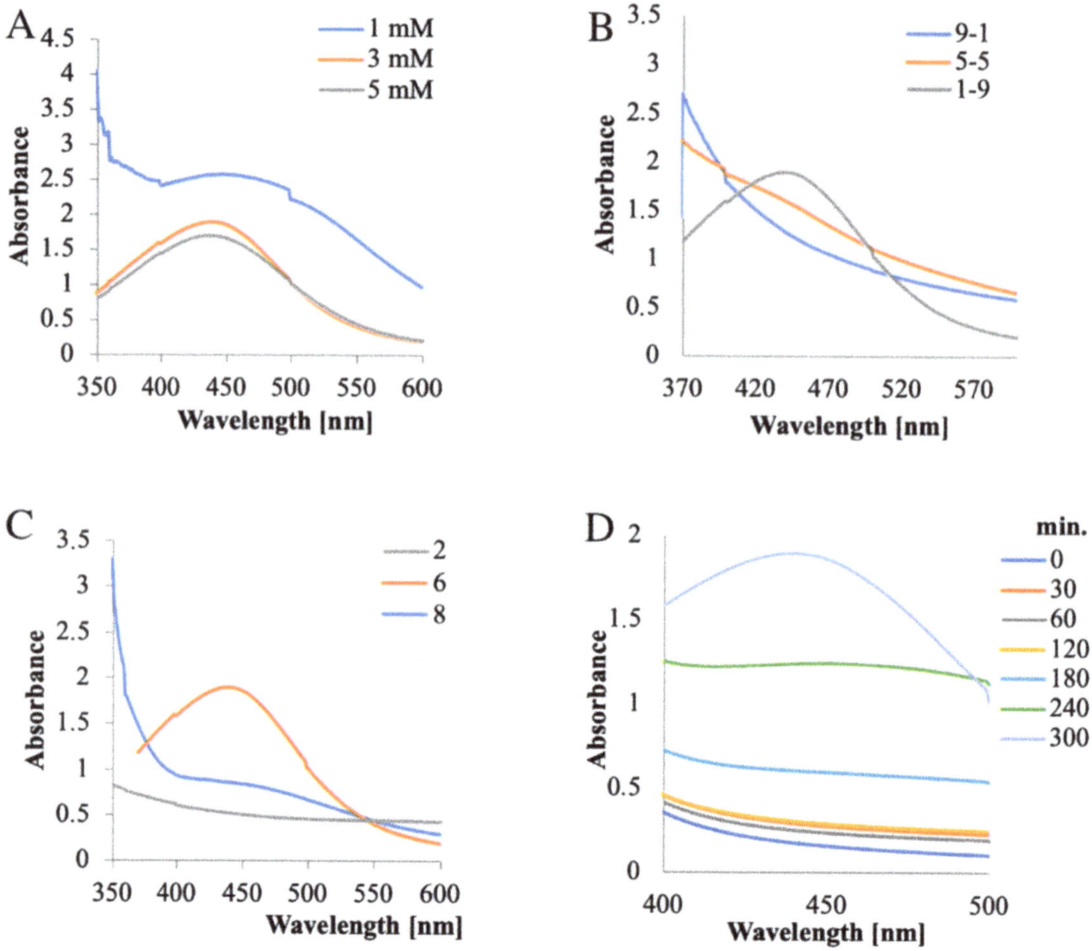

Figure 3. Optimization of parameters for AgNP synthesis: (**A**) $AgNO_3$ concentration; (**B**) extract:$AgNO_3$ volume ratio; (**C**) pH; (**D**) time.

Several extract:$AgNO_3$ volume ratios (9:1, 5:5, and 1:9) were examined using constant extract (10 g%) and silver salt (3 mM) concentrations (Figure 3B). For the 9:1 and 5:5 extract:$AgNO_3$ (*v:v*) ratios, no peak was observed, but the peak appeared at a 1:9 ratio; hence, this proportion was considered optimum for synthesis.

When optimizing the pH required for the reaction, it was found that, at pH 2, the synthesis was suppressed, while at pH values of 6 and 8, AgNPs were obtained. However, at pH 8, the peak was lower compared to that obtained at pH 6. Consequently, the last value was considered optimal for the synthesis (Figure 3C).

The optimal time required for synthesis was also examined. Initially, the reaction was slow; the peak appeared after 240 min and increased up to 300 min, after which there was no further significant increase, thus it can be considered that the synthesis was completed in 300 min (Figure 3D).

Therefore, the optimal conditions considered for the AgNP synthesis were: 3 mM $AgNO_3$ concentration, 1:9 (v/v) extract:$AgNO_3$ ratio, a pH of 6, and a 300 min synthesis time.

3.2. DLS Characterization, Zeta Potential, and EDX Analysis

The Zeta sizer showed an average hydrodynamic diameter of 136 nm with a polydispersity index of 1.116 for the prepared AgNPs. The obtained Zeta potential value was −23.99 (Figure 4A).

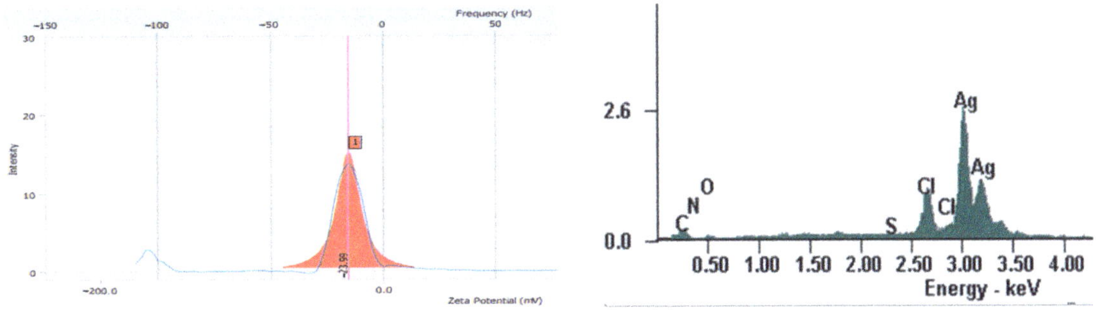

(A) (B)

Figure 4. (**A**) Zeta potential; (**B**) EDX spectrum. 1—Zeta potential value.

The EDX qualitative analysis of the AgNPs pointed out the characteristic peak for metallic silver at 3 keV (Figure 4B). The EDX quantitative analysis showed that the AgNPs contained 74.75% silver, 11.56% carbon, 1.2% nitrogen, and 3.66% (m%) oxygen.

3.3. Transmission Electron Microscopy (TEM) Analysis

The morphology of the nanoparticles was studied with TEM microscopy after solvent removal by evaporation at room temperature. TEM micrographs show nearly spherically shaped nanoparticles with a uniform morphology distribution without the appearance of edges, corners, or accentuation of an ovoid shape. In the dry state, the nanoparticles highlight the dimensions of the metallic Ag core, with an average diameter of 25 nm (Figure 5A). It seems that the vegetable extract effectively stabilizes the aqueous dispersion of the nanoparticles, considering that, in the TEM images at high magnification, there is no noticeable tendency for agglomeration after the removal of the solvent (Figure 5B).

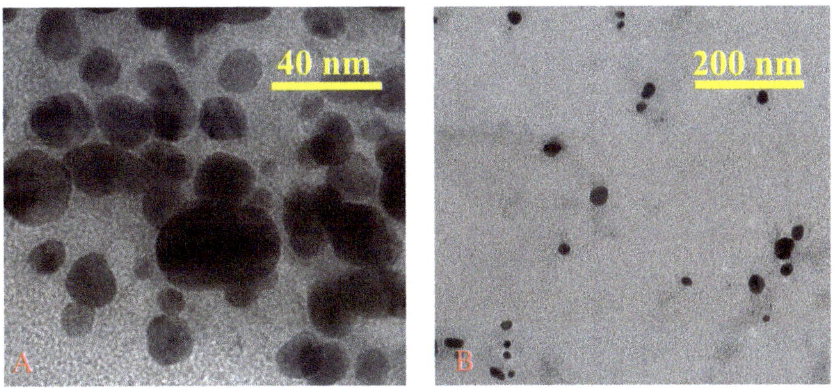

Figure 5. TEM micrographs of AgNPs with different magnifications (**A**,**B**).

3.4. FTIR Analysis

In order to determine the functional groups that participate in the synthesis and stabilization of AgNPs, FTIR spectra of the extract and AgNPs were recorded.

The extract spectrum showed significant absorbance bands at: 3442 cm^{-1}, corresponding to O-H stretching intermolecular hydrogen bonding from alcohols or phenol groups; 2925 cm^{-1} and 2356 cm^{-1}, representing C–H stretching of CH_3 and CH_2 (alkanes); 1591 cm^{-1}, corresponding to C=O, C–N (amide I), and COO– stretching vibrations; 1406 cm^{-1}, C–O (amide) stretching vibrations and C–C stretching vibrations of phenyl groups; COO symmetric stretching vibrations and CH_2 bond vibrations; 1261 cm^{-1}, C–O stretching vibrations of alcohols, ethers, esters, and carboxylic acids; 1116 cm^{-1}, C–O and C–C stretching vibrations from carbohydrates (Figure 6).

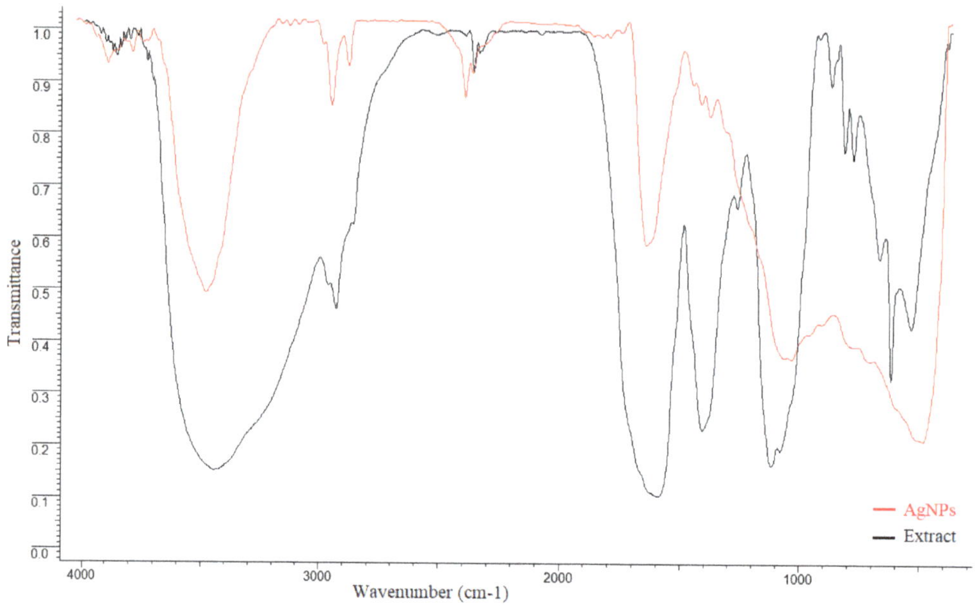

Figure 6. Comparative FTIR spectra of extract and AgNPs.

3.5. Microbiological Assay

When analyzing the obtained mean CFU values in subgroups 1, it can be observed that control subgroup A1 reached the highest number of colonies, at 237.8 ± 136.2, while subgroup C1 recorded the lowest CFU value of 83.9 ± 38.2 (Table 1).

Table 1. Mean number and standard deviation (SD) of *Streptococcus mutans* colonies (CFU) on 5% sheep blood agar.

Groups	Group A (0% wt. AgNPs)	Group B (0.5% wt. AgNPs)	Group C (1% wt. AgNPs)	Group D (1.5% wt. AgNPs)
CFU *Streptococcus mutans* -subgroups 1-	237.8 ± 136.2	145.3 ± 89.1	83.9 ± 38.2	91.7 ± 70.8
CFU *Streptococcus mutans* (samples soaked in 0.01 M NaOH solution) -subgroups 2-	201.7 ± 112.3	169.8 ± 113.4	125.9 ± 88.1	142.2 ± 105.5

The statistical analysis showed significant differences between control subgroup A1 and subgroup C1, with a significance level of $p = 0.012$, and between control subgroup A1 and subgroup D1, with a p value of 0.025. In subgroups 2, the peak was reached by control subgroup A2, with a mean of 201.7 ± 112.3, and the lowest value was recorded by subgroup C2, with a value of 125.9 ± 88.1. The statistical analysis showed no significant differences between the study subgroups and the control subgroup ($p > 0.05$).

3.6. Vickers Hardness Test Results

In subgroups 1, the highest value was recorded by subgroup C1 (Table 2), with a mean VHN of 60.62 ± 0.32, and the samples in subgroup B1 recorded the lowest value (59.88 ± 0.4). In subgroups 2, the peak was reached by subgroup D2 (58.8 ± 0.63), while the lowest value was recorded by subgroup B2 (56.91 ± 0.59). Within both subgroups, no statistically significant difference was recorded ($p > 0.05$).

Table 2. Mean Vickers hardness number (VHN) and standard deviation (SD) for each group recorded at surface and at base.

Groups	VHN -Subgroups 1-	VHN (Samples Soaked in 0.01 M NaOH Solution) -Subgroups 2-
A	60.14 ± 0.22	57.33 ± 0.76
B	59.88 ± 0.40	56.91 ± 0.59
C	60.62 ± 0.32	57.97 ± 0.88
D	60.48 ± 0.66	58.80 ± 0.63

4. Discussion

The aim of the study was to evaluate the antibacterial efficiency and surface hardness of a microhybrid composite resin loaded with silver nanoparticles, synthesized using an *Equisetum sylvaticum* extract. Different parts of plants, such as leaves, fruit, or flowers, were used for the green synthesis of AgNPs. The advantages of this technique are its low cost, short synthesis time, and the possibility of processing at a large scale [1]. Species from the Equisetum genus are perennial plants found in the wild flora of Romania and are traditionally used as treatments against cardiovascular, neurodegenerative, or infectious diseases [18]. AgNPs were synthesized using an eco-friendly method and further characterized by UV-Vis, EDX, FTIR spectroscopy, and TEM analysis. UV-Vis comparative spectra of the $AgNO_3$ extract solution and final mixture, as well as the color modification assessment at two different moments, initial and final, were performed. Initially, no absorbance peak was observed for the extract or $AgNO_3$ in the 400–500 nm range, but a peak appeared at 440 nm for the mixture; thus, the modification of color and the appearance of the peak can be explained by surface plasmon resonance. The peak corresponding to the colloidal solution was large, indicating that the AgNPs solution was polydisperse. This is explained by the variety of biomolecules found in the extract with different potentials to reduce Ag^+, which influences the nucleation and synthesis of AgNPs [24]. The negative Zeta potential value demonstrates that biomolecules found on the AgNPs' surfaces are negatively charged, implying a repulsion between nanoparticles, with the colloidal solution being stable [25–27]. The results obtained by EDX analysis demonstrate the presence of metallic silver as well as the presence of other elements that are found in the biomolecules responsible for the capping of nanoparticles [28]. The comparative FTIR analysis showed that, in the case of AgNPs, there were some shifts or disappearances of some bands. Moreover, the same analysis highlighted that several groups of compounds are involved in the synthesis and capping of AgNPs, with the functional groups belonging to classes of compounds such as flavonoids, proteins, amino acids, sterols, carbohydrates, and phenols found in the extract [29–31].

The green synthesized AgNPs were added in different concentrations (0%, 0.5%, 1%, and 1.5%) by mixing in a commercial microhybrid composite resin. The obtained results were consistent with the results of other studies that found antibacterial efficacy of AgNPs against *Streptococcus mutans* in concentrations of 1% or more [32,33]. Unlike the study conducted by Jenabi et al., in which, by adding 0.5% AgNPs, the number of colonies was significantly reduced, in our study there were no changes in CFU values for the same AgNP concentration [34]. Other reports concluded that even lower concentrations of 0.025% or 0.05% AgNP can have antibacterial effects against *S. mutans*, but our study showed no significant antibacterial effect for AgNPs concentrations lower than 1% [35]. This can be explained by the different antibacterial capacities of the products according to the size, shape, method of incorporation of the particles, or size of the samples [34]. Previous studies have reported that smaller particles show a stronger bactericidal effect due to higher surface/volume ratios [35,36].

Yassaei et al. consider a 1% concentration of AgNPs to be the maximum concentration that can be added to the composite resin, since higher concentrations can present increased toxicity and alter the aesthetic properties. The same author considers that a concentration of 1% is the maximum added AgNP concentration that can present antibacterial activity and, at the same time, not affect the mechanical properties of the composite material [33]. Nonetheless, other studies have used higher concentrations of AgNPs to test the antibacterial and mechanical properties of experimental composites [37,38].

In our study, AgNPs loaded in a commercial composite resin in 1% and 1.5% concentrations showed an antibacterial effect against *Streptococcus mutans*, while the hardness of the material showed no significant changes irrespective of AgNP concentration. These conclusions agree with the results of a study conducted by Bapat et al. [39].

Silver-based nanoparticles are efficient antibacterial agents, and their effectiveness increases as the surface/volume ratio increases [33]. Previous studies have demonstrated an increased cytotoxic effect of silver oxide on pathogenic microorganisms such as gram-positive or gram-negative bacteria, affecting their multiplication potential and adherence [33,40]. Silver ions can interact with the peptidoglycan cell wall by damaging the transmembrane transport of electrons; they can interfere with bacterial proteins and the plasma membrane; or they can lyse the bacterial wall, leading to cell death [35,41,42]. An undesirable effect of silver is its increased toxicity, which limits its use in human medicine [33], but its presentation in the form of nanoparticles considerably reduces the toxicity and increases its antibacterial efficiency [41]. Therefore, in reduced quantities, silver is a non-toxic metal to animal cells, but it can be very toxic to bacterial cells [42,43].

The discovery of materials with antibacterial potential has been one of the main concerns of researchers in the past decades. Regarding silver nanoparticles, the studies carried out in this direction reported an antibacterial efficiency of composites loaded with 0.5% and 1% AgNPs on *Streptococcus mutans* after 15 days, but when evaluated after 30 days, this effect was considerably reduced [44,45]. Antibacterial property maintenance over time is necessary because composite resin restorations are long-term treatments. In our study, the experimentally obtained composite material was soaked for 7 days in a 0.01 M NaOH solution at a constant temperature of 60 °C, according to the protocol described by Prakk et al. [22]. The results showed that the antibacterial efficiency of the silver nanoparticles loaded in the composite resin at concentrations of 1% and 1.5% was considerably reduced after soaking in 0.01 M NaOH solution; these findings agree with the results of another previous study [46]. NaOH acts on the composite resin by chemical degradation, accelerating the hydrolysis process [47]. The Vickers hardness of the experimental composite resin was not affected by submersion in NaOH solution for 7 days. Other studies have reported significant changes in physical and mechanical properties after this aging procedure [47–49].

Nonetheless, one important limitation of the present in vitro study is the impossibility of achieving a complex environment such as that of the oral cavity. The study may also be limited by the use of Herculite XRV composite resin as a tested material due to the fact that it was launched onto the market more than 25 years ago and, as resin-based materials evolved in recent years, it has required improvements in its composition. Consequently, further in vivo or other in vitro studies that replicate the oral environment conditions considering the presence of saliva, variations of the salivary flow, enzyme activity, and thermal or pH variations are recommended to confirm the antibacterial potential and mechanical behavior of the studied material.

5. Conclusions

The use of 1% wt. and 1.5% wt. silver nanoparticles synthesized from *Equisetum sylvaticum* as a composite resin filler reduced the activity of *Streptococcus mutans*. Soaking of the obtained experimental composite resin in a 0.01 M NaOH solution decreased the antibacterial efficacy. The loading of a commercial microhybrid composite resin with silver nanoparticles in concentrations of 0.5% wt., 1% wt., and 1.5% wt. did not influence the surface hardness.

Author Contributions: Conceptualization, I.T., A.F.B., A.C. and S.S.; methodology, M.H., C.M., I.N., A.C.T.-G. and S.A.; validation, D.B.-M. and A.F.; formal analysis, A.C., M.H., C.M. and S.A.; investigation, I.T., A.F.B., I.N. and A.C.T.-G.; resources, I.T., A.F.B. and A.C.; writing—original draft preparation, I.T., A.F.B., A.C., S.S. and A.C.T.-G.; writing—review and editing, A.F., D.B.-M., M.H., C.M. and S.A.; visualization, S.S. and I.N.; supervision, A.C., M.H., and S.A.; project administration, I.T. and A.F.B.; funding acquisition, I.T., A.F.B. and A.C. All authors have read and agreed to the published version of the manuscript.

Funding: This article was published with the support of the project Net4SCIENCE: Applied doctoral and postdoctoral research network in the fields of smart specialization Health and Bioeconomy, project code POCU/993/6/13/154722.

Institutional Review Board Statement: The study was conducted in accordance with the Declaration of Helsinki, and approved by the Ethics Committee) of "Grigore T. Popa" University of Medicine and Pharmacy Iași (no. 291/10.04.2023).

Informed Consent Statement: Not applicable.

Data Availability Statement: All the data presented in this study are available within the article.

Conflicts of Interest: The authors declare no conflict of interest.

References

1. Enan, E.T.; Ashour, A.A.; Basha, S.; Felemban, N.H.; El-Rab, S.M. Antimicrobial activity of biosynthesized silver nanoparticles, amoxicillin, and glass-ionomer cement against *Streptococcus mutans* and *Staphylococcus aureus*. *Nanotechnology* **2021**, *32*, 215101.
2. Teodorescu, A.C.; Teslaru, S.; Solomon, S.M.; Zetu, L.; Luchian, I.; Sioustis, I.A.; Martu, M.A.; Vasiliu, B.; Martu, S. Assessment of Bacterial Associations Involved in Periodontal Disease Using Crevicular Fluid. *Rev. Chim.* **2019**, *70*, 2145–2149. [CrossRef]
3. Ahmed, O.; Sibuyi, N.R.; Fadaka, A.O.; Madiehe, M.A.; Maboza, E.; Meyer, M.; Geerts, G. Plant extract-synthesized silver nanoparticles for application in dental therapy. *Pharmaceutics* **2022**, *14*, 380. [PubMed]
4. Pancu, G.; Andrian, S.; Iovan, G.; Gheorghe, A.; Nica, I.; Topoliceanu, C.; Antonia, M.; Tofan, N.; Stoleriu, S. The Assessment of Microbiology of the Bacterial Biofilm Adjacent to Different Types of the Materials Used for the Sealing of the Dental Occlusal Fissures. *Rom. J. Oral. Rehab.* **2016**, *8*, 67–71.
5. Valm, A.M. The Structure of Dental plaque microbial communities in the transition from health to dental caries and periodontal disease. *J. Mol. Biol.* **2019**, *431*, 2957–2969. [PubMed]
6. Noronha, V.T.; Paula, A.J.; Durán, G.; Galembeck, A.; Cogo-Müller, K.; Franz-Montan, M.; Durán, N. Silver nanoparticles in dentistry. *Dent. Mater.* **2017**, *33*, 1110–1126. [CrossRef]
7. Rai, M.; Ingle, A.P.; Gade, A.K.; Duarte, M.C.; Duran, N. Three *Phoma* spp. synthesised novel silver nanoparticles that possess excellent antimicrobial efficacy. *IET Nanobiotechnol.* **2015**, *9*, 280–287.
8. Padovani, G.C.; Feitosa, V.P.; Sauro, S.; Tay, F.R.; Durán, G.; Paula, A.J.; Durán, N. Advances in dental materials through nanotechnology: Facts, perspectives and toxicological aspects. *Trends Biotechnol.* **2015**, *33*, 621–636. [CrossRef] [PubMed]
9. IMARC Group. Market Research Report: Silver Nanoparticles Market: Global Industry Trends, Share, Size, Growth, Opportunity and Forecast 2023–2028. 2023. Available online: https://www.imarcgroup.com/silver-nanoparticles-market#:~:text=Market%20Overview%3A,16%25%20during%202023%2D2028 (accessed on 15 July 2023).
10. Corrêa, J.M.; Mori, M.; Sanches, H.L.; Cruz, A.D.; Poiate, E.; Poiate, I.A. Silver nanoparticles in dental biomaterials. *Int. J. Biomater.* **2015**, *2015*, 485275. [CrossRef]
11. Yin, I.X.; Zhang, J.; Zhao, I.S.; Mei, M.L.; Li, Q.; Chu, C.H. The antibacterial mechanism of silver nanoparticles and its application in dentistry. *Int. J. Nanomed.* **2020**, *15*, 2555–2562. [CrossRef]
12. Gudkov, S.V.; Serov, D.A.; Astashev, M.E.; Semenova, A.A.; Lisitsyn, A.B. Ag_2O nanoparticles as a candidate for antimicrobial compounds of the new generation. *Pharmaceuticals* **2022**, *15*, 968.

13. Quinteros, M.A.; Viviana, C.A.; Onnainty, R.; Mary, V.S.; Theumer, M.G.; Granero, G.E.; Paraje, M.G.; Páez, P.L. Biosynthesized silver nanoparticles: Decoding their mechanism of action in *Staphylococcus aureus* and *Escherichia coli*. *Int. J. Biochem.* **2018**, *104*, 87–93. [CrossRef]
14. Yin, I.X.; Zhao, I.S.; Mei, M.L.; Lo, E.C.; Tang, J.; Li, Q.; So, L.Y.; Chu, C.H. Synthesis and characterization of fluoridated silver nanoparticles and their potential as a non-staining anti-caries agent. *Int. J. Nanomed.* **2020**, *15*, 3207–3215. [CrossRef]
15. Azarsina, M.; Kasraei, S.; Yousefi-Mashouf, R.; Dehghani, N.; Shirinzad, M. The antibacterial properties of composite resin containing nanosilver against *Streptococcus mutans* and *Lactobacillus*. *J. Contem. Dent.* **2013**, *14*, 1014. [CrossRef]
16. Panpaliya, N.P.; Dahake, P.T.; Kale, Y.J.; Dadpe, M.V.; Kendre, S.B.; Siddiqi, A.G.; Maggavi, U.R. In vitro evaluation of antimicrobial property of silver nanoparticles and chlorhexidine against five different oral pathogenic bacteria. *Saudi Dent. J.* **2019**, *31*, 76–83.
17. Saliminasab, M.; Jabbari, H.; Farahmand, H.; Asadi, M.; Soleimani, M.; Fathi, A. Study of antibacterial performance of synthesized silver nanoparticles on *Streptococcus mutans* bacteria. *Nanomed. Res. J.* **2022**, *7*, 391–396.
18. Batir-Marin, D.; Boev, M.; Cioanca, O.; Mircea, C.; Burlec, A.F.; Beppe, G.J.; Spac, A.; Corciova, A.; Hritcu, L.; Hancianu, M. Neuroprotective and antioxidant enhancing properties of selective *Equisetum* extracts. *Molecules* **2021**, *26*, 2565.
19. Halliwell, B.; Gutteridge, J.M.C. *Free Radicals in Biology and Medicine*; Oxford University Press: Oxford, UK, 2015.
20. Elufioye, T.O.; Berida, T.I.; Habtemariam, S. Plants-Derived Neuroprotective Agents: Cutting the Cycle of Cell Death through Multiple Mechanisms. *Evid.-Based Complement. Altern. Med.* **2017**, *2017*, 3574012. [CrossRef]
21. Hritcu, L.; Foyet, H.S.; Stefan, M.; Mihasan, M.; Asongalem, A.E.; Kamtchouing, P. Neuroprotective effect of the methanolic extract of Hibiscus asper leaves in 6-hydroxydopamine-lesioned rat model of Parkinson's disease. *J. Ethnopharmacol.* **2011**, *137*, 585–591. [CrossRef]
22. Prakki, A.; Cilli, R.; Mondelli, R.F.; Kalachandra, S.; Pereira, J.C. Influence of pH environment on polymer based dental material properties. *J. Dent.* **2005**, *33*, 91–98. [CrossRef]
23. Kasraei, S.; Sami, L.; Hendi, S.; AliKhani, M.Y.; Rezaei-Soufi, L.; Khamverdi, Z. Antibacterial properties of composite resins incorporating silver and zinc oxide nanoparticles on *Streptococcus mutans* and *Lactobacillus*. *Restor. Dent. Endod.* **2014**, *39*, 109–114.
24. Veerasamy, R.; Xin, T.Z.; Gunasagaran, S.; Xiang, T.F.; Yang, E.F.; Jeyakumar, N.; Dhanaraj, S.A. Biosynthesis of silver nanoparticles using mangosteen leaf extract and evaluation of their antimicrobial activities. *J. Saudi Chem. Soc.* **2011**, *15*, 113–120. [CrossRef]
25. Varadavenkatesan, T.; Selvaraj, R.; Vinayagam, R. Phyto-synthesis of silver nanoparticles from *Mussaenda erythrophylla* leaf extract and their application in catalytic degradation of methyl orange dye. *J. Mol. Liq.* **2016**, *221*, 1063–1070. [CrossRef]
26. Padalia, H.; Moteriya, P.; Chanda, S. Green synthesis of silver nanoparticles from marigold flower and its synergistic antimicrobial potential. *Arab. J. Chem.* **2015**, *8*, 732–741. [CrossRef]
27. Patil, S.; Chaudhari, G.; Paradeshi, J.; Mahajan, R.; Chaudhari, B.L. Instant green synthesis of silver-based herbo-metallic colloidal nanosuspension in *Terminalia bellirica* fruit aqueous extract for catalytic and antibacterial applications. *3 Biotech.* **2017**, *7*, 36.
28. Jyoti, K.; Singh, A. Green synthesis of nanostructured silver particles and their catalytic application in dye degradation. *J. Genet Eng. Biotechnol.* **2016**, *14*, 311–317.
29. Sinha, S.N. In vitro antibacterial activity of ethanolic extract of *Equisetum arvense* L. *Indian J. Pharm. Biol. Res.* **2012**, *3*, 19–21.
30. Sandhu, N.S.; Kaur, S.A.; Chopra, D.I. Equisetum arvense: Pharmacology and phytochemistry-a review. *Asian J. Pharm. Clin. Res.* **2010**, *3*, 146–150.
31. Mimica-Dukic, N.; Simin, N.; Cvejic, J.; Jovin, E.; Orcic, D.; Bozin, B. Phenolic compounds in field horsetail (*Equisetum arvense* L.) as natural antioxidants. *Molecules* **2008**, *13*, 1455–1464. [CrossRef]
32. Do Amaral, G.S.; Negrini, T.; Maltz, M.; Arthur, R.A. Restorative materials containing antimicrobial agents: Is there evidence for their antimicrobial and anticaries effects? A systematic review. *Aus. Dent. J.* **2016**, *61*, 6–15. [CrossRef]
33. Yassaei, S.; Nasr, A.; Zandi, H.; Motallaei, M.N. Comparison of antibacterial effects of orthodontic composites containing different nanoparticles on *Streptococcus mutans* at different times. *Dental Press J. Orthod.* **2020**, *25*, 52–60. [CrossRef] [PubMed]
34. Jenabi, N.; Sadeghian, S.; Karimzadeh, F.; Pour, M.S.; Rakhshan, V. Antibacterial activity and shear bond strength of fiber-reinforced composites and bonding agents containing 0.5%, 1%, 2.5%, and 5% silver nanoparticles. *Dent. Res. J.* **2022**, *20*, 1–10.
35. Ahn, S.J.; Lee, S.J.; Kook, J.K.; Lim, B.S. Experimental antimicrobial orthodontic adhesives using nanofillers and silver nanoparticles. *Dent. Mater.* **2009**, *25*, 206–213. [CrossRef] [PubMed]
36. Wijnhoven, S.W.; Peijnenburg, W.J.; Herberts, C.A.; Hagens, W.I.; Oomen, A.G.; Heugens, E.H.; Roszek, B.; Bisschops, J.; Gosens, I.; Van De Meent, D.; et al. Nano-silver–a review of available data and knowledge gaps in human and environmental risk assessment. *Nanotoxicology* **2009**, *3*, 109–138.
37. Spencer, C.G.; Campbell, P.M.; Buschang, P.H.; Cai, J.; Honeyman, A.L. Antimicrobial effects of zinc oxide in an orthodontic bonding agent. *Angle Orthod.* **2009**, *79*, 317–322. [CrossRef] [PubMed]
38. Yoshida, K.; Tanagawa, M.; Matsumoto, S.; Yamada, T.; Atsuta, M. Antibacterial activity of resin composites with silver-containing materials. *Eur. J. Oral. Sci.* **1999**, *107*, 290–296.
39. Bapat, R.A.; Chaubal, T.V.; Joshi, C.P.; Bapat, P.R.; Choudhury, H.; Pandey, M.; Gorain, B.; Kesharwani, P. An overview of application of silver nanoparticles for biomaterials in dentistry. *Mater. Sci. Eng. C. Mater. Biol. Appl.* **2018**, *91*, 881–898.
40. Andrian, S.; Munteanu, B.; Tărăboanţă, I.; Negraia, D.; Nica, P.E.; Stoleriu, S.; Nica, I. Surface roughness after finishing and polishing of a restorative nanocomposite material. In *E-Health and Bioengineering Conference (EHB)*; IEEE: Piscataway, NJ, USA, 2017; pp. 101–104.

41. Monteiro, D.R.; Gorup, L.F.; Takamiya, A.S.; Ruvollo-Filho, A.C.; Camargo, E.R.; Barbosa, D.B. The growing importance of materials that prevent microbial adhesion: Antimicrobial effect of medical devices containing silver. *Int. J. Antimicrob. Agents* **2009**, *34*, 103–110.
42. Arif, W.; Rana, N.F.; Saleem, I.; Tanweer, T.; Khan, M.J.; Alshareef, S.A.; Sheikh, H.M.; Alaryani, F.S.; Al-Kattan, M.O.; Alatawi, H.A.; et al. Antibacterial Activity of Dental Composite with Ciprofloxacin Loaded Silver Nanoparticles. *Molecules* **2022**, *27*, 7182.
43. Smirnova, V.V.; Chausov, D.N.; Serov, D.A.; Kozlov, V.A.; Ivashkin, P.I.; Pishchalnikov, R.Y.; Uvarov, O.V.; Vedunova, M.V.; Semenova, A.A.; Lisitsyn, A.B.; et al. A Novel Biodegradable Composite Polymer Material Based on PLGA and Silver Oxide Nanoparticles with Unique Physicochemical Properties and Biocompatibility with Mammalian Cells. *Materials* **2021**, *14*, 6915.
44. Kim, J.S.; Shin, D.H. Inhibitory effect on *Streptococcus mutans* and mechanical properties of the chitosan containing composite resin. *Restor. Dent. Endod.* **2013**, *38*, 36–42. [CrossRef] [PubMed]
45. Argueta-Figueroa, L.; Scougall-Vilchis, R.J.; Morales-Luckie, R.A.; Olea-Mejia, O.F. An evaluation of the antibacterial properties and shear bond strength of copper nanoparticles as a nanofiller in orthodontic adhesive. *Aust. Orthod. J.* **2015**, *31*, 42–48. [CrossRef] [PubMed]
46. Poosti, M.; Ramazanzadeh, B.; Zebarjad, M.; Javadzadeh, P.; Naderinasab, M.; Shakeri, M.T. Shear bond strength and antibacterial effects of orthodontic composite containing TiO_2 nanoparticles. *Eur. J. Orthod.* **2013**, *35*, 676–679. [CrossRef]
47. Szczesio-Wlodarczyk, A.; Fronczek, M.; Ranoszek-Soliwoda, K.; Grobelny, J.; Sokolowski, J.; Bociong, K. The first step in standardizing an artificial aging protocol for dental composites—Evaluation of basic protocols. *Molecules* **2022**, *27*, 3511. [CrossRef]
48. Krüger, J.; Maletz, R.; Ottl, P.; Warkentin, M. In vitro aging behavior of dental composites considering the influence of filler content, storage media and incubation time. *PLoS ONE* **2018**, *13*, e0195160. [CrossRef]
49. Tărăboanță, I.; Stoleriu, S.; Gurlui, S.; Nica, I.; Tărăboanță-Gamen, A.C.; Iovan, A.; Andrian, S. The Influence of Abrasive and Acidic Aggressions on the Surface Condition of Flowable Composite Resin. *Materials* **2022**, *15*, 1000. [CrossRef]

Journal of
Functional
Biomaterials

MDPI

Article

Composite Hydrogels of Ultrasound-Assisted-Digested Formic Acid-Decellularized Extracellular Matrix and Sacchachitin Nanofibers Incorporated with Platelet-Rich Plasma for Diabetic Wound Treatment

Chien-Ju Lin [1,†], Hong-Liang Lin [1,†], Wen-Chen You [2], Hsiu-O Ho [2], Ming-Thau Sheu [2], Ling-Chun Chen [3,*] and Wei-Jie Cheng [2,*]

[1] School of Pharmacy, College of Pharmacy, Kaohsiung Medical University, Kaohsiung 80708, Taiwan; mistylin@kmu.edu.tw (C.-J.L.); hlglin@kmu.edu.tw (H.-L.L.)
[2] School of Pharmacy, College of Pharmacy, Taipei Medical University, Taipei 11031, Taiwan; hsiuoho@tmu.edu.tw (H.-O.H.); mingsheu@tmu.edu.tw (M.-T.S.)
[3] Department of Biotechnology and Pharmaceutical Technology, Yuanpei University of Medical Technology, Hsinchu 30015, Taiwan
* Correspondence: d8801004@mail.ypu.edu.tw (L.-C.C.); d301107001@tmu.edu.tw (W.-J.C.)
† These authors contributed equally to this work.

Abstract: In this study, an ultrasound-assisted digestion method of a formic acid-decellularized extracellular matrix (dECM) of porcine skin was developed and optimized to form UdECM hydrogels for diabetic wound healing. Results demonstrated that ultrasonication improved the extraction rate of collagen from dECM samples, preserved the collagen content of dECM, reduced residual cells, and extracted greater DNA contents. Scanning electron microscope (SEM) analyses were performed, which demonstrated the optimal porosity on the surface and density of the cross-section in the hydrogel structure, which could control the release of growth factors embedded in UdECM hydrogels at desirable rates to boost wound healing. A wound-healing study was conducted with six different composite hydrogels, both empty materials and materials enriched with rat platelet-rich plasma (R-PRP), sacchachitin nanofibers (SCNFs), and TEMPO-oxidized sacchachitin in diabetic rats. The assessment based on scars stained with hematoxylin and eosin (H&E), Masson's trichrome (MT), and a cluster of differentiation 31 (CD31) staining showed that the UdECM/SC/R-PRP treatment group had the most significant efficacy of promoting healing and even recovery of diabetic wounds to normal tissues. UdECM/R-PRP and UdECM/SCNFs demonstrated better healing rates than UdECM hydrogel scaffolds, which had only recovered 50% resemblance to normal skin. Treatment with both UdECM/TEMPO 050 and UdECM/TEMPO 050/R-PRP hydrogel scaffolds was ranked last, with even poorer efficacy than UdECM hydrogels. In summary, formulated UdECM and SCNF hydrogels loaded with PRP showed synergistic effects of accelerating wound healing and ultimately stimulating the wound to recover as functional tissues. This newly UdECM/SCNF composite hydrogel has promising potential for healing and regenerating diabetic wounds.

Keywords: ultrasound; decellularized extracellular matrix; platelet-rich plasma; sacchachitin; wound dressing; hydrogel

Citation: Lin, C.-J.; Lin, H.-L.; You, W.-C.; Ho, H.-O.; Sheu, M.-T.; Chen, L.-C.; Cheng, W.-J. Composite Hydrogels of Ultrasound-Assisted-Digested Formic Acid-Decellularized Extracellular Matrix and Sacchachitin Nanofibers Incorporated with Platelet-Rich Plasma for Diabetic Wound Treatment. *J. Funct. Biomater.* **2023**, *14*, 423. https://doi.org/10.3390/jfb14080423

Academic Editors: Christie Ying Kei Lung and Zhen Zheng

Received: 26 June 2023
Revised: 8 August 2023
Accepted: 10 August 2023
Published: 11 August 2023

1. Introduction

Extracellular matrices (ECMs) obtained from decellularized tissues have emerged as a highly promising approach for tissue engineering scaffolds and acellular regenerative strategies [1–5]. Notably, decellularized (d)ECM can undergo enzymatic digestion to form inducible hydrogels, facilitating minimally invasive delivery through injection to target tissues [6–9]. These degradable dECM hydrogels are prosurvival, immunomodulatory, and vascularized, among a host of other pro-regenerative functions [2,10–12]. The most prevalent method to produce dECM hydrogels is via pepsin-mediated (stirred in pepsin with dilute hydrochloric acid) solubilization from the comminuted (powder) form of dECM (also called "ECM digestion") as reported by Freytes et al. (designated herein as the "Freytes method") [6]. Another method involves the use of 0.5 M acetic acid instead of 0.1 M HCl as the base medium for the pepsin enzyme (designated the "Voytik-Harbin method") [13]. Alternatives include an extraction process to solubilize and form ECM hydrogels from soft tissues (designated the "Uriel method") [14,15]. Recently, a new method was reported by Hussey et al. to prepare hydrogels from ECM bioscaffolds by rapid ultrasonic (US) cavitation without acidic or alkaline solutions, protease digestion, or chemical extraction and dialysis [16]. The US cavitation method described herein claimed to produce marked improvements in rheological properties and processing time over traditional enzymatic methods.

In order to obtain dECM for dECM hydrogels, a simple and effective decellularization technique was developed and optimized to attain dECM from porcine skin in our previous study [17]. dECM hydrogels were further fabricated by digesting the so-obtained dECM powder with pepsin in various acidic solutions and then treatment with pH-controlled neutralization and a temperature-controlled gelation procedure. The study on wound healing in diabetic patients and the histological examinations demonstrated that the combination of resultant dECM hydrogels and sacchachitin nanofibers (SCNFs) in composite hydrogels effectively expedited the healing process of diabetic wounds. Additionally, this synergistic approach stimulated the regeneration of hair follicles and sweat glands, resulting in the restoration of fully functional tissues. Although formic acid-decellularized pepsin-soluble ECM hydrogels were effective for treating diabetic wounds, the process still was time-consuming and less effective for retaining collagen and glycosaminoglycan (GAG). Overall, there is an unmet need to develop simple, effective, and optimized decellularization techniques and fabrication to obtain dECM hydrogels from porcine skin for wound-healing medical applications.

Studies have shown that US can increase collagen production by up to 124% and significantly reduce extraction times compared to traditional pepsin isolation methods. Further, the triple helix structure of collagen remains intact with US extraction. Finally, because US can promote the dispersion of large enzyme aggregates and widen the contact area with collagen fibers for hydrolysis, it further improves the digestion of enzyme molecules only on the surface of collagen fibrils [18]. In conclusion, gentle US for collagen extraction can optimize the extraction capacity without affecting the collagen content.

A previous study showed that SCNFs generated by NanoLyzer® and TEMPO-oxidation possessed a gel-forming property to act as a biomaterial with ideal characteristics [19]. For diabetic wound-healing applications, both mechanically disintegrated SCNFs (SCN5) and TEMPO-oxidized sacchachitin nanofibers (T050SC; TEMPO 050) accelerated wound-healing and formed nearly identical normal tissues. As for T050SC/H, it further boosted the growth of sweat glands and hair follicles and had the ability to rebuild wounds into functional tissues. Furthermore, multifunctional biomaterials based on SCNFs might have great potential for future clinical applications. Some studies also showed that wound healing can be accelerated with platelet-rich plasma (PRP), which is enriched from blood and usually has 3–7 times the average platelet concentration of whole blood. After activation, greater amounts of growth factors, such as platelet-derived growth factor (PDGF), vascular endothelial growth factor (VEGF), transforming growth factor (TGF)-β, and epidermal growth factor (EGF), can be released. It not only stops bleeding but also inhibits cytokines and inflammatory symptoms, promotes the formation of microvessels, and accelerates the epithelialization of chronic wounds, thereby stimulating wound healing and tissue regeneration [20,21].

In this study, we hypothesized that the combination of the acidic addition and ultrasonication could improve the preparation of dECM hydrogel. This approach aims to reduce decellularization and fabrication time while increasing the extraction rate of collagen and related glycosaminoglycans (GAGs). Furthermore, we incorporated SCNFs or TEMPO

050 to form a composite hydrogel, which carried PRP and was expected to demonstrate a synergistic wound-healing effect on diabetic wounds.

2. Materials and Methods

2.1. Materials

The protein marker was purchased from BioTools (New Taipei, Taiwan). Tetramethylethylenediamine (TEMED), ammonium persulfate (APS),30% acrylamide/bis solution (37.5:1), and glycine were purchased from Bio-Rad (Hercules, CA, USA). Sodium chloride (NaCl) and sodium hydroxide (NaOH) were purchased from Showa Chemical (Tokyo, Japan). Trypsin, pepsin, formic acid, 10% formaldehyde, papain, Trizma® base, L-cysteine hydrochloride, sodium EDTA, sodium dodecyl sulfate (SDS), acetic acid, glycerol, bromophenol blue, 2-mercaptoethanol, methanol, Coomassie blue R-250, and nicotinamide (Vitamin B3) were bought from Sigma-Aldrich (St. Louis, MO, USA). The glycosaminoglycan kit and hydroxyproline kit were obtained from Chondrex (Redmond, WA, USA). Penicillin/streptomycin (PS) and phosphate-buffered saline (PBS) were obtained from Corning (Corning, NY, USA). Triethanolamine (TEA) and hydrochloric acid (HCl) were purchased from Merck (Darmstadt, Germany). Porcine type I collagen standard-FlexiCol® was bought from Advanced BioMatrix (San Diego, CA, USA). SC (2% *w*/*v* hydrogel) prepared by the mechanical disintegration of SCNFs in double-distilled (dd)H_2O, was previously reported [19]. Fetal bovine serum (FBS), minimum essential medium (MEM), and a PicoGreen quantification kit were purchased from Thermo Fisher Scientific (Waltham, MA, USA). Streptozotocin (STZ) was bought from ChemCruz® (Santa Cruz, CA, USA). An anti-cluster of differentiation 31 (CD31) antibody was obtained from Abcam (Cambridge, UK). Rat (R)-PRP was provided by the laboratory of Prof. Wu Yina, Department of Medicine, Fu Jen University (New Taipei City, Taiwan). Fresh white-fur pigskin was purchased from Wuxing Traditional Market (Taipei, Taiwan). Attane (isoflurane) was supplied by Panion & BF Biotech (New Taipei City, Taiwan). A mouse TGF-β1 DuoSet enzyme-linked immunosorbent assay (ELISA) (DY1679-05) and Mouse/Rat PDGF-AB DuoSet ELISA (DY8460-05) were purchased from R&D Systems (Minneapolis, MN, USA). A Quant-iT™ PicoGreen™ dsDNA assay kit (Molecular Probes, P7589) was provided by Thermo Fisher Scientific.

2.2. Methods

2.2.1. Preparation of US-Assisted Pepsin-Solubilized dECM Hydrogels

Pretreatment of porcine skin to obtain dECM fragments followed the method developed by Hsieh et al. [17]. After ground dECM fragments were passed through a #40 screen, 250 mg of dECM plus 1 mg of pepsin (250 U/mg) were extracted in a 50 mL centrifuge tube with 10 mL of 0.5 M acetic acid. US was performed to assist pepsin solubilization in a Q Sonica (Q700) US breaker with a 1/8″ probe immersed at the 4 cm position in the centrifuge tube. The operational parameters were set with a 2 s on-time followed by a 3 s off-time. The frequency and power settings were 24 kHz and 150 W, respectively. Over a duration of 5 min, the cumulative energy delivered amounted to 2300 J. After extraction, 1.0 M NaCl was added to salt out for 12 h, then it was centrifuged to remove the supernatant, and the precipitate was redissolved in 10 mL of a 0.5 M acetic acid solution. Afterward, the solution was loaded onto a dialysis membrane with a molecular weight (MW) cutoff of 3000 and subjected to dialysis against a 0.1 M acetic acid solution for a duration of 2 days. Finally, ddH_2O was added for the final dialysis until neutral, and the solution was lyophilized to obtain UdECM. Then, 25, 50, and 75 mg of UdECM were weighed in 5 mL centrifuge tubes, and 0.5 M acetic acid was added. After being completely dissolved, the pH value was adjusted to neutral (pH 7.0) with 2 N NaOH to obtain UdECM hydrogels, which were respectively designated UdECM25, UdECM50, and UdECM75.

2.2.2. Biochemical Characterization of UdECM Hydrogels

Qualitative Analysis of Collagen

In order to characterize the collagen in the UdECM hydrogels, SDS-polyacrylamide gel electrophoresis (PAGE) was conducted using 5% (*w*/*v*) polyacrylamide for the separating gel and 4% (*w*/*v*) polyacrylamide for the stacking gel. UdECM hydrogels and porcine type I collagen standard were prepared in 0.5 M Tris-HCl buffer (pH 6.8) containing 30% glycerol, 10% SDS, 1% 2-mercaptoethanol, and 0.02% bromophenol blue. The samples were heated to 95 °C for 5 min to ensure proper denaturation. Subsequently, 8 μL of a protein marker, 30 μL of the standard, and 30 μL of hydrogels were loaded into the gel. The electrophoresis was conducted at 85–100 V on vertical slab gels until the bromophenol blue front reached the bottom of the gel. The polyacrylamide gels were stained for 2 h with 0.1% Coomassie blue R-250 in acetic acid/methanol/water (2:5:5, *v*/*v*/*v*). The gels were then destained in a solution containing 7.5% acetic acid/15% methanol. Following the staining process, the stained polyacrylamide gels were recorded using an iPhone camera (Apple, Cupertino, CA, USA)

Quantitative Analysis of Collagen

Collagen contents in UdECM were determined with a Hydroxyproline assay kit (#6017, Chondrex, Woodinville, WA, USA). Samples were prepared according to the instructions with the assay kit as follows: UdECM hydrogels were hydrolyzed with 10 N hydrochloric acid and reacted with chloramine-T for 20 min. Next, a p-dimethylaminobenzaldehyde solution was added, and the resultant mixture was heated to 60 °C for 30 min. After reacting, the absorbance value was measured at a wavelength of 530 nm, and the amount of hydroxyproline was calculated by an automated imaging system and multifunctional optical detector (Cytation 3™, Cell Imaging Multi-Mode Reader, BioTek, Winooski, VT, USA). The total amount of collagen was calculated based on the following formula with the so-obtained hydroxyproline level:

$$\text{Hydroxyproline level}\left(\frac{\mu g}{mg}\right) \times \frac{100}{13.5} = \text{Collagen level}\left(\frac{\mu g}{mg}\right) \quad (1)$$

DNA Quantitative Analysis

In order to quantify the amount of residue DNA in UdECM, the PicoGreen dsDNA detection kit was used. Sample preparation was performed according to the manufacturer's instructions. Subsequently, the fluorescence intensity was measured using the Cytation 3 cell imaging multi-mode reader (BioTek) with an excitation wavelength of 480 nm and an emission wavelength of 520 nm. To subtract the background fluorescence, the fluorescence of the DNA-free blank was conducted and subtracted from the fluorescence of the experimental groups, ensuring accurate measurement of the sample's fluorescence signal.

2.2.3. Physical Characterizations of UdECM Hydrogels

Rheological Studies of UdECM Hydrogels

We evaluated the rheological characteristics of different concentrations (25, 50, and 75 mg/mL) of UdECM (UdECM25, UdECM50, and UdECM75) at 25 and 37 °C as previously reported [22]. We tested the oscillatory shear stress by the same procedure as previously reported [23], and the Amplitude Sweep program in the intelligent advanced rheometer (MCR 102e Anton-Paar, St. Albans Hertfordshire, UK) was used to analyze the UdECM25, UdECM50, and UdECM75 hydrogels, which yielded values for G' (storage modulus) and G'' (loss modulus). The elastic modulus was measured by placing 0.3 mL of a hydrogel sample on rheometer plates at 25 and 37 °C with a fixed amplitude frequency of 1 Hz and a stress (strain) range of 0.1–100 Pa. Plotting G' on the X-axis as and G'' on the Y-axis, the linear viscoelastic region (LVR) was obtained to numerically analyze the colloidal stability of the sample.

Morphological Observations

Freeze-dried UdECM25, UdECM50, UdECM75, and UdECM/SC sponges were mounted with conductive carbon tape, which was sputter-coated with gold (Hitachi IB-2, Tokyo, Japan). The images were assessed by Hitachi SU3500 scanning electron microscopy (SEM) at a 5 mm working distance and an accelerating voltage of 2.5 kV. The surface morphology and internal pores were observed for different concentrations of those UdECM hydrogels.

2.2.4. R-PRP Preparation and Determination of PDGF-AB and TGF-β1

In order to improve the ability of wound-healing, R-PRP provided by the laboratory of Prof. Wu Yina of the Department of Medicine of Fu Jen University (New Taipei City, Taiwan) was incorporated with composite hydrogels as an adjuvant for treatment of diabetic wounds in animals. In this experiment, the commercially available enzyme immunosorbent assay kits (#DY1679-05, DY8460-05, R&D Systems) were used in an ELISA to determine the contents of PDGF-AB and TGF-β1 in R-PRP. The preparation method of the standard and test product was based on the attached instructions. Furthermore, the automated imaging system and multifunction optical detector (Cytation™ 3 Cell Imaging Multi-Mode Reader, BioTek) measured the absorbance at a wavelength of 450 nm, and PDGF-AB and TGF-β1 contents were calculated. Each test had two replicates, and an average value was obtained.

2.2.5. In Vivo Studies of Diabetic Wound-Healing

Preparation of Wound Dressings for Animal Studies

Wound dressings were formulated in the proportions shown in Table 1. Formulation A (5% UdECM hydrogel, UdECM50) contained 50 mg of UdECM completely dissolved in 1.0 mL of 0.5 M acetic acid, and the pH was adjusted to neutral (pH 7.0) with 0.2 N NaOH. Formulation B (UdECM/PRP) had 800 pg of R-PRP added to 1.0 mL of formulation A and evenly mixed; it was then placed in a refrigerator at 4 °C for later use. Formulation C (UdECM/SCNF hydrogel) consisted of 50 mg of SCNFs dispersed in 1.0 mL of formulation A and mixed until it had formed a hydrogel. Formulation D (UdECM/SCNF/R-PRP) had 800 pg of R-PRP added to 1.0 mL of formulation C and evenly mixed; it was then placed in a refrigerator at 4 °C for later use. Formulation E (UdECM/TEMPO 050) had 50 mg Temp50 dispersed in 1.0 mL of formulation A and mixed until it had formed a hydrogel. Formulation F (UdECM/TEMPO 050/PRP) had 800 pg of R-PRP added to 1.0 mL of formulation E and evenly mixed; it was then placed in a refrigerator at 4 °C for later use.

Table 1. Formulations of wound dressings.

	A	B	C	D	E	F
Wound Dressing	**UdECM**	**UdECM/PRP**	**UdECM/SCNFs**	**UdECM/SCNFs/PRP**	**UdECM/TEMPO 050**	**UdECM/TEMPO 050/PRP**
UdECM (mg)	50	50	50	50	50	50
SCNFs (mg)			50	50		
TEMPO 050 (mg)					50	50
R-PRP (pg)		800		800		800

Abbreviations: UdECM, ultrasound-assisted pepsin-solubilized decellularized extracellular matrix; SCNFs, sacchachitin nanofibers; TEMPO 050, TEMPO-oxidized sacchachitin nanofibers; R-PRP, rat platelet-rich plasma.

Laboratory Animals

This animal experiment and experimental protocol were approved by the Institutional Animal Care and Use Committee of Taipei Medical University (Approval No.: LAC-2021-0024) in compliance with the Taiwanese Animal Welfare Act. Sprague-Dawley (SD) male rats at approximately 280–290 g and 8 weeks of age were purchased from BioLASCO (Taipei, Taiwan). Experimental rats were kept at 22–25 °C and a humidity of 60–70% with free intake of aseptic standard feed and drinking water under 12 h of automatic light and dark control. The experimental animals were given sufficient time to acclimate to the environment, allowing their physiological and psychological states to stabilize before conducting the experiment.

Evaluation of Induction of Type 1 Diabetes of the Animal Model

In this experiment, rats were raised for 1 week to achieve more than 300 g in weight, and then 180 mg/kg of a 50% nicotinamide solution (nicotinamide, NAM, vitamin B3) was given by an intraperitoneal (IP) injection. After 15 min, the animal model of type 1 diabetes was established by an IP injection of streptozotocin (STZ) (65 mg/kg) dissolved in 0.1 M citrate buffer (pH 4.5). Because STZ can cause large amounts of pancreatic insulin to be released and lead to fatal hypoglycemia, it is important to pay attention to any adverse effects and deaths over the next 24 h. The weight was measured, and blood glucose was detected by ACCU-CHEK® (Performa #06870244, Roche, Basel, Switzerland) on days 3, 5, and 7 after induction. Rats exhibiting moderate hyperglycemia (i.e., blood glucose concentration of >250 mg/dL) and typical symptoms of diabetes (eating more, drinking more, frequent urination, and weight loss) were selected for the experiments [24].

Experimental Design and Surgical Methods

Following the protocol shown in Scheme 1, rats with successful induction of diabetes were divided into six groups (n = 5 or 6) and IP injected with 10 mg/kg of Zoletil® to anesthetize them. The back hair was removed, and an 8 mm tissue sampler was used to create a wound. Each rat had four holes drilled into its back, and a silicone sheet was glued with instant glue around the wound. Next, the silicone sheet was fixed onto the skin with nylon thread sutures. Then, 0.28 g of different wound dressings (UdECM, UdECM/PRP, UdECM/SCNFs, UdECM/SCNFs/PRP, UdECM/TEMPO 050, UdECM/TEMPO 050/PRP) were placed on the wound, followed by breathable film fixation, and finally being dressed in elastic bandages.

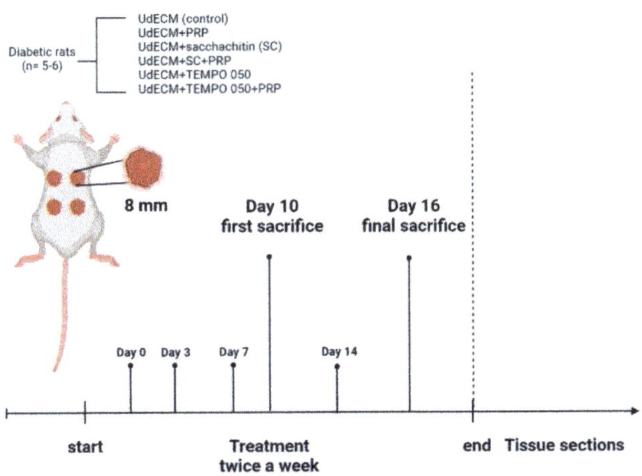

Scheme 1. Experimental schedule of the animal experiment of diabetic rats.

Analysis of Wound-Healing

After wound establishment, we changed the hydrogel dressings about every 3 days (on days 3, 7, 10, 14, 16, and 20). Furthermore, we took images of the wound, monitored blood glucose, and recorded the weight every day. ImageJ software (National Institutes of Health, Bethesda, MD, USA) was used to calculate the wound area and assess wound healing.

Histopathological Biopsy Evaluation

On days 10 and 16 post-treatment, rats were euthanized with excess CO_2. Then, the skin of the wound was cut and fixed with 10% formaldehyde solution for 24 h. Paraffin-embedding and sectioning treatment were entrusted to Toson Technology (Zhubei City, Hsinchu County, Taiwan). Tissue sections were stained with hematoxylin and eosin (H&E) and Masson's trichrome (MT) stain to assess collagen regeneration and, with CD31 endothelial cell markers (Abcam, ab182981), to evaluate the degree of angiogenesis. After sealing the film, skin tissue recovery was observed by light microscopy (BX43, Olympus, Center Valley, PA, USA).

2.2.6. Statistical Analysis

Statistical analyses were performed using one-way analysis of variance (ANOVA) along with Bartlett's statistical correction. A significance level of $p < 0.05$ was considered to indicate a statistically significant difference. The significance levels $p < 0.05$, $p < 0.01$, and $p < 0.005$ were denoted by *, **, and ***, respectively.

3. Results and Discussion

3.1. Preparation and Characterization of UdECM Hydrogels

3.1.1. UdECM Process Development

dECM was decellularized with 30% formic acid and made into physically crosslinked UdECM hydrogels with US. The collagen content was analyzed, and then rheology and morphology were used to examine the UdECM colloidal type. In a previous experiment, we found that dECM solubilized by citric acid at various concentrations was unable to form UdECM hydrogels above 25 °C. Therefore, we solubilized dECM with acetic acid. We sieved granular dECM through a #40 screen and dissolved it in 1 mg/mL pepsin (250 U/mg in a 0.5 M acetic acid solution); also, the US setup illustrated in Figure 1A was optimized to form UdECM hydrogels at 25 mg/mL dECM in a 0.5% acetic acid solution. After US and neutralization, as shown in Figure 1B,C, the UdECM solution could transform into UdECM hydrogels at room temperature. As reported previously, the collagen fiber could self-assemble into bundled fibers that physically crosslink to form the hydrogel at neutral pH [25]. Additionally, hydrolyzed collagen was also found to form hydrogels when cooled to temperatures below 30 °C. The phenomenon is similar to the thermomechanical properties of UdECM hydrogels treated by sonication [26]. Therefore, this result demonstrated that the application of ultrasonication could effectively extract collagen proteins and subsequently facilitate their self-assembly to form hydrogels.

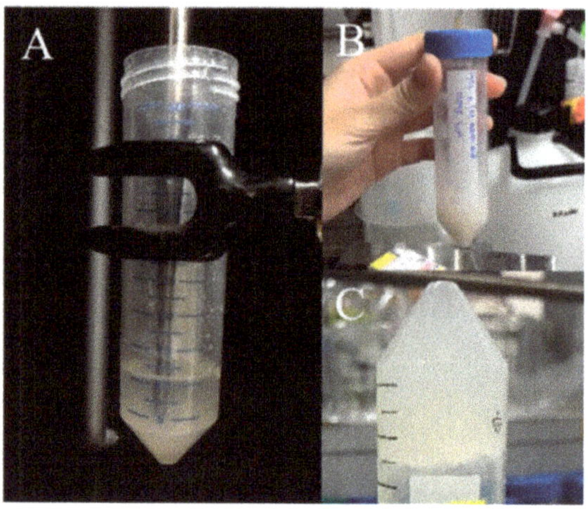

Figure 1. (**A**) The 1/8″ probe tip was dipped into the 4 cm position of the 50 mL centrifuge tube for ultrasound (US) treatment. (**B**) US-assisted pepsin-solubilized decellularized extracellular matrix (UdECM) was ground and sifted through a #40 mesh screen, then US processed into UdECM hydrogels at low temperature. (**C**) Inversion of a UdECM hydrogel in a centrifuge tube to confirm hydrogel formation.

3.1.2. Qualitative and Qualitative Analysis of Collagen in UdECM

SDS-PAGE analysis was used to evaluate collagen qualities for three groups of UdECM hydrogels (UdECM25: 25 mg/mL, UdECM50: 50 mg/mL, and UdECM75: 75 mg/mL), and results are shown in Figure 2. Typically, after heating the UdECM with SDS to unfold and linearize the protein structures by denaturation, type I collagen should show two different α-chains (two α1-chains and one α2-chain), two different β-chains (with crosslinks between two α1-chains or one α1-chain and one α2-chain), and a single γ-chain (with crosslinks between three α-chains). Different combinations of α-chains (two α1-chains and one α2-chain) comprised type I collagen fibers. As illustrated in Figure 2, the leftmost lane A of SDS-PAGE is a protein marker with molecular weights of 245, 180, 135, 100, 75, and 63 kDa from top to bottom, while lane B is a type I collagen standard, and lanes C to E are UdECM25, UdECM50, and UdECM75 collagen samples, respectively. In Figure 2, the type I collagen standard and three UdECM collagen samples all showed a band at 245 kDa, which represents the β-chain, and two bands at 135 kDa (α1 and α2) with the formula $[\alpha1(I)]_2[\alpha2(I)]$. This meant that the typical triple helix structure of collagen was similar to the type I collagen standard and the three UdECM samples. In summary, US-treated UdECM hydrogels were rich in type I collagen.

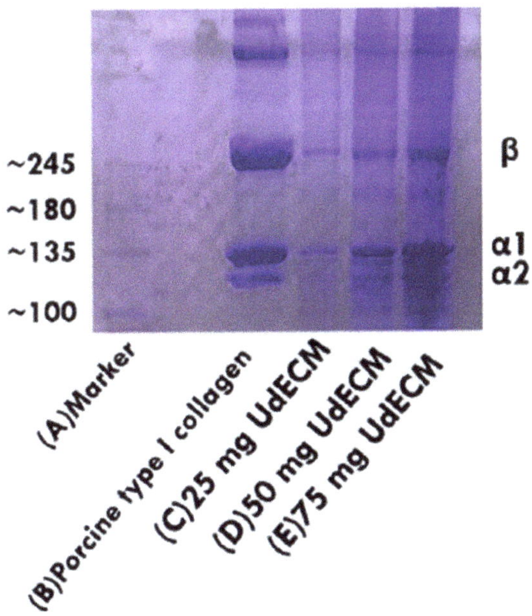

Figure 2. SDS-PAGE electrophoresis patterns. A, Marker; B, porcine type I collagen standard; C, ultrasound-assisted pepsin-solubilized decellularized extracellular matrix at 25 mg/mL (UdECM25); D, UdECM50; E, UdECM75.

Collagen in UdECM samples was quantitatively analyzed with a hydroxyproline assay kit. Lyophilized-dried hydrogels of UdECM25, UdECM50, and UdECM75 were first hydrolyzed with a strong acid (10 N HCl), and we measured the total collagen content compared to those for a $dECM_{HCl}$ hydrogel (25 mg/mL) (pepsin-mediated solubilization of dECM in a dilute HCl solution) and fresh porcine skin tissue (porcine) as control groups [17]. Results showed that the total content of collagen for the control group, $dECM_{HCl}$ (25 mg/mL), UdECM25, UdECM50, and UdECM75 were 715.4 ± 7.34, 556.01 ± 5.9, 668.0 ± 0.8, 721.0 ± 0.7, and 757.0 ± 0.1 μg/mg dry weight, respectively. Collagen contents for the control group, UdECM50, and UdECM75 were in a similar range of 700–750 μg/mg dry weight, followed by the UdECM25 group with a collagen content of 650–700 μg/mg dry weight, while the $dECM_{HCl}$ (25 mg/mL), extracted using pepsin solubilization in an HCl solution for two days, contained the least amount of collagen. This confirmed that US could produce cavitation between the solvent and sample, thereby increasing the dispersion and contact area of the collagen sample with enzymes in the extraction solvent. As a result, the extraction rate of collagen from the dECM sample is improved by the addition of ultrasonication.

3.1.3. DNA Quantitative Analysis

DNA contents of those decellularized samples were measured compared to fresh porcine skin tissue, which was the control group, by following a previously reported procedure [17,27]. The results showed that following a 72 h decellularization process in 30% formic acid, the DNA contents of both dECM and UdECM were consistently reduced to 46.49 ± 0.79 and 42.95 ± 0.73 ng/mg dry weight, respectively. In comparison, fresh porcine skin soaked in PBS exhibited a much higher DNA content of 349.77 ± 8.33 ng/mg dry weight, resulting in a removal efficiency of nearly 80%. With a residual DNA content of less than the minimum allowance of 50 ng/mg dry weight, both dECM and UdECM had reached the minimum standard for decellularization. Furthermore, US was able to destroy residual cells to extract more DNA contents and reduce the DNA content of 46.49 ± 0.79 ng/mg dry weight for dECM to that of 42.95 ± 0.73 ng/mg dry weight for UdECM with statistical significance.

3.2. *Rheological Studies*

The Amplitude Sweep program in the intelligent advanced rheometer (MCR 102e Anton-Paar) was used to evaluate the rheological characteristics of these hydrogels, including UdECM25, UdECM50, and UdECM75. Changes in G' and G'' in the range of 0.1–100 Pa were measured at 25 and 37 °C. If $G' > G''$, the sample shows a gel-like or solid structure and can be termed a viscoelastic solid material. However, if $G'' > G'$, the sample displays fluid characteristics and can be termed a viscoelastic liquid. Results are presented as the diagram in which the storage modulus, G', and loss modulus, G'', were respectively plotted on a logarithmic scale as the X-axis and Y-axis. The limit of the linear viscoelastic region (LVR) determined the stress range in which the test could be carried out without destroying the sample structure. A wide LVR means that the hydrogel has good strength with structural stability to resist high stress.

As shown by Figure 3A, when the temperature was 25 °C, G'' for UdECM25 was greater than G', indicating a viscoelastic liquid state. However, at 37 °C, G' surpassed G'', demonstrating a gel-like or solid structure. Nonetheless, the LVR was short, and the hydrogel structure was not strong enough to withstand high stresses. On the contrary, both UdECM50 (Figure 3B) and UdECM75 (Figure 3C) exhibited a gel-like or solid structure throughout the temperature range from 25 to 37 °C. Moreover, the hydrogel strength at 37 °C was higher than that at 25 °C, and the LVR was longer at 37 °C. These results revealed that with higher dECM concentrations than 25 mg/mL, the gel-like structure remained unchanged as the temperature increased from 25 to 37 °C as evidenced by $G' >> G''$ being significantly greater than G''. Importantly, it was observed that a higher concentration of dECM contributed to a larger LVR and higher G, leading to a greater stiffness and structure stability of the hydrogel. However, other rheologic properties, including oscillatory deformation and steady shear deformation, were not evaluated in this study and need to be further investigated.

To the present, the reason for US being capable of gelling dECM is still unclear. Based on our research, it was assumed that collagen proteins were not the only self-assembling molecules in the UdECM mixture. Interactions among various components dissolved in UdECM might include other self-assembling molecules such as collagen, laminin, and proteoglycans. The above results were similar to a reference that ECM subjected to US could induce rapid hydrogelation below 25 °C [16].

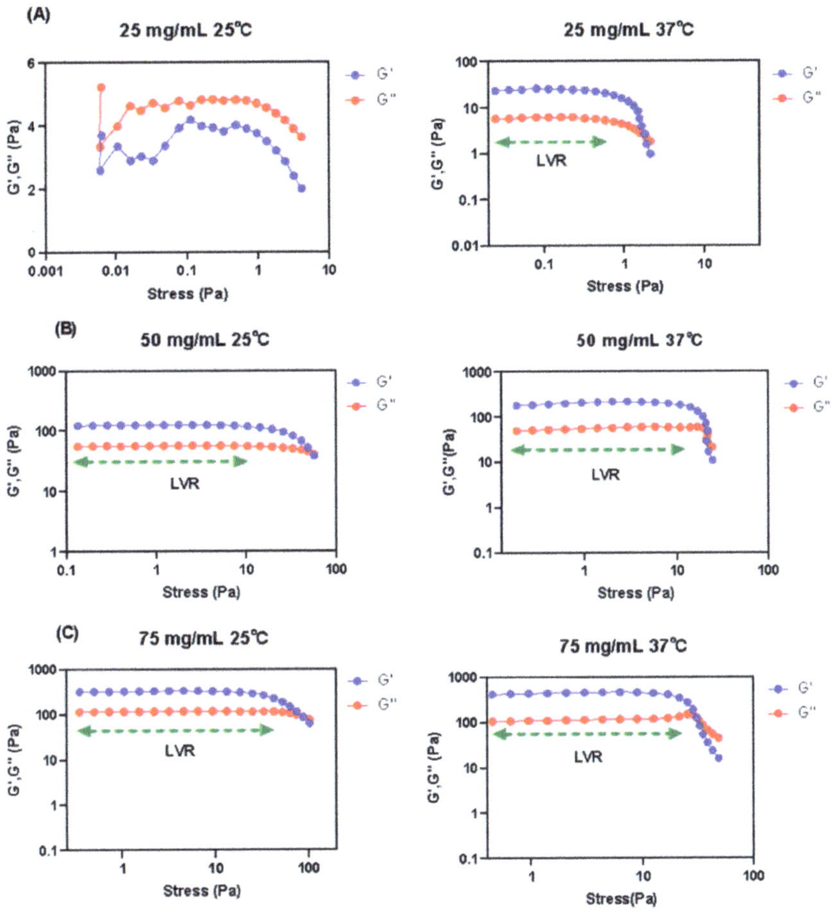

Figure 3. Viscoelasticity characterization of (**A**) ultrasound-assisted pepsin-solubilized decellularized extracellular matrix at 25 mg/mL (UdECM25), (**B**) UdECM50 (50 mg/mL), and (**C**) UdECM75 (75 mg/mL). Representative graphs of the UdECM hydrogel gelation kinetic profiles at two temperatures for three different concentrations of UdECM.

Morphological Studies

The UdECM25, UdECM50, and UdECM75 hydrogels were freeze-dried to form a spongy matrix. Furthermore, the density, porosity, and cross-section were observed by SEM. Figure 4 shows images of the surface (top) and cross-section (bottom) of (A) UdECM25, (B) UdECM50, and (C) UdECM75 spongy matrices. SEM demonstrated that all three spongy matrices were porous on the surface and had dense, rough, layered sheet structures on the cross-section. The denser cross-section with a less-porous surface increased the concentration of UdECM in the spongy matrix. The optimal porosity and density of the cross-section are critical to controlling the release of growth factors embedded in the UdECM hydrogels at desirable rates to promote utmost wound healing.

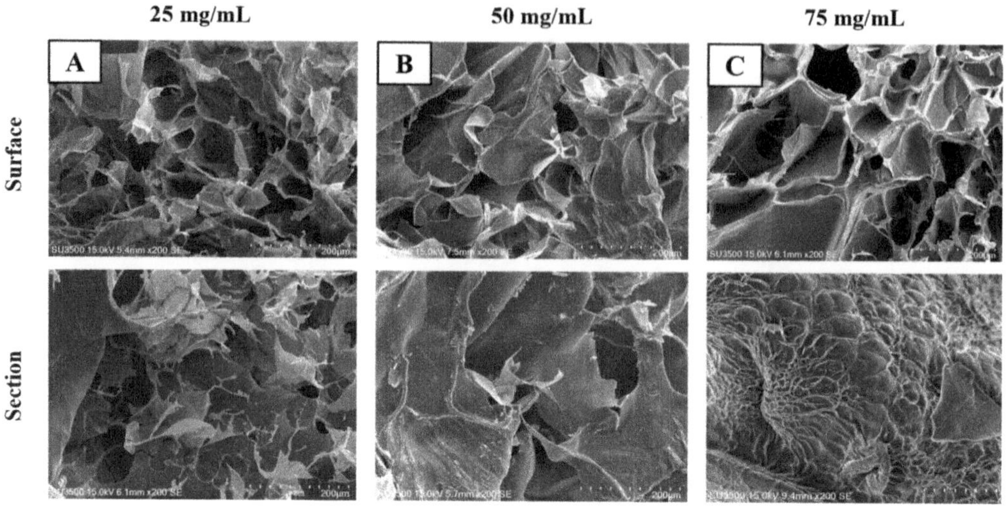

Figure 4. SEM images of the surface (top) and the middle section (bottom) of (**A**) ultrasound-assisted pepsin-solubilized decellularized extracellular matrix at 25 mg/mL (UdECM25), (**B**) UdECM50, and (**C**) UdECM75. Samples were imaged at 200× magnification (scale bars = 200 μm).

3.3. Contents of PDGF-AB and TGF-β1 in R-PRP

PRP, as a plasma product, was found in clinical applications to have large quantities of autologous growth factors, which are released after activation. These autologous growth factors play vital roles in cell growth and differentiation processes associated with wound healing [28]. Among them, TGF-β, VEGF, and PDGF-AB have the highest concentrations and can stimulate collagen synthesis, callus formation, and tissue regeneration. Thus, the PDGF-AB and TGF-β contents in PRP were used as indicators of the bioactivity of R-PRP to promote wound healing. After activation, the contents of PDGF-AB and TGF-β1 in R-PRP were 1103 ± 6.7 and 104 ± 5.0 ng/mL, while those for inactivated R-PRP were 30.2 and 23.3 ng/mL, respectively [29]. In conclusion, activation of PRP can boost the release of PDGF-AB and TGF-β1 growth factors from α-granules stored in platelet cells, and thus, the higher concentrations of both can enhance the repair ability and reduce wound-healing times.

3.4. In Vivo Testing on Diabetic Animals

In diabetic animal experiments, 180 mg/kg of a 50% nicotinamide (Vitamin B3) solution was IP administered to rats in the various groups, and after 15 min, 65 mg/kg body weight (BW) of STZ was IP injected to produce an animal model close to type 1 diabetes. Figure 5 shows the blood glucose values and BWs on days zero, three, five, and seven after injections of Vitamin B3 and STZ. BWs slightly decreased on the third day during the induction period, and BWs were maintained at 300–400 g from the third to seventh days. After the drug was given, blood glucose values sharply rose on the third day and reached a peak on the seventh day. Because STZ can destroy β cells of animal islets, most of the pancreatic islet β cells in the rats were destroyed, which resulted in insufficient insulin secretion, and energy could not be converted after eating. Due to insufficient insulin, glycogen could not be inhibited. As the BW decreased, the blood glucose value continued to rise. Furthermore, these rats were found to have polyphagia, polydipsia, and frequent urination. Therefore, this experiment successfully established a rat model of diabetes.

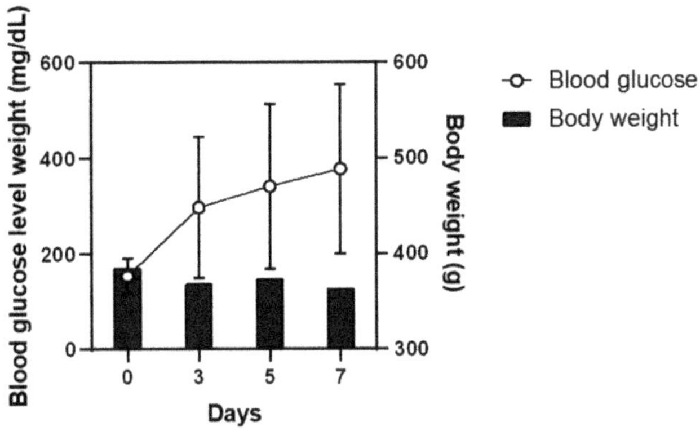

Figure 5. Blood glucose levels and body weights of a rat model of vitamin B3/streptozotocin (STZ)-induced diabetics. Values are presented as the mean ± SD (n = 13).

In order to explore the application of UdECM as a scaffold for diabetic wound healing, an 8 mm tissue sampler was used to establish a wound on the back of diabetic rats, and a silicone sheet was sutured near the wound to avoid false wound-healing and wound measurement errors. Then six different hydrogel dressings of UdECM, UdECM/R-PRP, UdECM/SCNFs, UdECM/SCNFs/R-PRP, UdECM/TEMPO 050, and UdECM/TEMPO 050/R-PRP were randomly assigned to each wound of diabetic rats. The hydrogel dressings were changed on days 0, 3, 7, 10, 14, 16, and 20. Pictures were taken to record the wound healing, and we continuously monitored blood glucose levels and BWs.

Figure 6 shows assessments of the wound area after 20 days of treatment by the wound appearance, and healing profiles were also calculated using the ImageJ image-processing analytical software. Figure 6A demonstrates that all wounds gradually healed without redness or ulceration. Among them, the UdECM/SCNFs/R-PRP group had completely healed by day 14, followed by the UdECM/SCNFs group (without R-PRP) that had healed by day 16, and the other groups (UdECM, UdECM/R-PRP, UdECM/TEMPO 050 and UdECM/TEMPO 050/R-PRP) which had healed by day 20. UdECM, UdECM/R-PRP, UdECM/TEMPO 050, and UdECM/TEMPO 050/R-PRP had effectively healed by day 20, which indicated that UdECM itself was relatively rich in original proteins and retained various cytokines after US, such as fibroblast growth factor and TGF, that could effectively help tissues regenerate. Among them, UdECM/SCNFs/R-PRP-treated wounds healed rapidly by the 14th day and revealed that its SC antibacterial efficacy was the same as that of UdECM, which was a mixed wound dressing with an external antimicrobial microenvironment and tissue growth factors.

As illustrated in Figure 6B, rapid healing began on day seven in all groups, and the degree of wound-healing for UdECM/SCNFs/R-PRP was 58% ± 0.6% compared to day 0, whereas that of UdECM/TEMPO 050/R-PRP was only 45% ± 0.1%, and there was a significant difference ($p < 0.05$) between these two groups. Furthermore, the UdECM/SCNFs/R-PRP group showed wound-healing on day 10, which had reached nearly 90% (89% ± 0.7%), whereas the UdECM/TEMPO 050/R-PRP group was only at 59% ± 0.7% healing ($p < 0.005$). This indicates that SCNFs were more effective than TEMPO 050. UdECM and R-PRP with rich growth factors can be combined to form homogeneous hydrogels to help enhance therapeutic effects.

Blood glucose levels and BWs were monitored during the treatment period to ensure that the rats continued to maintain diabetic disease patterns. As shown in Figure 7, during the follow-up period, blood glucose levels of diabetic rats continually remained higher than 350 mg/dL, and BWs remained at approximately 200 g, which confirmed that the rats were still hyperglycemic during the experiment.

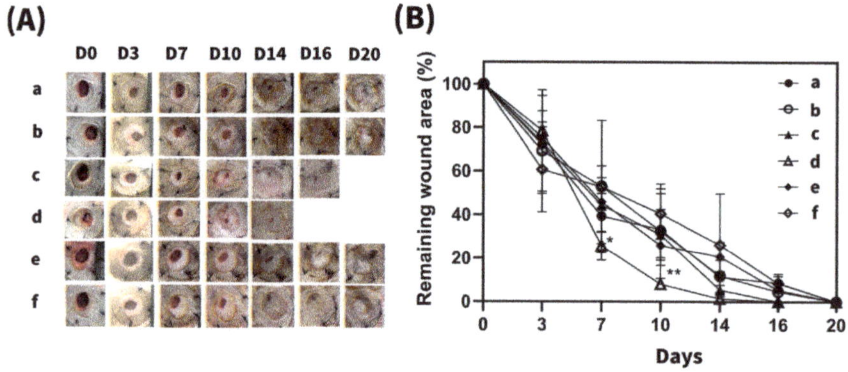

Figure 6. Wound-healing studies (n = 5 or 6) in diabetic rats. Patterns of the healing process vs. time (**A**) and the remaining wound area vs. time profiles (**B**) for six kinds of hydrogel dressings including (a) ultrasound-assisted pepsin-solubilized decellularized extracellular matrix (UdECM), (b) UdECM/rat platelet-rich plasma (R-PRP), (c) UdECM/sacchachitin nanofibers (SCNFs), (d) UdECM/SCNFs/R-PRP, (e) UdECM/TEMPO 050, and (f) UdECM/TEMPO 050/R-PRP. Values are presented as the mean ± SD (n = 5 or 6). * $p < 0.05$, ** $p < 0.01$ compared with UdECM/TEMPO 050/R-PRP group.

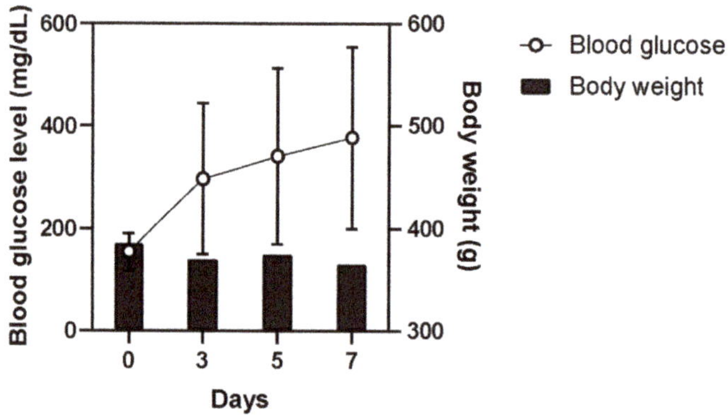

Figure 7. Determination of blood glucose levels and body weights in diabetic rats for 20 days. Values are presented as the mean ± standard deviation (n = 9–13).

The wound-healing test on euthanized diabetic rats was conducted on days 10 and 16. After removal of UdECM, UdECM/R-PRP, UdECM/SCNFs, UdECM/SCNFs/R-PRP, UdECM/TEMPO 050, and UdECM/TEMPO 050/R-PRP hydrogel scaffolds, skin tissues of the wound area were sectioned and stained with H&E, MT, and platelet endothelial cell adhesion molecule (CD31) for histopathological biopsies. H&E stain is a dual staining method for tissues that visualizes the distribution of nuclei (purple) and intracellular proteins (red). MT staining is used to observe the differential staining of collagen fibers and muscle fibers in connective tissue. Black indicates the nucleus, red the cytoplasm, keratin, or muscle fibers, and blue indicates collagen. CD31 is a surface marker for vascular endothelial cells, which are present in platelets, neutrophils, monocytes, and endothelial cells. Further, CD31 is commonly used to understand the distribution of neovascular growth.

Figure 8 illustrates H&E staining of skin tissues sampled on days 10 (Figure 8A) and 16 (Figure 8) after treatment with UdECM, UdECM/R-PRP, UdECM/SCNFs, UdECM/SCNFs/R-PRP, UdECM/TEMPO 050, and UdECM/TEMPO 050/R-PRP hydrogel scaffolds. Figure 8A shows that the newly grown tissue was in a red-dotted frame, and large numbers of cells were found to have accumulated around the wound in all treatment groups by day 10. This indicates that the wounds were gradually healing.

Figure 8B shows the great improvement in wound healing with insignificant inflammation on day 16 in all treatment groups. As for dermal restoration, purple fibroblasts stained with H&E were uniformly dispersed in the dermis, which represented the transition of wound healing from the proliferation phase to the final tissue-remodeling stage. Compared to UdECM, UdECM/TEMPO 050, and UdECM/TEMPO 050/R-PRP hydrogel treatment groups, UdECM/R-PRP, UdECM/SCNFs, and UdECM/SCNFs/R-PRP treatment groups showed regeneration of new skin tissues in the epidermal layer. Obviously, epithelial tissue cells were neat, the blood vessel density of the wound had returned to normal, and the presence of hair follicles and sweat glands hinted at the significantly better healing efficacy of the latter three groups than the former three groups.

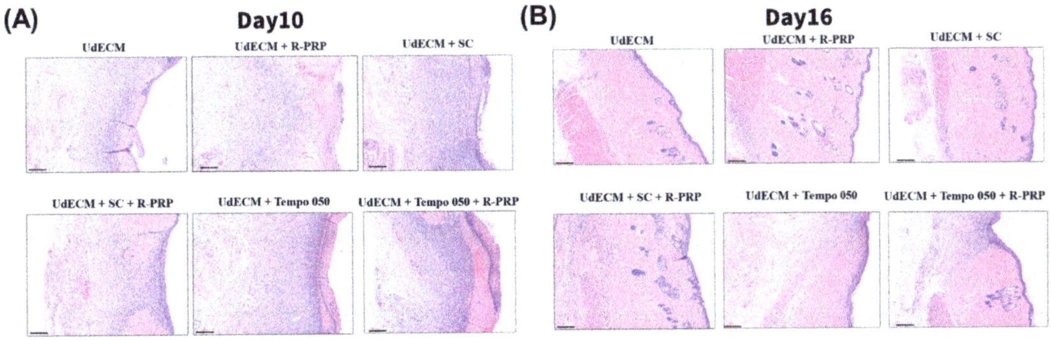

Figure 8. Hematoxylin and eosin (H&E)-stained images of ultrasound-assisted pepsin-solubilized decellularized extracellular matrix (UdECM), UdECM/rat platelet-rich plasma (R-PRP), UdECM/sacchachitin nanofibers (SCNFs), UdECM/SCNFs/R-PRP, UdECM/TEMPO 050 and UdECM/TEMPO 050/R-PRP hydrogel-treated wounds on days 10 (**A**) and 16 (**B**). (4×, scale bar = 300 μm).

During the tissue-remodeling period, excess microvessels degenerated and became atrophied, the vascular density of the wound gradually returned to normal, collagen tissue was elongated and neatly arranged, and scars became flat and faded. To further assess wound healing, MT staining was used to confirm collagen recovery in regenerated tissues. As shown in Figure 9, MT staining for skin tissues sampled on days 10 (Figure 9A) and 16 (Figure 9B) are illustrated after treatment with UdECM, UdECM/R-PRP, UdECM/SCNFs, UdECM/SCNFs/R-PRP, UdECM/TEMPO 050, and UdECM/TEMPO 050/R-PRP hydrogel scaffolds. In Figure 9A, newly grown tissue appeared inside the black dotted frame, and there was a little collagen formed, which was irregularly arranged for the UdECM, UdECM/R-PRP, UdECM/SCNFs, UdECM/TEMPO 050, and UdECM/TEMPO 050/R-PRP treatment groups and which indicated that the wound had not completely healed. Exceptionally, the UdECM/SCNFs/R-PRP group was the only arm that showed collagen deposition and had gradually entered the tissue-repair phase.

Figure 9B illustrates MT three-color staining graphs for tissues sampled on day 16. From the area inside the black dotted frame, it was found that all treatment groups had large amounts of collagen deposition (blue-violet), but the UdECM, UdECM/SCNFs, UdECM/TEMPO 050, and UdECM/TEMPO 050/R-PRP treatment groups exhibited irregular arrangements of collagen in the dermal layer, which indicated that the tissues were still undergoing regeneration during the tissue-remodeling period. Obviously, the UdECM/SCNFs/R-PRP and UdECM/R-PRP hydrogel scaffolds demonstrated that the collagen in the dermal layer was neatly arranged and densely uniform, while UdECM/SCNFs/R-PRP showed the best wound-healing effect, as the tissue was completely repaired and remodeled by 16 days after treatment. Figure 9C illustrates the quantification results of collagen in tissue sections by image analytical software. The collagen content of the UdECM/SCNF/R-PRP treatment group had increasingly been deposited to 30% from days 10 to 16, whereas those for UdECM, UdECM/R-PRP and UdECM/SCNFs had moderately increased to 20–25%, and those for UdECM/TEMPO 050 and UdECM/TEMPO 050/R-PRP had slightly increased to less than 20%. This indicates that the use of collagen-rich

UdECM could provide a good microenvironment for long-term wound repair. Furthermore UdECM with the addition of either SCNFs (UdECM/SCNFs) and/or R-PRP (UdECM/R-PRP UdECM/SCNFs/R-PRP) further synergistically enhanced diabetic wound-healing with greater collagen deposition.

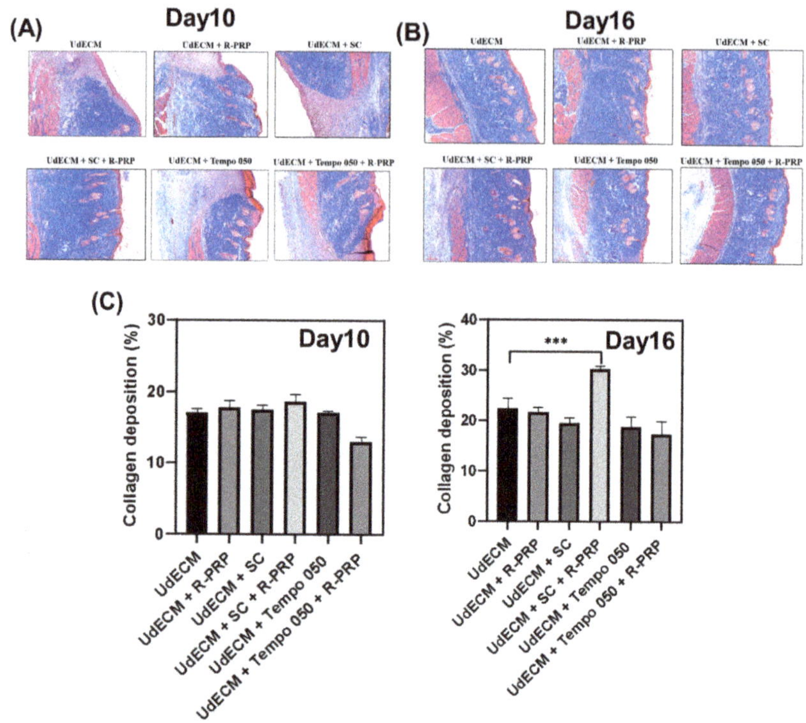

Figure 9. Masson's trichrome (MT)-stained images of ultrasound-assisted pepsin-solubilized decellularized extracellular matrix (UdECM), UdECM/rat platelet-rich plasma (R-PRP), UdECM/sacchachitin nanofibers (SCNFs), UdECM/SCNFs/R-PRP, UdECM/TEMPO 050, and UdECM/TEMPO 050/R-PRP hydrogel-treated wounds on days 10 (**A**) and 16 (**B**). (4×, scale bar = 300 μm). Quantitative collagen deposition (**C**) based on MT staining of skin tissues sampled on days 10 and 16. *** $p < 0.005$ compared to the UdECM group.

In immunohistochemical (IHC) staining, CD31 acts as the surface marker for vascular endothelial cells and is mainly used to indicate the presence of endothelial cell tissues and the distribution of neovascularization. As shown in Figure 10, CD31 was stained with brown dots or strips and was found to be evenly distributed in the dermal layer of skin tissues sampled on days 10 (Figure 10A) and 16 (Figure 10B). The brown dots or strips, as highlighted by red triangles in the figure, were new microvessels. On day 10, there was a large amount of CD31 expression in the UdECM/SCNFs/R-PRP group, which obviously indicated the formation of many new blood vessels. Figure 10B shows that expression in all groups significantly increased and was uniformly distributed from days 10 to 16. Furthermore, IHC staining of CD31 for the UdECM/SCNF/R-PRP and UdECM/R-PRP hydrogel treatment groups exhibited a more uniform and larger amount compared to the UdECM, UdECM/TEMPO 050, and UdECM/TEMPO 050/R-PRP groups. This indicated that more neovascularization had occurred in the dermis and also that excess microvessels were degenerating and atrophying to normalize the density of blood vessels in the wound. Thus, it was confirmed that the UdECM/SCNF/R-PRP and UdECM/R-PRP hydrogels greatly promoted the regeneration of blood vessels and provided sufficient nutrients to increase the ability of tissue repair and achieve wound healing.

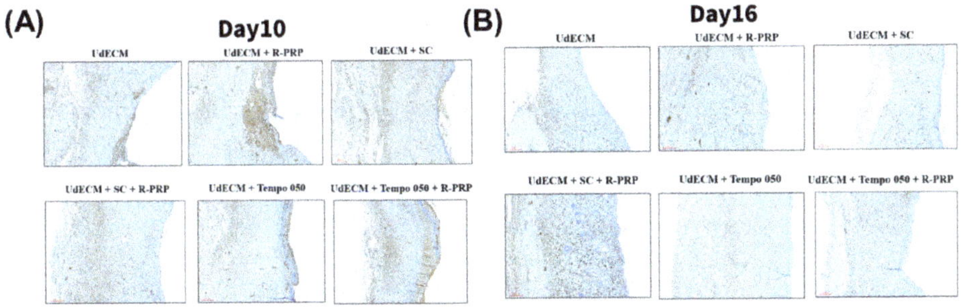

Figure 10. CD31 immunohistochemically stained (CD31) images of ultrasound-assisted pepsin-solubilized decellularized extracellular matrix (UdECM), UdECM/rat platelet-rich plasma (R-PRP), UdECM/sacchachitin nanofibers (SCNFs), UdECM/SCNFs/R-PRP, UdECM/TEMPO 050, and UdECM/TEMPO 050/R-PRP hydrogel-treated wounds on days 10 (**A**) and 16 (**B**). (4×, scale bar = 300 μm).

In the comprehensive evaluation, wounds treated with different hydrogel scaffolds were scored based on scarring and the results obtained from H&E, MT, and CD31 staining on day 16. The detailed outcomes are presented in Table 2. The extent of wound healing was assessed using a grading scale ranging from 1 to 5, depending on the level of tissue remodeling observed (5: closest to normal skin appearance after treatment; 1: least resemblance to normal skin). The total score for normal skin was 20, indicating the highest level of recovery and resemblance to normal skin. Results in Table 2 show that the UdECM/SCNFs/R-PRP treatment group had the most significant efficacy for promoting the diabetic wound-healing process with a total score of 18, which meant the recovery was closest to normal tissues, followed by the UdECM/R-PRP and UdECM/SCNFs groups with similar total scores of 13–14. Additionally, UdECM hydrogel scaffolds with a total score of nine seemed to have only recovered to a 50% resemblance of normal skin. Treatment with the UdECM/TEMPO 050 and UdECM/TEMPO 050/R-PRP hydrogel scaffolds that combined UdECM with TEMPO 050, ranked the last with total scores of 5–7, which was even poorer than the UdECM hydrogel.

Table 2. Overall assessment of the in vivo wound-healing study on day 16. The degree of wound healing was graded from 1 to 5.

Wound Dressing	Scarring	H&E	MT	CD31	Sum
UdECM	1	2	4	2	9
UdECM/R-PRP	3	3	4	4	14
UdECM/SCNFs	3	4	3	3	13
UdECM/SCNFs/R-PRP	4	5	5	4	18
UdECM/TEMPO 050	1	1	2	1	5
UdECM/TEMPO 050/R-PRP	1	2	2	2	7

Abbreviations: UdECM, ultrasound-assisted pepsin-solubilized decellularized extracellular matrix; R-PRP, rat platelet-rich plasma; SCNFs, sacchachitin nanofibers; TEMPO 050, TEMPO-oxidized sacchachitin nanofibers; H&E, hematoxylin and eosin; MT, Masson trichrome; CD31, cluster of differentiation 31.

Several studies have demonstrated that ECM hydrogels effectively facilitate cell infiltration, with a particular emphasis on macrophages and progenitor cells. Furthermore, these hydrogels also promote neovascularization and exhibit positive functional remodeling, further highlighting their potential in tissue repair and regenerative applications [3]. This is the reason that UdECM hydrogels can moderately promote diabetic wound healing. Further, composite hydrogels composed of UdECM with SCNFs (UdECM/SCNFs) or R-PRP (UdECM/R-PRP) accelerated wound-healing faster than only UdECM hydrogels (UdECM). The observed potential mechanism underlying UdECM/SCNF hydrogels' efficacy may be attributed to two main factors: rapid re-epithelialization and normal ECM deposition. These hydrogels create a moist microenvironment that fosters enhanced cellular movements, promoting an accelerated healing process. Furthermore, the inclusion of SCNFs in the hydrogel formulation may contribute to a chemotactic effect on inflammatory cells, facilitating angiogenesis and promoting the formation of granulation tissue Ultimately, this synergistic effect leads to the formation of new tissue, thereby contributing to the overall improvement in diabetic wound healing [30,31]. Obviously, combination therapy of UdECM hydrogels with R-PRP (UdECM/R-PRP) accelerated diabetic wound-healing because of greater amounts of growth factors, such as PDGF, VEGF, TGF-β, and EGF, which can inhibit inflammatory symptoms and promote the formation of microvessels and epithelialization of chronic wounds.

Obviously, UdECM/SCNFs/R-PRP showed the greatest potency for diabetic wound healing. Notably, the diabetic wounds treated with UdECM/SCNF/R-PRP hydrogels achieved complete healing within 14 days, which closely resembles the healing timeline observed in our prior study where diabetic wounds were treated with aECMHCl,25/SC hydrogels and also fully healed within 14 days [17]. These findings underscore the remarkable therapeutic potential of UdECM/SCNF/R-PRP hydrogels and their efficacy in promoting efficient diabetic wound closure. However, Hung et al. reported that skin wounds covered with an SC membrane needed 21 days to heal [32]. This discrepancy might be attributed to synergistic promotion by the composite hydrogel and the incorporation of enriched amounts of several growth factors. Furthermore, compared to SC membranes, the design of UdECM/SCNF/R-PRP hydrogels aims to create a moist environment that fosters optimal wound-healing conditions. With a high moisture content exceeding 90% water, UdECM/SCNFs/R-PRP hydrogels effectively contribute to impeding bacteria from accessing the wound site, acting as a protective barrier against potential infections. This unique characteristic enhances the hydrogel's ability to support the healing process and leads to improved wound closure outcomes [33]. It was concluded that UdECM hydrogels could be prepared by US digestion, which has the advantage of rapid and effective preparation for type I collagen-rich hydrogels from formic acid-decellularized ECM (dECM). In in vivo tests, UdECM/SC composite hydrogels loaded with PRP could heal diabetic wounds in 14 days. In histopathological sections, UdECM/SCNFs/R-PRP effectively promoted diabetic wound healing and angiogenesis. Overall, UdECM/SCNF/R-PRP hydrogels controlled the release rate of platelet-derived growth factors and retained the antibacterial ability of SC, thereby shortening wound-healing times.

4. Conclusions

In this study, ultrasound-assisted digestion of decellularized extracellular matrix (UdECM) was successfully developed. Composite hydrogels composed of UdECM hydrogels and SCNF hydrogels or UdECM hydrogels incorporated with R-PRP improved diabetic wound healing faster than UdECM hydrogels. UdECM/SCNFs/R-PRP demonstrated the greatest enhancement of diabetic wound healing. In conclusion, UdECM hydrogels developed in this study combined with SCNF hydrogels and/or incorporated with R-PRP became a convenient and effective hydrophilic dressing, which ultimately has a wide range of applications for wound-healing and regenerative medicine.

Author Contributions: Conceptualization, writing—review and editing and project administration: C.-J.L. and H.-L.L.; methodology, software, validation, formal analysis and writing—original draft preparation: W.-J.C. and L.-C.C.; investigation, data curation and visualization: W.-C.Y.; resources, supervision and funding acquisition: H.-O.H. and M.-T.S. All authors have read and agreed to the published version of the manuscript.

Funding: This study was partially supported by research grants from the National Science and Technology Council, Taiwan (MOST 111-2221-E-264 -002 -MY2).

Data Availability Statement: Source data are available from the corresponding author upon reasonable request.

Acknowledgments: The authors acknowledge the technical support provided by TMU Core Facility.

Conflicts of Interest: The funders had no role in the design of the study; in the collection, analyses, or interpretation of data; in the writing of the manuscript; or in the decision to publish the results.

References

1. Badylak, S.F. The extracellular matrix as a biologic scaffold material. *Biomaterials* **2007**, *28*, 3587–3593. [CrossRef] [PubMed]
2. Saldin, L.T.; Cramer, M.C.; Velankar, S.S.; White, L.J.; Badylak, S.F. Extracellular matrix hydrogels from decellularized tissues: Structure and function. *Acta Biomater.* **2017**, *49*, 1–15. [CrossRef] [PubMed]
3. Spang, M.T.; Christman, K.L. Extracellular matrix hydrogel therapies: In vivo applications and development. *Acta Biomater.* **2018**, *68*, 1–14. [CrossRef] [PubMed]
4. Badylak, S.F.; Freytes, D.O.; Gilbert, T.W. Extracellular matrix as a biological scaffold material: Structure and function. *Acta Biomater.* **2009**, *5*, 1–13. [CrossRef] [PubMed]
5. Christman, K.L. Biomaterials for tissue repair. *Science* **2019**, *363*, 340–341. [CrossRef] [PubMed]
6. Freytes, D.O.; Martin, J.; Velankar, S.S.; Lee, A.S.; Badylak, S.F. Preparation and rheological characterization of a gel form of the porcine urinary bladder matrix. *Biomaterials* **2008**, *29*, 1630–1637. [CrossRef] [PubMed]
7. Singelyn, J.M.; DeQuach, J.A.; Seif-Naraghi, S.B.; Littlefield, R.B.; Schup-Magoffin, P.J.; Christman, K.L. Naturally derived myocardial matrix as an injectable scaffold for cardiac tissue engineering. *Biomaterials* **2009**, *30*, 5409–5416. [CrossRef] [PubMed]
8. Singelyn, J.M.; Sundaramurthy, P.; Johnson, T.D.; Schup-Magoffin, P.J.; Hu, D.P.; Faulk, D.M.; Wang, J.; Mayle, K.M.; Bartels, K.; Salvatore, M.; et al. Catheter-deliverable hydrogel derived from decellularized ventricular extracellular matrix increases endogenous cardiomyocytes and preserves cardiac function post-myocardial infarction. *J. Am. Coll. Cardiol.* **2012**, *59*, 751–763. [CrossRef]
9. Seif-Naraghi, S.B.; Singelyn, J.M.; Salvatore, M.A.; Osborn, K.G.; Wang, J.J.; Sampat, U.; Kwan, O.L.; Strachan, G.M.; Wong, J.; Schup-Magoffin, P.J.; et al. Safety and efficacy of an injectable extracellular matrix hydrogel for treating myocardial infarction. *Sci. Transl. Med.* **2013**, *5*, 173ra125. [CrossRef]
10. Wassenaar, J.W.; Gaetani, R.; Garcia, J.J.; Braden, R.L.; Luo, C.G.; Huang, D.; DeMaria, A.N.; Omens, J.H.; Christman, K.L. Evidence for Mechanisms Underlying the Functional Benefits of a Myocardial Matrix Hydrogel for Post-MI Treatment. *J. Am. Coll. Cardiol.* **2016**, *67*, 1074–1086. [CrossRef]
11. Dziki, J.L.; Badylak, S.F. Immunomodulatory biomaterials. *Curr. Opin. Biomed. Eng.* **2018**, *6*, 51–57. [CrossRef]
12. Sadtler, K.; Sommerfeld, S.D.; Wolf, M.T.; Wang, X.; Majumdar, S.; Chung, L.; Kelkar, D.S.; Pandey, A.; Elisseeff, J.H. Proteomic composition and immunomodulatory properties of urinary bladder matrix scaffolds in homeostasis and injury. *Semin. Immunol.* **2017**, *29*, 14–23. [CrossRef] [PubMed]
13. Voytik-Harbin, S.L.; Brightman, A.O.; Waisner, B.Z.; Robinson, J.P.; Lamar, C.H. Small intestinal submucosa: A tissue-derived extracellular matrix that promotes tissue-specific growth and differentiation of cells in vitro. *Tissue Eng.* **1998**, *4*, 157–174.
14. Uriel, S.; Huang, J.J.; Moya, M.L.; Francis, M.E.; Wang, R.; Chang, S.Y.; Cheng, M.H.; Brey, E.M. The role of adipose protein derived hydrogels in adipogenesis. *Biomaterials* **2008**, *29*, 3712–3719. [CrossRef]
15. Uriel, S.; Labay, E.; Francis-Sedlak, M.; Moya, M.L.; Weichselbaum, R.R.; Ervin, N.; Cankova, Z.; Brey, E.M. Extraction and assembly of tissue-derived gels for cell culture and tissue engineering. *Tissue Eng. Part C Methods* **2009**, *15*, 309–321. [CrossRef] [PubMed]
16. Hussey, G.S.; Nascari, D.G.; Saldin, L.T.; Kolich, B.; Lee, Y.C.; Crum, R.J.; El-Mossier, S.O.; D'Angelo, W.; Dziki, J.L.; Badylak, S.F. Ultrasonic cavitation to prepare ECM hydrogels. *Acta Biomater.* **2020**, *108*, 77–86. [CrossRef] [PubMed]
17. Hsieh, C.-M.; Wang, W.; Chen, Y.-H.; Wei, P.-S.; Liu, Y.-H.; Sheu, M.-T.; Ho, H.-O. A Novel Composite Hydrogel Composed of Formic Acid-Decellularized Pepsin-Soluble Extracellular Matrix Hydrogel and Sacchachitin Hydrogel as Wound Dressing to Synergistically Accelerate Diabetic Wound Healing. *Pharmaceutics* **2020**, *12*, 538. [CrossRef]
18. Li, D.; Mu, C.; Cai, S.; Lin, W. Ultrasonic irradiation in the enzymatic extraction of collagen. *Ultrason. Sonochem.* **2009**, *16*, 605–609. [CrossRef]
19. Chao, F.C.; Wu, M.H.; Chen, L.C.; Lin, H.L.; Liu, D.Z.; Ho, H.O.; Sheu, M.T. Preparation and characterization of chemically TEMPO-oxidized and mechanically disintegrated sacchachitin nanofibers (SCNF) for enhanced diabetic wound healing. *Carbohydr. Polym.* **2020**, *229*, 115507. [CrossRef]
20. Van Buul, G.M.; Koevoet, W.L.; Kops, N.; Bos, P.K.; Verhaar, J.A.; Weinans, H.; Bernsen, M.R.; van Osch, G.J. Platelet-rich plasma releasate inhibits inflammatory processes in osteoarthritic chondrocytes. *Am. J. Sports Med.* **2011**, *39*, 2362–2370. [CrossRef]
21. Qian, Z.; Wang, H.; Bai, Y.; Wang, Y.; Tao, L.; Wei, Y.; Fan, Y.; Guo, X.; Liu, H. Improving Chronic Diabetic Wound Healing through an Injectable and Self-Healing Hydrogel with Platelet-Rich Plasma Release. *ACS Appl. Mater. Interfaces* **2020**, *12*, 55659–55674. [CrossRef] [PubMed]
22. Jhan, H.-J.; Liu, J.-J.; Chen, Y.-C.; Liu, D.-Z.; Sheu, M.-T.; Ho, H.-O. Novel injectable thermosensitive hydrogels for delivering hyaluronic acid–doxorubicin nanocomplexes to locally treat tumors. *Nanomedicine* **2015**, *10*, 1263–1274. [CrossRef] [PubMed]
23. Ho, H.-O.; Lin, L.-H.; Sheu, M.-T. Characterization of collagen isolation and application of collagen gel as a drug carrier. *J. Control. Release* **1997**, *44*, 103–112. [CrossRef]
24. Palsamy, P.; Subramanian, S. Resveratrol, a natural phytoalexin, normalizes hyperglycemia in streptozotocin-nicotinamide induced experimental diabetic rats. *Biomed. Pharmacother.* **2008**, *62*, 598–605. [CrossRef] [PubMed]
25. Antoine, E.E.; Vlachos, P.P.; Rylander, M.N. Review of collagen I hydrogels for bioengineered tissue microenvironments: Characterization of mechanics, structure, and transport. *Tissue Eng. Part B Rev.* **2014**, *20*, 683–696. [CrossRef] [PubMed]

26. Tosh, S.M.; Marangoni, A.G. Determination of the maximum gelation temperature in gelatin gels. *Appl. Phys. Lett.* **2004**, *84*, 4242–4244. [CrossRef]
27. Lin, H.J.; Wang, T.J.; Li, T.W.; Chang, Y.Y.; Sheu, M.T.; Huang, Y.Y.; Liu, D.Z. Development of Decellularized Cornea by Organic Acid Treatment for Corneal Regeneration. *Tissue Eng. Part A* **2019**, *25*, 652–662. [CrossRef]
28. Lai, F.; Kakudo, N.; Morimoto, N.; Taketani, S.; Hara, T.; Ogawa, T.; Kusumoto, K. Platelet-rich plasma enhances the proliferation of human adipose stem cells through multiple signaling pathways. *Stem Cell Res. Ther.* **2018**, *9*, 107. [CrossRef]
29. Van den Dolder, J.; Mooren, R.; Vloon, A.P.; Stoelinga, P.J.; Jansen, J.A. Platelet-rich plasma: Quantification of growth factor levels and the effect on growth and differentiation of rat bone marrow cells. *Tissue Eng.* **2006**, *12*, 3067–3073. [CrossRef]
30. Guo, X.; Mu, D.; Gao, F. Efficacy and safety of acellular dermal matrix in diabetic foot ulcer treatment: A systematic review and meta-analysis. *Int. J. Surg.* **2017**, *40*, 1–7. [CrossRef]
31. Lin, Y.C.; Tan, F.J.; Marra, K.G.; Jan, S.S.; Liu, D.C. Synthesis and characterization of collagen/hyaluronan/chitosan composite sponges for potential biomedical applications. *Acta Biomater.* **2009**, *5*, 2591–2600. [CrossRef]
32. Hung, W.S.; Fang, C.L.; Su, C.H.; Lai, W.F.; Chang, Y.C.; Tsai, Y.H. Cytotoxicity and immunogenicity of SACCHACHITIN and its mechanism of action on skin wound healing. *J. Biomed. Mater. Res.* **2001**, *56*, 93–100. [CrossRef]
33. Solanki, D.; Vinchhi, P.; Patel, M.M. Design Considerations, Formulation Approaches, and Strategic Advances of Hydrogel Dressings for Chronic Wound Management. *ACS Omega* **2023**, *8*, 8172–8189. [CrossRef]

Journal of
*Functional
Biomaterials*

MDPI

Article

Hydroxyapatite-Based Coatings on Silicon Wafers and Printed Zirconia

Antoine Chauvin [1], Marie-Rose Garda [1], Nathan Snyder [2], Bai Cui [2], Nicolas Delpouve [1,*] and Li Tan [2]

1 Groupe de Physique des Matériaux UMR 6634, CNRS, Université de Rouen Normandie, INSA Rouen Normandie, F-76000 Rouen, France; marie-rose.garda@univ-rouen.fr (M.-R.G.)
2 Department of Mechanical and Materials Engineering, University of Nebraska, Lincoln, NE 68588, USA; bcui@unl.edu (B.C.); ltan4@unl.edu (L.T.)
* Correspondence: nicolas.delpouve1@univ-rouen.fr; Tel.: +33-(0)232955165

Abstract: Dental surgery needs a biocompatible implant design that can ensure both osseointegration and soft tissue integration. This study aims to investigate the behavior of a hydroxyapatite-based coating, specifically designed to be deposited onto a zirconia substrate that was intentionally made porous through additive manufacturing for the purpose of reducing the cost of material. Layers were made via sol–gel dip coating by immersing the porous substrates into solutions of hydroxyapatite that were mixed with polyethyleneimine to improve the adhesion of hydroxyapatite to the substrate. The microstructure was determined by using X-ray diffraction, which showed the adhesion of hydroxyapatite; and atomic force microscopy was used to highlight the homogeneity of the coating repartition. Thermogravimetric analysis, differential scanning calorimetry, and Fourier transform infrared spectroscopy showed successful, selective removal of the polymer and a preserved hydroxyapatite coating. Finally, scanning electron microscopy pictures of the printed zirconia ceramics, which were obtained through the digital light processing additive manufacturing method, revealed that the mixed coating leads to a thicker, more uniform layer in comparison with a pure hydroxyapatite coating. Therefore, homogeneous coatings can be added to porous zirconia by combining polyethyleneimine with hydroxyapatite. This result has implications for improving global access to dental care.

Keywords: HAp; additive manufacturing; thermal properties; microscopy; sol–gel dip coating

Citation: Chauvin, A.; Garda, M.-R.; Snyder, N.; Cui, B.; Delpouve, N.; Tan, L. Hydroxyapatite-Based Coatings on Silicon Wafers and Printed Zirconia. *J. Funct. Biomater.* **2024**, *15*, 11. https://doi.org/10.3390/jfb15010011

Academic Editor: Christie Ying Kei Lung

Received: 23 October 2023
Revised: 12 December 2023
Accepted: 13 December 2023
Published: 27 December 2023

1. Introduction

The last few decades have seen a growing demand for dental surgery as the population increases and ages, and the market for dentistry implants is expected to significantly increase [1]. The most notable current targets of improvement for dental reconstruction are related to implant design and stability [2–10], material choice optimization [11–17], osseointegration [9,10,18–20], biocompatibility [13,14,21–23], soft tissue integration [24–30], and healing activity [31,32]. Titanium and zirconia are the most frequently used scaffolds. Zirconia is the more stable candidate mechanically and chemically, and it also exhibits aesthetic advantages (as it has the color of a tooth). Moreover, zirconia-based implants can compete against titanium in terms of bacteria resistance [33,34]. Although zirconia is a favored candidate, its biological inertness limits osseointegration [35,36]. Hydroxyapatite (HAp), the main mineral constituent of dentine and tooth enamel, is commonly used as a coating component, notably for overlaying titanium- or ceramic-based implants. HAp can be successfully added to zirconia to improve dental reconstruction [36]. In brief, HAp has a bone-like mineral composition, and it contributes to osseointegration when it dissolves. The released calcium and phosphorous ions can be used by the surrounding tissues to consolidate the dental implant [37].

Ideal HAp coatings for orthopedic implants are thin (<50 μm), have high cohesive strength to prevent delamination, have a high hardness scale rating to reduce the wear, and are rough with sufficient porosity to promote the ingrowth of bone tissues [38–40] Sol–gel coating [41], pulsed laser deposition [42], pulsed electrochemical deposition [43], pulsed plasma deposition [44], atmospheric plasma spraying [45], and suspension plasma spraying [46] are some examples of deposition techniques. Atmospheric plasma spraying is the most common thermal spray process for fabricating HAp coatings due to its relatively low cost and ability to coat large areas of complex shapes [40]. On the other hand, it was reported that the design of thin coating via atmospheric plasma spraying leads to phase decomposition and poor adhesion strength [38,40]. Moreover, this process could introduce local heating, which can degrade HAp or induce phase transformation [36,47].

Sol–gel dip coating [48–50] is a low-temperature alternative that exhibits advantages in terms of flexibility and biocompatibility. Some of its merits include a simple setup, flexibility with irregularly shaped substrates, and easy control of both thickness and porosity. Because it requires minimal consumption of materials, it is easy to coat a wide range of substrates with bioactive and biocompatible materials [51]. The substrate is simply dipped into a starting colloidal solution (the 'sol') and is then withdrawn at a constant speed to enable solution drainage and instantaneous gelation. The thickness and morphology of coatings can be adjusted by varying the substrate withdrawal speed and viscosity of the solution [50].

This present study considers a zirconia scaffold that is manufactured via additive manufacturing (3D printing). This process can realize complex designs, which enables precise control over porosity for the benefits of osseointegration [40] while reducing the cost of materials. Unfortunately, some complications emerge with the coating of HAp, especially regarding dewetting and the formation of a discontinuous layer. These issues are, for the first time, investigated by blending a sol–gel solution with polyethyleneimine (PEI). PEI is a versatile polymer that exists in a linear and branched form and can act as a ligand with several molecules. It has attracted attention from the paper industry [52], from the electronic domain [53], and in heat exchange membranes [54]. It is used biomedically as a transfection agent for gene delivery [55] and stabilizer for HAP nanoparticles for drug delivery [56]. On the other hand, PEI is not completely harmless to the human body: it is cytotoxic at high concentrations and can cause cell apoptosis [55].

Therefore, several tests were attempted to remove the excess PEI. The first tests were performed on HAp-coated wafers of silicon, which are less expensive than zirconia. Following this procedure, thermogravimetric analysis (TGA), differential scanning calorimetry (DSC), and Fourier transform infrared spectroscopy (FTIR) were performed on the annealed materials to record the possible residual signatures of PEI. Then, the HAp coating microstructure and morphology were characterized via X-ray diffraction (XRD) and atomic force microscopy (AFM). Finally, the proper coating of HAp on zirconia porous ceramics was verified using scanning electron microscopy (SEM). To the authors' knowledge, this is the first time that this strategy has been proposed for the coating of HAp, that is, combining the design of porous zirconia scaffolds, which are made using digital light processing (DLP), with an adapted deposition method based on mineral-organic interactions. Thus, the aim of this study is to develop the PEI-HAp coating to improve the adhesion of hydroxyapatite to zirconia.

2. Materials and Methods

2.1. Preparation of Zirconia Substrate

The preparation of printed supports was performed via digital light processing (DLP). Its technology is similar to stereolithography, but it uses a UV projector as a light source, which enables it to cure a complete layer of resin at a time. This process significantly increases the rate of manufacturing. The practice of applying this technology to zirconia ceramic printing has been widely presented in the literature, including for applications in the dental industry [57–61]. The protocol followed in this study was reported by Snyder [62]. Oleic acid dispersant and ceramic powder were mixed in ethanol before the solvent was removed. For the DLP method, a slurry, or semi-liquid mixture, must be created to make the ceramic green body (GB). The GB is the ceramic body that results after printing. It is composed of ceramic particles that are held together by a cured polymer, and it must be heat treated further to create a dense ceramic part.

The build plate descends into the slurry, and a UV light projector shines up from below, projecting the curing layer image. After lifting, the build plate waits for resin to refill, then returns to the vat with the adjusted layer thickness and cures the next layer. The subsequent debinding process, the initial heating stage post-sintering, removes the polymer binder, leaving ceramic particles in the body. Finally, the system undergoes sintering, with the local temperature increased to 1500 °C.

2.2. HAp Solution

To prepare the HAp solution, 0.015 mol (2.494 g) of triethylphosphite (TEP, Thermo Scientific Alfa Aesar, Ward Hill, MA, USA) was added to a mixed solution of 23.92 mL ethanol and 1.08 mL deionized water. Then, 0.0251 mol (4.1102 g) of calcium nitrate tetrahydrate ($Ca(NO_3)_2 \cdot 4H_2O$, Thermo Scientific Alfa Aesar, USA) was dissolved in 25 mL ethanol. These solutions were mixed at room temperature for 3 days and aged at 37 °C for 1 day with continuous stirring to make the HAp sol. The theoretical concentration of the final HAp sol was 0.05 M when it was assumed that all reactants completely reacted.

2.3. PEI-HAp Solution

First, 0.625×10^{-3} mol (500 mg) of branched PEI (Sigma Aldrich 408727, Saint Louis, MO, USA, CAS number = 9002–98–6), was added to a solution of 50 mL of deionized water. This solution was stirred at 80 °C for one minute. Then, this solution was mixed with HAp solution under a ratio of 1:1 in a 10 mL plastic test tube.

2.4. PEI-HAp Coating on Zirconia

The PEI-HAp was coated with a laboratory dip-coating equipment. The dip-coating rate was 0.25 mm/s and the experiments were performed thrice. The sol was dried for 5 min in the oven at 37 °C between each dipping step. The annealing was performed after the coating was completed. The PEI-HAp-coated zirconia sample was oven-dried at 70 °C overnight and annealed at 230 °C for 1 h (to degrade the PEI) then at 800 °C for 1 h. For both ramps, the heating rate was 5 °C/min. Before any characterization, the coating adhesion was tested by sticking a piece of tape on the substrate, then removing it from several angles to mimic tearing solicitations. Only samples without coating damage were kept for investigation. An example is provided in Supplementary Materials (Figure S1).

2.5. Characterization of the PEI-HAp Coating

A Dimension 3100 SPM System along with a Digital Instrument Nanoscope IIIa was used for the atomic force microscopy (AFM). This system was set to contact mode, applying a constant force of 0.1 N/m. The measurement was performed with a resolution of 0.5 nm and an accuracy of 1 nm.

Thermogravimetric analyses (TGAs) were carried out using a TGA Discovery instrument from TA Instruments®, Guyancourt, France. The analyses were carried out under nitrogen atmosphere at a 25 mL min^{-1} flow rate, in the temperature range of 30–800 °C and a scanning rate of 10 K min^{-1} on 2 mg samples. The apparatus calibration procedure included a baseline, mass, and temperature calibration. The temperature calibration was performed using the Curie point of nickel as reference. The mass calibration was performed using standard masses, and the calibration in mass loss was performed using calcium oxalate as the reference.

The differential scanning calorimetry (DSC) experiments were conducted on a DSC Q100 from TA Instruments®, Guyancourt, France. The experiments were conducted on samples of 2 mg mass at 10 K/min, and under nitrogen atmosphere at a 50 mL min^{-1} flow rate. The calibration procedure includes a baseline with an empty furnace. To calibrate the temperature and energy, a standard sample of indium (T_m = 156.60 °C and ΔH_m = 28.38 J g^{-1} corresponding to melting temperature and enthalpy respectively was used.

The infrared spectroscopic details were recorded on a Thermo Fischer® (Montigny-le-Bretonneux, France) Nicolet iS10 in transmission mode in the 425–4000 cm^{-1} range, with a resolution of 4 cm^{-1}. A background correction was performed prior to the experiment. The analyses of IR spectra were performed using the apparatus Omnic 9.3. software.

The X-ray diffraction (XRD) analyses were performed using a Rigaku® Multiflex Diffractometer (Austin, TX, USA). The Rigaku Multiflex consists of a 2 kW copper tube and a θ/θ goniometer. The sample holder remains horizontal during the scan and therefore, there is no need to use adhesive substances to mount samples on to a low background sample holder plate. It is also configured in focusing geometry where a secondary monochromator removes the scattered signal except that corresponding to Cu Kalpha wavelength. The samples analyzed via XRD were under the form of powders that were obtained by heating the HAp solutions at 80 °C to boil ethanol.

The scanning electron microscopy (SEM) pictures were acquired using a Hitachi® (Schaumburg, IL, USA) S4700 field-emission SEM with a high-resolution digital processing system. The work distance was in the order of 20 nm. The sample was fractured in liquid nitrogen, then polished and fixed on the support with a conductive coating.

3. Results and Discussion

3.1. Coating Topography

The surface relief is an important element to validate the dip-coating experimental conditions. A first glance at the impact of sol–gel deposition on the coating morphology is provided on Figure 1. It shows images produced by AFM measurements of the coating, which was made by mixing HAp with PEI after being overlayed onto the silica wafer and annealed. The AFM probe provides a precise nanoscale measurement of the different cliffs and clusters. Figure 1a provides the topography (the relief depth is ±125 nm on the AFM picture) and Figure 1b provides the deflection signal. A bright point in each image indicates a high value while a dark point indicates a low value. One can observe from the topography images that slight agglomerations of brighter light are dispersed throughout the sample. They are attributed to the coating domain with a higher thickness, while the black parties should indicate the absence of coating in the probed area. One can consider, despite slight irregularities, that the coating is fairly homogeneous. Moreover, one can observe that the surface does not exhibit any cracks.

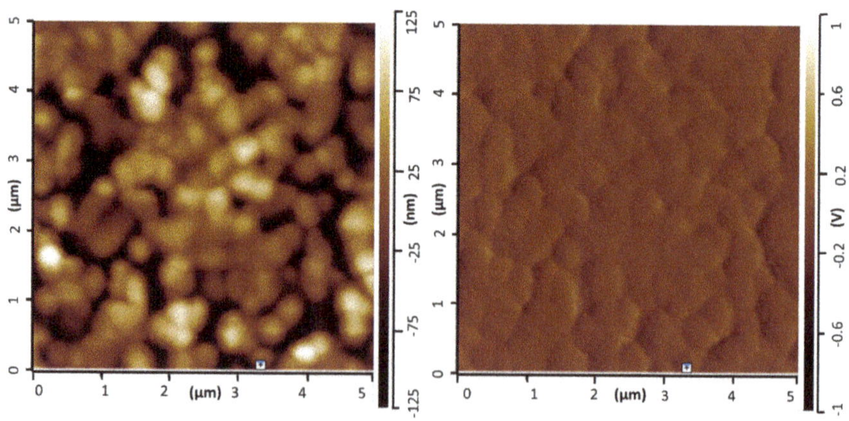

(a) (b)

Figure 1. AFM picture of the silica wafer surface that was coated using HAp-PEI, then annealed: (**a**) topography relief of the coating; (**b**) deflection signal related to the AFM measurement.

3.2. Coating Characterization

To ensure that the annealing procedure did not damage or alter the microstructure of the HAp coating, X-ray analyses were performed. The results were compared with those obtained for pure HAp. These results are shown in Figure 2A for HAp alone and in Figure 2B for the coating on the silica wafer. Thus, according to the spectra, the signature of the coating is very similar to the one of the non-treated HAp and is in agreement with the literature [63]. The peaks of the diffracting planes (200) and (111) are found between 2Theta = $2\theta = 20°$ and $2\theta = 25°$, with 2θ the angle of deviation between the incoming beam and the reflected beam. The peaks of the highest intensity I are observed between $2\theta = 25°$ and $2\theta = 35°$. They characterize the diffracting planes (002), (210), (211), (112), (300), and (202). Finally, the peaks corresponding to the diffracting planes (301), (212), and (310) are recorded between $2\theta = 35°$ and $2\theta = 40°$. This reflects that the annealing did not degrade the HAp coating.

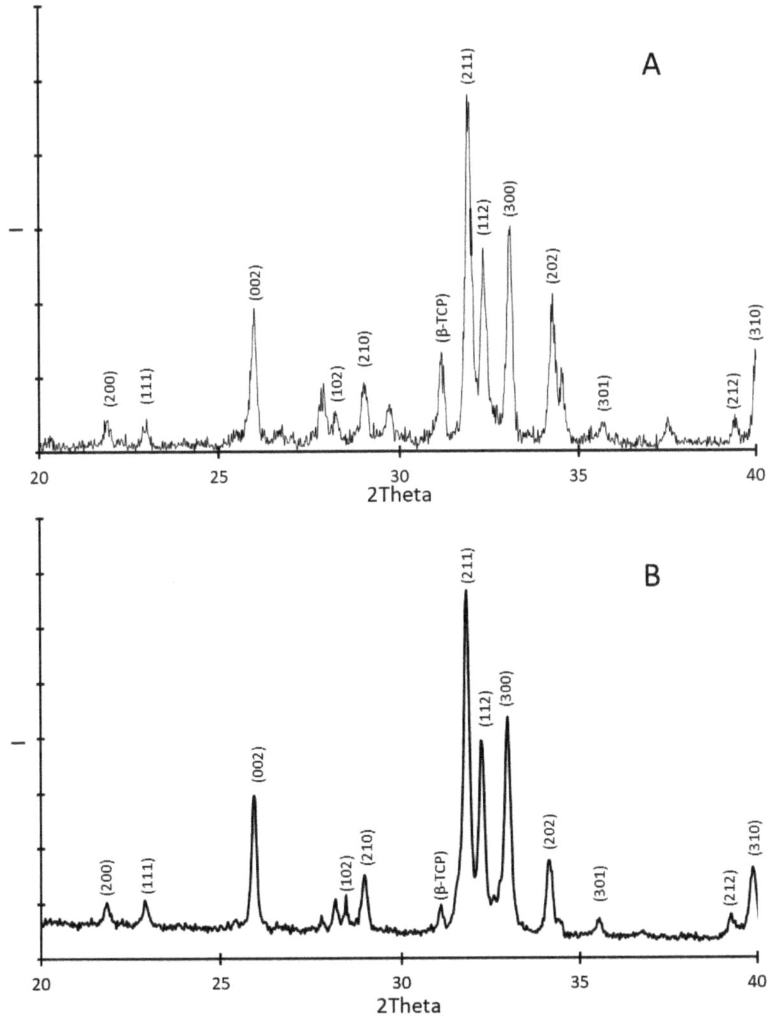

Figure 2. XRD analyses on (**A**) synthesized hydroxyapatite; (**B**) the coating made from the mix of HAp and PEI and deposited on the silica wafer.

To check whether its composition and crystalline form was preserved after the annealing, it was necessary to track for possible singularities between the two spectra. The most significant difference that can be observed lies in the peak situated around 2θ = 31°, which is characteristic of the presence of β-tricalcium phosphate (β-TCP) [64], one of the most used and potent synthetic bone graft substitutes [65]. This corresponds to the rhombohedral crystalline form of tricalcium phosphate (TCP) [66]. This is the densest TCP polymorph but it is not a stable crystalline form at high temperatures, contrary to both monoclinic α-TCP and hexagonal α′-TCP. The β-TCP peak is clearly visible in the non-treated HAp and barely detectable in the coating. This suggests that the annealing performed up to 800 °C partially or totally decomposes β-TCP. It is worth mentioning that there is no evidence of the presence of α-TCP either. Since the removal or the transformation of β-TCP is expected to impact the residual properties of the implant [67], different annealing procedures should be evaluated to determine which one leads to the most interesting macroscopic behavior regarding the targeted dental applications.

While previous results attest to the coating survival consecutive to the annealing procedure, the removal of PEI, which is expected to degrade at about 300 °C, remains to be verified. According to the TGA results, presented in Figure 3, the process of heating silicon wafers coated with the HAp and PEI mix under a nitrogen atmosphere does not reveal significant loss of mass from the ambient temperature up to 900 °C. Only above 600 °C could some artefacts of measurements be observed, which are probably linked to the impact of high temperature on the sensors. The absence of any degradation signature might indicate that PEI was properly decomposed during the previous annealing step. Nevertheless, it is worth mentioning that data collected from the literature do not necessarily agree concerning the behaviour of HAp when submitted to a temperature ramp.

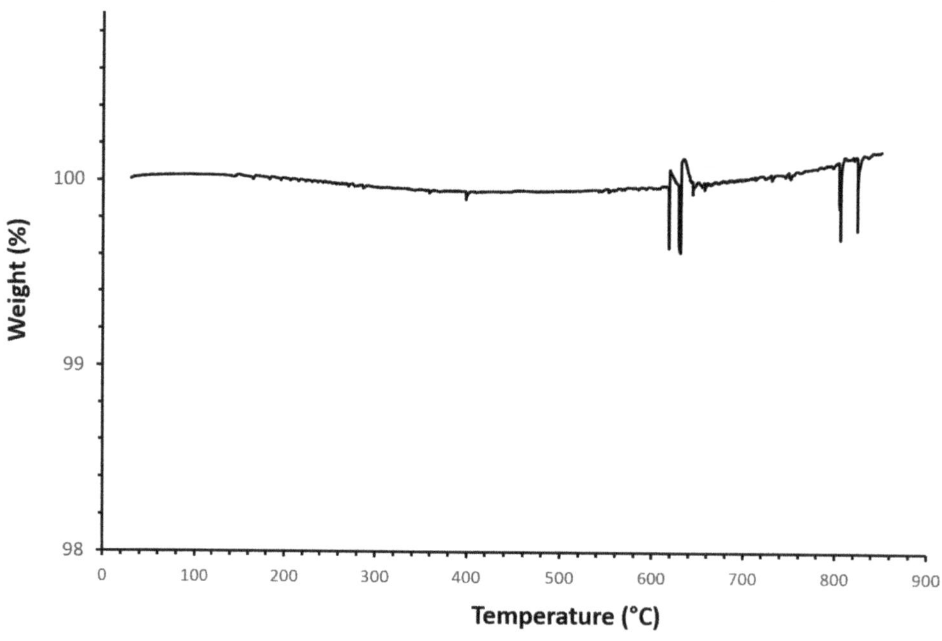

Figure 3. Evolution of the sample mass percentage as a function of temperature from thermogravimetric analyses performed on the silica wafers that were coated by the mix of HAp and PEI, then annealed.

According to Phuong et al. [68], the total weight loss of HAp when subjected to harsh conditions, that is, heated under air up to 800 °C, is insignificant in comparison to what has previously been reported by other authors [69–71], as the results of this study show

HAp residue values that were close to 99.5% weight. This slight weight loss was essentially attributed to the evaporation of residual water. On the other hand, the studies performed by the aforementioned authors [69–71] reveal more significant traces of degradation of the HAp structure. Lazić et al. [69] identified three stages of HAp degradation. Up to 200 °C, the weight loss is imputed to water evaporation. From 200 to 650 °C, the decomposition of hydrogen phosphate occurs, which can be combined with the loss of interstitial water. Above 650 °C, reactions involving the decomposition of diphosphates take place. The final loss of weight is estimated to be about 7%, but it is already higher than 3% at the end of the first stage, which does not correspond to the results presented there, since it suggests that a more significant amount of water is present in the material before the TGA run. Jagadale et al. [70] reported an almost full HAp degradation into nano-porous powders with a very prominent first stage, but they did not interpret the high-temperature behaviour. Safarzadeh et al. [71] separated the degradation of carbonated Hap into two stages. The first stage is essentially caused by the liberation of absorbed and lattice water, which accounts for weight loss up to 12%. The second stage is less pronounced and was assumed to be linked to the decomposition of carbonates.

These strong inconsistencies between reports in terms of interpretations and results lead us to two possible assumptions regarding our results. First, it assumed that the sample weight remains constant throughout the analysis, as could be expected from any scan performed on HAp-coated silicon. In this case, annealing at 800 °C should have completely removed PEI from the substrate. In addition, the absence of water loss below 200 °C can be easily explained by the absence of residual water into the material, due to the annealing step performed prior to analysis. Nevertheless, this is not the only case that could be considered. The apparatus may not have detected degradation due to resolution issues. In this latter case, the absence of weight loss would not prove that PEI effectively disappeared during the annealing step that was performed before TGA. Although this interpretation seems less likely, other characterisations were performed to consider additional elements.

The DSC results, shown in Figure 4, lead to similar interrogations. According to the heat flow variations with temperature, there is no evidence of any transition in the temperature range covered by the analysis. The events of interest for this analysis are the PEI glass transition and its onset of decomposition. Typically, the glass transition of PEI is expected to be around −25 °C [72], while the thermal decomposition has been reported to start at about 250 °C [73]. However, the obtained DSC response is identical to the signal obtained when performing the baseline step during the calibration procedure. This observation is consistent with the results reported by Feng et al. [74] for a plasma-sprayed Ti-6Al-4V/HAp composite. In this study, no calorimetric event was recorded before 700 °C, that is, before the phase transformation of HAp. The absence of any characteristic thermal event is also consistent with TGA results. Thus, it is reasonable to assume that there is no residual PEI on the sol–gel dip coated support, due to the annealing procedure. Nevertheless, it is worth reminding that the quantity of PEI mixed with HAp is very small, as its role lies essentially in facilitating the adhesion of HAp on the support. Thus, the detection of residual traces can be particularly challenging and possibly beyond the resolution limits of both thermal analysis apparatuses.

Therefore, the results obtained from thermogravimetric and calorimetric investigations were complemented with infrared spectroscopy analyses. The resulting spectrum is presented in Figure 5. The PEI infrared signature [75] is classically characterized by the N–H stretching band at 3300 cm^{-1}, the C–H stretching bands around 3000 cm^{-1}, the C–H stretching bands around 3000 cm^{-1}, the bending vibration of the CH2 group at 1500 cm^{-1}, and the C–N stretching band around 1100 cm^{-1}. No peak was recorded at these characteristic wavenumbers. On the other hand, the obtained spectrum is in good agreement with references from the literature describing the infrared signature of hydroxyapatite [76,77], which confirms the adhesion of the coating, as shown in Figure 2. At about 1050 cm^{-1}, a main intense peak is recorded, which can be attributed to the asymmetric stretching vibration of phosphate P–O bonds. Around 1400 cm^{-1}, one can find the broad signature of C–O bond asymmetric stretching from carbonate ions. The small peak above 1600 cm^{-1} belongs to residual water according to Prekajski et al. [78]. Finally, the broad peak located between 3200 and 3600 cm^{-1} is linked to O–H bonds, but its intensity remains weak, probably due to the strong annealing procedure performed prior to the material characterization. The unambiguous infrared signature of hydroxyapatite indicates that the infrared spectroscopy is adequate to investigate the composition of the coating. Due the absence of band characteristics of PEI, one can think that the annealing treatment successfully removed any trace of PEI.

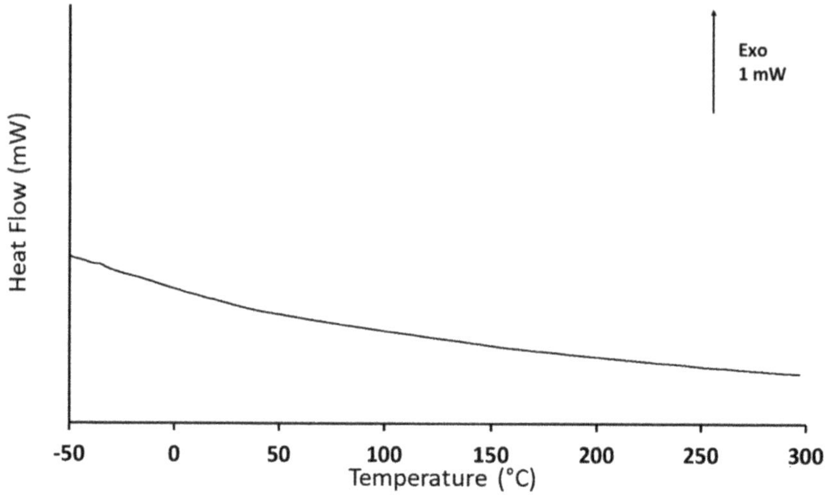

Figure 4. Differential scanning calorimetry analyses performed on the silica wafers that were coated by the mix of HAp and PEI, then annealed: evolution of the heat flow between sample and empty furnace as a function of temperature.

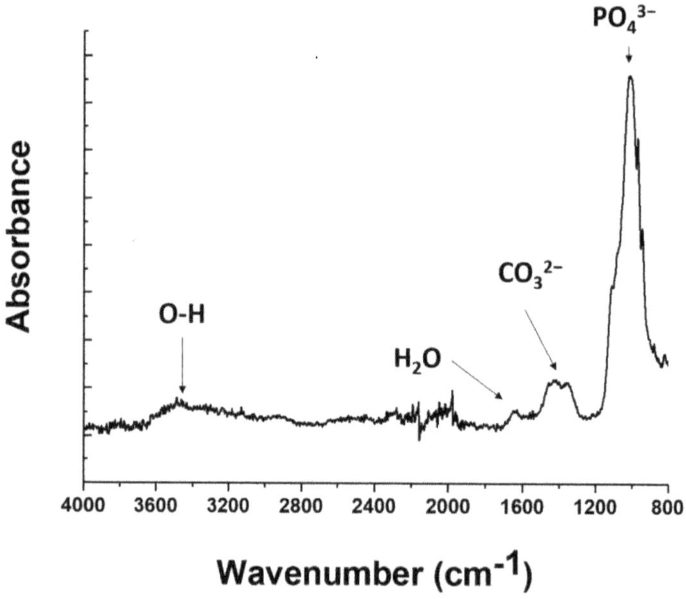

Figure 5. Infrared spectrum analysis on the silica wafers that were coated by the mix of HAp and PEI.

3.3. Coating Morphology on Printed Zirconia

The HAp coating was seemingly preserved after the removal of PEI. A conceptual test was performed by comparing coatings of HAp alone and HAp mixed with PEI on 3D-printed zirconia. Zirconia is a commonly used material in dentistry implants, and its compatibility with this HAp coating is a main requirement for future applications. SEM pictures of both systems are presented in Figure 6. These describe the coating morphology and adhesion for each case. The picture presented in Figure 6a was obtained after analysing 3D-printed zirconia coated with the HAp and PEI mixture. It shows the cross-section between a coated and an uncoated part after the sample was cut in two. The coating is clearly visible, and it apparently takes the form of the porous structure obtained by the rapid prototyping procedure. It is not perfectly uniform, in comparison with reports from literature regarding dense systems, for example the report by Jiang et al. [79]. However, a clear improvement is noticeable when compared with the HAp coating alone (Figure 6b), which exhibits a much-disorganised repartition of the coating with insulated portions of HAp facing many wide empty spaces. Moreover, the bare-coated sample also exhibits a craze that may indicate a more fragile structure in the absence of ligand polymer. Therefore, a clear improvement of the coating adhesion is assessed when HAp is coupled with PEI before being deposited. In other terms, HAp coating alone is not sustainable on 3D-printed zirconia; thus, HAp-PEI is proposed as a coating instead. It is worth mentioning that the HAp-PEI does not fully cover the surface of the substrate either. This could be related to the quantity of material being too low. It is expected that a higher grafted quantity of material would lead to a more continuous coating.

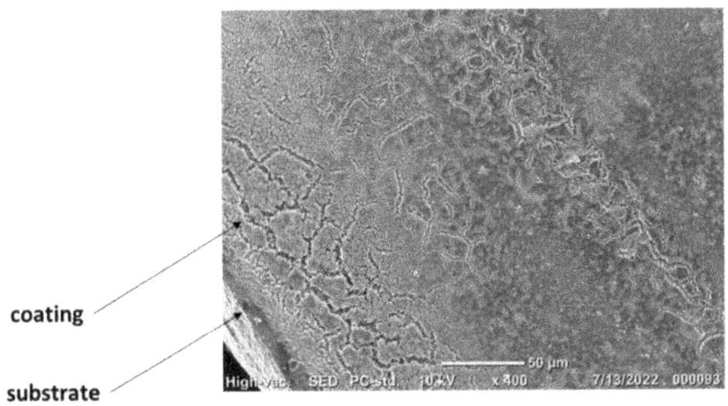

(a)

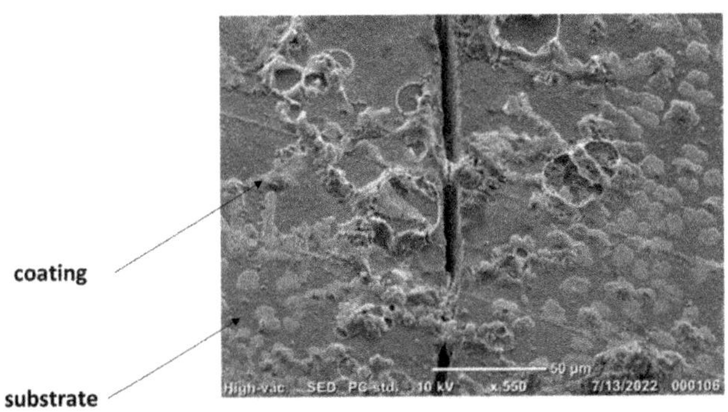

(b)

Figure 6. SEM pictures of coated porous zirconia ceramic obtained by additive manufacturing: (**a**) cross-sectional view of HAp/PEI coating on zirconia annealed at 800 °C; (**b**) bare HAp on printed zirconia.

3.4. Benefits and Drawbacks of DLP and Organic–Mineral Interactions for HAp Coating

The long-term objective behind this work is to facilitate the access to dental care by proposing a protocol that is transferable to the industrial scale that compensates the technological cost related to the scaffold design by the reduction in global content of matter. Therefore, it is important to critically evaluate the properties of HAP coatings on DLP porous substrates. The process also includes the mixing of HAp with PEI to improve the homogeneity and the sustainability of the coating. This step, based on the interactions between mineral and organic compounds, has been shown to be of value as it significantly improves the continuity of the coating. In Table 1, the main characteristics of this HAp coating are summarized and compared with others from the literature, either on plain or porous scaffolding.

Table 1. Features of the designed coating procedure in comparison with other methodologies from previous studies.

Methodology	Objectives	Results
Pulsed laser deposition [42]	- Controlled thickness - Controlled transfer at a stoichiometric level of the composition of the material - Good adherence	- Coating morphology tailorable by varying the incident energy - Microhardness values of the hydroxyapatite coating increase with the incident energy - Good stoichiometry
Pulsed electrochemical deposition [43]	- Improved compositional control - Coating uniformity - Versatility - Low-cost procedure	- High control in the amount, purity, and homogeneity of the coating - Morphology influenced by current density
Pulsed plasma deposition [44]	- Improved tribological properties - Wear resistance	- Design of coated spheres - Sub-micrometric grain size distribution - Low roughness - Lower wear rate in comparison with uncoated systems
Atmospheric plasma spraying [45]	- Design of bioactive glass - Improved bond strength by combining bioactive glass with HAp	- Improvement of adhesion strength by thermal treatment - Coating properties influenced by the amorphous/crystal ratio
Suspension plasma spraying [46]	- Improved mechanical properties	- Uniform microstructure - Low porosity - Enhanced mechanical properties
Axial suspension plasma spraying [40]	- Improved adhesion strength - Limited phase degradation	- Successful design of coatings with high adhesion strength - The lower solid load content and lower mean solute particle size in the suspension was found to be beneficial to achieve porous and rougher coatings
In-situ sol–gel process [41]	- Improved corrosion behaviour - Improved in vitro activity	- Homogenous deposition - Compact and crack-free coating - Surface roughness, adhesion strength and hydrophilicity are tailorable - Good performances in terms of corrosion and bioactivity
DLP + dip-coating [this study]	- Rapid substrate design - Design of a porous scaffold to facilitate the osseointegration - Reducing of the material cost	- Successful design of porous zirconia scaffold - HAp alone does not adhere on the substrate - Adhesion improved by PEI - Sustainability of the coating post-annealing

4. Conclusions

In this study, a dip-coating procedure effectively deposited PEI-HAp onto the surface of 3D-printed zirconia. The coating of HAp on porous substrates is not uniform and its adhesion is poor. The results are significantly better when HAp is mixed with PEI. The adhesion of HAp mixed with PEI is possible on several surfaces including silica and zirconia in the present case. A subsequent annealing procedure successfully removes PEI. It does not damage the HAp coating.

The interest of this procedure lies in the relative conservation under high-temperature annealing of the HAp structural integrity while destructing PEI. Consecutively to the annealing, the hydroxyapatite signature is indeed clearly visible via X-ray diffraction and infrared spectroscopy whereas the signature of the polyethyleneimine is not recorded by TGA, DSC, or IR. Therefore, not only does PEI improve the adhesion of HAp, but it can also be successfully removed without significantly damaging the HAp coating.

Several immediate conclusions result from this work:

- Zirconia substrate obtained via digital light processing (DLP) can be used as a scaffold for coating deposition.
- The coating of HAp on porous substrates is not uniform and its adhesion is poor. The results are significantly better when Hap is mixed with PEI.
- The adhesion of HAp mixed with PEI is possible on several surfaces including silicon and zirconia in the present case.
- A subsequent annealing procedure successfully removes PEI. It does not damage the HAp coating.

It is worth mentioning that the possibility of interactions between HAp and PEI remains to be investigated. Future works should also consider further examination of the removal of polyethyleneimine, adhesion tests, and toxicity controls.

Supplementary Materials: The following supporting information can be downloaded at: https://www.mdpi.com/article/10.3390/jfb15010011/s1, Figure S1, Surface state consecutive to the removal of a piece of tape stuck on the coating.

Author Contributions: Conceptualization, L.T.; validation, N.D.; investigation, A.C., M.-R.G., and N.S.; original draft preparation, A.C.; writing—review and editing, N.D.; supervision, N.D. and L.T.; project administration, N.D. and L.T.; funding acquisition and resources, N.D., L.T., and B.C. All authors have read and agreed to the published version of the manuscript.

Funding: This research was primarily supported by the Erasmus+/ICM program between the University of Nebraska-Lincoln (USA) and the University of Rouen Normandy (France). The authors are grateful for the University of Nebraska Collaboration Initiative Seed Grant, which supported DLP acquisition.

Data Availability Statement: Data are available in this article.

Conflicts of Interest: The authors declare no conflict of interest.

References

1. Londoño, J.J.; Ramos, A.M.; Correa, S.A.; Mesnard, M. Review of expandable dental implants. *Brit. J. Oral Max. Surg.* **2021**, *59*, 546–554. [CrossRef]
2. De Stefano, M.; Lanza, A.; Faia, E.; Ruggiero, A. A novel ultrashort dental implant design for the reduction of the bone stress/strain: A comparative numerical investigation. *Biomed. Eng. Adv.* **2023**, *5*, 100077. [CrossRef]
3. Panchal, M.; Khare, S.; Khamkar, P.; Bhole, K.S. Dental implants: A review of types, design analysis, materials, additive manufacturing methods, and future scope. *Mater. Today Proc.* **2022**, *68*, 1860–1867. [CrossRef]
4. Matsko, A.; França, R. Design, manufacturing, and clinical outcomes for additively manufactured titanium dental implants: A systematic review. *Dent. Rev.* **2022**, *2*, 100041. [CrossRef]
5. Yin, S.; Zhang, W.; Tang, Y.; Yang, G.; Wu, X.; Lin, S.; Liu, X.; Cao, H.; Jiang, X. Preservation of alveolar ridge height through mechanical memory: A novel dental implant design. *Bioact. Mater.* **2021**, *6*, 75–83. [CrossRef]
6. Moreira, A.; Madeira, S.; Buciumeanu, M.; Fialho, J.; Carvalho, A.; Silva, F.; Monteiro, F.J.; Caramês, J. Design and surface characterization of micropatterned silica coatings for zirconia dental implants. *J. Mech. Behav. Biomed.* **2022**, *126*, 105060. [CrossRef]

7. Dantas, T.A.; Pinto, P.; Vaz, P.C.S.; Silva, F.S. Design and optimization of zirconia functional surfaces for dental implants applications. *Ceram. Int.* **2020**, *46*, 16328–16336. [CrossRef]
8. Zanetti, E.M.; Pascoletti, G.; Cali, M.; Bignardi, G.; Franceschini, G. Clinical Assessment of Dental Implant Stability During Follow-Up: What Is Actually Measured, and Perspectives. *Biosensors* **2018**, *8*, 68. [CrossRef]
9. Kittur, N.; Oak, R.; Dekate, D.; Jadhav, S.; Dhatrak, P. Dental implant stability and its measurements to improve osseointegration at the bone-implant interface: A review. *Mater. Today Proc.* **2021**, *43*, 1064–1070. [CrossRef]
10. Qin, J.; Yang, D.; Maher, S.; Lima-Marques, L.; Zhou, Y.; Chen, Y.; Atkins, G.J.; Losic, D. Micro- and nano-structured 3D printed titanium implants with a hydroxyapatite coating for improved osseointegration. *J. Mater. Chem. B* **2018**, *6*, 3136–3144. [CrossRef]
11. Olander, J.; Ruud, A.; Wennerberg, A.; Stenport, V.F. Wear particle release at the interface of dental implant components: Effects of different material combinations. An in vitro study. *Dent. Mater.* **2022**, *38*, 508–516. [CrossRef] [PubMed]
12. Das, I.; Chattopadhyay, S.; Mahato, A.; Kundu, B.; De, G. Fabrication of a cubic zirconia nanocoating on a titanium dental implant with excellent adhesion, hardness and biocompatibility. *RSC Adv.* **2016**, *6*, 59030–59038. [CrossRef]
13. Bose, S.; Koski, C.; Vu, A.A. Additive manufacturing of natural biopolymers and composites for bone tissue engineering. *Mater. Horiz.* **2020**, *7*, 2011–2027. [CrossRef]
14. Zou, R.; Bi, L.; Huang, Y.; Wang, Y.; Wang, Y.; Li, L.; Liu, J.; Feng, L.; Jiang, X.; Deng, B. A biocompatible silicon nitride dental implant material prepared by digital light processing technology. *J. Mech. Behav. Biomed.* **2023**, *141*, 105756. [CrossRef]
15. Hasan, J.; Bright, R.; Hayles, A.; Palms, D.; Zilm, P.; Barker, D.; Vasilev, K. Preventing peri-implantitis: The quest for a next generation of titanium dental implants. *ACS Biomater. Sci. Eng.* **2022**, *8*, 4697–4737. [CrossRef]
16. Gallegos, S.I.; Parsaei, S.; Siddiqui, D.A.; Biguetti, C.C.; Palmer, K.L.; Rodrigues, D.C. Can dental cement composition affect dental implant success? *ACS Biomater. Sci. Eng.* **2019**, *5*, 5116–5127. [CrossRef]
17. Milone, D.; Fiorillo, L.; Alberti, F.; Cervino, G.; Filardi, V.; Pistone, A.; Cicciù, M.; Risitano, G. Stress distribution and failure analysis comparison between Zirconia and Titanium dental implants. *Procedia Struct. Integr.* **2022**, *41*, 680–691. [CrossRef]
18. Kitagawa, I.L.; Miyazaki, C.M.; Pitol-Palin, L.; Okamoto, R.; de Vasconcellos, L.M.R.; Constantiono, C.J.L.; Lisboa-Filho, P.N. Titanium-based alloy surface modification with TiO2 and poly(sodium 4-styrenesulfonate) multilayers for dental implants. *ACS Appl. Bio Mater.* **2021**, *4*, 3055–3066. [CrossRef] [PubMed]
19. Kim, W.-H.; Shin, Y.C.; Lee, S.-H.; Lee, S.-H.; Kang, M.S.; Lee, M.-S.; Lee, J.H.; Lee, J.-H.; Han, D.-W.; Kim, B. Dental implants with electrochemical nanopattern formation to increase osseointegration. *J. Ind. Eng. Chem.* **2022**, *116*, 543–555. [CrossRef]
20. Wu, Y.; Feng, F.; Xin, H.; Li, K.; Tang, Z.; Guo, Y.; Qin, D.; An, B.; Diao, X.; Dou, C. Fracture strength and osseointegration of an ultrafine-grained titanium mini dental implant after macromorphology optimization. *ACS Biomater. Sci. Eng.* **2019**, *5*, 4122–4130. [CrossRef]
21. Gautam, C.; Joyner, J.; Gautam, A.; Rao, J.; Vajtai, R. Zirconia based dental ceramics: Structure, mechanical properties, biocompatibility and applications. *Dalton Trans.* **2016**, *45*, 19194–19215. [CrossRef] [PubMed]
22. Ma, M.; Zhao, M.; Deng, H.; Liu, Z.; Wang, L.; Ge, L. Facile and versatile strategy for fabrication of highly bacteriostatic and biocompatible SLA-Ti surfaces with the regulation of Mg/Cu coimplantation ratio for dental implant applications. *Colloids Surf. B* **2023**, *223*, 113180. [CrossRef] [PubMed]
23. Yilmaz, E.; Çalişkan, F. A new functional graded dental implant design with biocompatible and antibacterial properties. *Mater. Chem. Phys.* **2022**, *277*, 125481. [CrossRef]
24. Jayasree, A.; Gómez-Cerezo, M.N.; Verron, E.; Ivanovski, S.; Gulati, K. Gallium-doped dual micro-nano titanium dental implants towards soft-tissue integration and bactericidal functions. *Mater. Today Adv.* **2022**, *16*, 100297. [CrossRef]
25. Ren, X.; van der Mei, H.C.; Ren, Y.; Busscher, H.J. Keratinocytes protect soft-tissue integration of dental implant materials against bacterial challenges in a 3D-tissue infection model. *Acta Biomater.* **2019**, *96*, 237–246. [CrossRef] [PubMed]
26. Mühl, A.; Szabó, P.; Krafcsik, O.; Aigner, Z.; Kopniczky, J.; Nagy, A.; Marada, G.; Turzó, K. Comparison of surface aspects of turned and anodized titanium dental implant, or abutment material for an optimal soft tissue integration. *Heliyon* **2022**, *8*, e10263. [CrossRef]
27. Kim, J.C.; Lee, M.; Yeo, I.-S.L. Three interfaces of the dental implant system and their clinical effects on hard and soft tissues. *Mater. Horiz.* **2022**, *9*, 1387–1411. [CrossRef]
28. Yang, Z.; Liu, M.; Yang, Y.; Zheng, M.; Yang, Y.; Liu, X.; Tan, J. Biofunctionalization of zirconia with cell-adhesion peptides via polydopamine crosslinking for soft tissue engineering: Effects on the biological behaviors of human gingival fibroblasts and oral bacteria. *RSC Adv.* **2020**, *10*, 6200–6212. [CrossRef]
29. Matter, M.T.; Maliqi, L.; Keevend, K.; Guimond, S.; Ng, J.; Armagan, E.; Rottmar, M.; Herrmann, I.K. One-step synthesis of versatile antimicrobial nano-architected implant coatings for hard and soft tissue healing. *ACS Appl. Mater. Interfaces* **2021**, *13*, 33300–33310. [CrossRef]
30. Fischer, N.G.; Moussa, D.G.; Skoe, E.P.; De Jong, D.A.; Aparicio, C. Keratinocyte-specific peptide-based surfaces for hemidesmosome upregulation and prevention of bacterial colonization. *ACS Biomater. Sci. Eng.* **2020**, *6*, 4929–4939. [CrossRef]
31. Fischer, N.G.; Munchow, E.A.; Tamerler, C.; Bottino, M.C.; Aparicio, C. Harnessing biomolecules for bioinspired dental biomaterials. *J. Mater. Chem. B* **2020**, *8*, 8713–8747. [CrossRef]
32. De Avila, E.D.; Castro, A.G.B.; Tagit, O.; Krom, B.P.; Lőwik, D.; van Well, A.A.; Bannenberg, L.J.; Vergani, C.E.; van den Beucken, J.J.J.P. Anti-bacterial efficacy via drug-delivery system from layer-by-layer coating for percutaneous dental implant components. *Appl. Surf. Sci.* **2019**, *488*, 194–204. [CrossRef]

33. Scarano, A.; Piattelli, M.; Caputi, S.; Favero, G.A.; Piattelli, A. Bacterial adhesion on commercially pure titanium and zirconium oxide disks: An in vivo human study. *J. Periodontol.* **2004**, *75*, 292–296. [CrossRef] [PubMed]
34. do Nascimento, C.; Pita, M.S.; Fernandes, F.H.N.C.; Pedrazzi, V.; de Albuquerque Junior, R.F.; Ribeiro, R.F. Bacterial adhesion on the titanium and zirconia abutment surfaces. *Clin. Oral Implant. Res.* **2014**, *25*, 337–343. [CrossRef] [PubMed]
35. Caravaca, C.; Shi, L.; Balvay, S.; Rivory, P.; Laurenceau, E.; Chevolot, Y.; Hartmann, D.; Gremillard, L.; Chevalier, J. Direct silanization of zirconia for increased biointegration. *Acta Biomater.* **2016**, *46*, 323–335. [CrossRef] [PubMed]
36. Seesala, V.S.; Rajasekaran, R.; Ojha, A.K.; Mahato, A.; Korrayi, R.R.; Das, B.; Venugopal, S.P.; Roy, S.; Dhara, S. A novel functional gradient hydroxyapatite coating for zirconia-based implants. *Surf. Coat. Technol.* **2023**, *469*, 129817. [CrossRef]
37. Gledhill, H.C.; Turner, I.G.; Doyle, C. In vitro dissolution behaviour of two morphologically different thermally sprayed hydroxyapatite coatings. *Biomaterials* **2001**, *22*, 695–700. [CrossRef]
38. Heimann, R.B. Plasma-sprayed hydroxylapatite-based coatings: Chemical, mechanical, microstructural, and biomedical properties. *J. Therm. Spray Technol.* **2016**, *25*, 827–850. [CrossRef]
39. Bøe, B.G.; Röhrl, S.M.; Heier, T.; Snorrason, F.; Nordsletten, L. A prospective randomized study comparing electrochemically deposited hydroxyapatite and plasma-sprayed hydroxyapatite on titanium stems. *Acta Orthop.* **2011**, *82*, 13–19. [CrossRef]
40. Ganvir, A.; Nagar, S.; Markocsan, N.; Balani, K. Deposition of hydroxyapatite coatings by axial plasma spraying: Influence of feedstock characteristics on coating microstructure, phase content and mechanical properties. *J. Eur. Ceram.* **2021**, *41*, 4637–4649. [CrossRef]
41. Ansari, Z.; Kalantar, M.; Kharaziha, M.; Ambrosio, L.; Raucci, M.G. Polycaprolactone/fluoride substituted-hydroxyapatite (PCL/FHA) nanocomposite coatings prepared by in-situ sol-gel process for dental implant applications. *Prog. Org. Coat.* **2020**, *147*, 105873. [CrossRef]
42. González-Estrada, O.A.; Pertuz Comas, A.D.; Ospina, R. Characterization of hydroxyapatite coatings produced by pulsed-laser deposition on additive manufacturing Ti6Al4V ELI. *Thin Solid Films* **2022**, *763*, 139592. [CrossRef]
43. Lissandrello, F.; Magagnin, L. Pulsed electrochemical deposition of calcium phosphate coatings for biomedical applications. *Trans. Inst. Mater. Finish.* **2023**, *101*, 173–178. [CrossRef]
44. Bianchi, M.; Lopomo, N.; Boi, M.; Gambardella, A.; Marchiori, G.; Berni, M.; Pavan, P.; Marcacci, M.; Russo, A. Ceramic thin films realized by means of pulsed plasma deposition technique: Applications for orthopedics. *J. Mech. Med. Biol.* **2015**, *15*, 1540002. [CrossRef]
45. Garrido, B.; Martin-Morata, A.; Dosta, S.; Cano, I.G. Improving the bond strength of bioactive glass coatings obtained by atmospheric plasma spraying. *Surf. Coat. Technol.* **2023**, *470*, 129837. [CrossRef]
46. Chen, X.; Zhang, B.; Gong, Y.; Zhou, P.; Li, H. Mechanical properties of nanodiamond-reinforced hydroxyapatite composite coatings deposited by suspension plasma spraying. *Appl. Surf. Sci.* **2018**, *439*, 60–65. [CrossRef]
47. Oyane, A.; Kakehata, M.; Sakamaki, I.; Pyatenko, A.; Yashiro, H.; Ito, A.; Torizuka, K. Biomimetic apatite coating on yttria-stabilized tetragonal zirconia utilizing femtosecond laser surface processing. *Surf. Coat. Technol.* **2016**, *296*, 88–95. [CrossRef]
48. Singh, P.P.; Dixit, K.; Sinha, N. A sol-gel based bioactive glass coating on laser textured 316L stainless steel substrate for enhanced biocompatibility and anti-corrosion properties. *Ceram. Int.* **2022**, *48*, 18704–18715. [CrossRef]
49. García-Arnáez, I.; Cerqueira, A.; Romero-Gavilán, F.; Elortza, F.; Azkargorta, M.; Iloro, I.; Suay, J.; Goñi, I.; Gurruchaga, M. Development and characterisation of strontium-doped sol-gel coatings to optimise the initial bone regeneration processes. *Mater. Today Commun.* **2022**, *33*, 104674. [CrossRef]
50. Catauro, M.; Bollino, F.; Giovanardi, R.; Veronesi, P. Modification of Ti6Al4V implant surfaces by biocompatible TiO_2/PCL hybrid layers prepared via sol-gel dip coating: Structural characterization, mechanical and corrosion behavior. *Mater. Sci. Eng. C* **2017**, *74*, 501–507. [CrossRef]
51. Alavi, S.E.; Panah, N.; Page, F.; Gholami, M.; Dastfal, A.; Sharma, L.A.; Shahmabadi, H.E. Hydrogel-based therapeutic coatings for dental implants. *Eur. Polym. J.* **2022**, *181*, 111652. [CrossRef]
52. Vieira, M.; Tavares, C.R.; Bergamasco, R.; Petrus, J.C.C. Application of ultrafiltration-complexation process for metal removal from pulp and paper industry wastewater. *J. Membr. Sci.* **2001**, *194*, 273–276. [CrossRef]
53. Lin, X.; Tran, D.T.; Song, M.-H.; Yun, Y.-S. Development of polyethyleneimine-starch fibers stable over the broad pH range for selective adsorption of gold from actual leachate solutions of waste electrical and electronic equipment. *J. Clean. Prod.* **2021**, *328*, 129545. [CrossRef]
54. Li, S.; Tu, L.; Lu, Y.; Lin, M.; Ma, J.; Bai, C.; Gao, C.; Xue, L. Polyethyleneimine (PEI) based thin film nanocomposite (TFN) total heat exchange membranes (THEMs) composed of shaped ZIF-8 crystalline micro-leaves (ZIF-L). *Sep. Purif. Technol.* **2023**, *324*, 124435. [CrossRef]
55. Zhang, C.; Wu, F.-G.; Hu, P.; Chen, Z. Interaction of polyethylenimine with model cell membranes studied by linear and nonlinear spectroscopic techniques. *J. Phys. Chem. C* **2014**, *118*, 12195–12205. [CrossRef]
56. Kong, L.; Mu, Z.; Yu, Y.; Zhang, L.; Hu, J. Polyethyleneimine-stabilized hydroxyapatite nanoparticles modified with hyaluronic acid for targeted drug delivery. *RSC Adv.* **2016**, *6*, 101790–101799. [CrossRef]
57. Li, X.; Zhong, H.; Zhang, J.; Duan, Y.; Bai, H.; Li, J.; Jiang, D. Dispersion and properties of zirconia suspensions for stereolithography. *Int. J. Appl. Ceram. Technol.* **2020**, *17*, 239–247. [CrossRef]
58. Borlaf, M.; Serra-Capdevila, A.; Colominas, C.; Graule, T. Development of UV-curable ZrO_2 slurries for additive manufacturing (LCM-DLP) technology. *J. Eur. Ceram. Soc.* **2019**, *39*, 3797–3803. [CrossRef]

59. Jang, K.J.; Kang, J.H.; Fisher, J.G.; Park, S.W. Effect of the volume fraction of zirconia suspensions on the microstructure and physical properties of products produced by additive manufacturing. *Dent. Mater.* **2019**, *35*, e97–e106. [CrossRef]
60. Coppola, B.; Schmitt, J.; Lacondemine, T.; Tardivat, C.; Montanaro, L.; Palmero, P. Digital light processing stereolithography of zirconia ceramics: Slurry elaboration and orientation-reliant mechanical properties. *J. Eur. Ceram. Soc.* **2022**, *42*, 2974–2982. [CrossRef]
61. Ji, S.H.; Da, S.K.; Park, M.S.; Ji, S.Y. Sintering process optimization for 3YSZ ceramic 3D-printed objects manufactured by stereolithography. *Nanomaterials* **2021**, *11*, 192. [CrossRef] [PubMed]
62. Snyder, N. Digital Light Processing (DLP) of Yttria-Stabilized-Zirconia (YSZ). Master's Thesis, The University of Nebraska Lincoln, Lincoln, NE, USA, May 2022.
63. Manafi, S.A.; Yazdani, B.; Rahimiopour, M.R.; Sadrnezhaad, S.K.; Amin, M.H.; Razavi, M. Synthesis of nano-hydroxyapatite under a sonochemical/hydrothermal condition. *Biomed. Mater.* **2008**, *3*, 025002. [CrossRef] [PubMed]
64. Lee, D.S.H.; Pai, Y.; Chang, S. Effect of thermal treatment of the hydroxyapatite powders on the micropore and microstructure of porous biphasic calcium phosphate composite granules. *J. Biomater. Nanobiotechnol.* **2013**, *4*, 114–118. [CrossRef]
65. Bohner, M.; Le Gars Santoni, B.; Döbelin, N. β-tricalcium phosphate for bone substitution: Synthesis and properties. *Acta Biomater.* **2020**, *113*, 23–41. [CrossRef] [PubMed]
66. Yashima, M.; Sakai, A.; Kamiyama, T.; Hoshikawa, A. Crystal structure analysis of β-tricalcium phosphate $Ca_3(PO_4)_2$ by neutron powder diffraction. *J. Solid State Chem.* **2003**, *175*, 272–277. [CrossRef]
67. Lin, H.K.; Pan, Y.H.; Salamanca, E.; Lin, Y.T.; Chang, W.J. Prevention of bone resorption by ha/β-TCP + collagen composite after tooth extraction: A case series. *Int. J. Environ. Res. Public Health* **2019**, *16*, 4616. [CrossRef] [PubMed]
68. Phuong, N.T.; Nam, N.H.; Hong, C.T.; Dac, D.V.Q.; Thu, L.P.; Hai, D.T.; Osial, M.; Giersig, M.; Thanh, D.T.M. Apatite Ore-based Nanostructures: Novel and Eco-friendly Sorbent for Efficient Removal of Wastewater Containing Pb^{2+} and Fe^{3+}. *Water Air Soil Pollut.* **2023**, *234*, 550. [CrossRef]
69. Lazić, S.; Zec, S.; Miljević, N.; Milonjić, S. The effect of temperature on the properties of hydroxyapatite precipitated from calcium hydroxide and phosphoric acid. *Thermochim. Acta* **2001**, *11*, 13–22. [CrossRef]
70. Jagadale, P.N.; Jagtap, P.P.; Joshi, M.G.; Bamane, S.R. A prototype synthesis and characterization of hydroxyapatite bioceramics nanocrystallites. *Adv. Mater. Lett.* **2016**, *7*, 325–329. [CrossRef]
71. Safarzadeh, M.; Ramesh, S.; Tan, C.Y.; Chandran, H.; Ching, Y.C.; Noor, A.F.M.; Krishnasamy, S.; Teng, W.D. Sintering behaviour of carbonated hydroxyapatite prepared at different carbonate and phosphate ratios. *Bol. Soc. Esp. Ceram. Vidr.* **2020**, *59*, 73–80. [CrossRef]
72. Román, F.; Colomer, P.; Calventus, Y.; Hutchinson, J.M. Study of hyperbranched poly(ethyleneimine) polymers of different molecular weight and their interaction with epoxy resin. *Materials* **2018**, *11*, 410. [CrossRef] [PubMed]
73. Grenda, K.; Idström, A.; Evenäs, L.; Persson, M.; Holmberg, K.; Bordes, R. An analytical approach to elucidate the architecture of polyethyleneimines. *J. Appl. Polym. Sci.* **2022**, *139*, 51657. [CrossRef]
74. Feng, C.F.; Khor, K.A.; Gu, Y.W.; Cheang, P. Analysis of phase changes in plasma-sprayed Ti-6Al-4V/hydroxyapatite composite coatings by DSC. *Mater. Lett.* **2001**, *51*, 88–93. [CrossRef]
75. Payne, M.E.; Lou, Y.; Zhang, X.; Sahiner, N.; Sandoval, N.R.; Shantz, D.F.; Grayson, S.M. Comparison of cross-linked branched and linear poly(ethylene imine) microgel microstructures and their impact in antimicrobial behavior, copper chelation, and carbon dioxide capture. *ACS Appl. Polym. Mater.* **2020**, *2*, 826–836. [CrossRef]
76. Chang, M.C.; Tanaka, J. FT-IR study for hydroxyapatite/collagen nanocomposite cross-linked by glutaraldehyde. *Biomaterials* **2002**, *23*, 4811–4818. [CrossRef] [PubMed]
77. Abidi, S.S.A.; Murtaza, Q. Synthesis and characterization of nano-hydroxyapatite powder using wet chemical precipitation reaction. *J. Mater. Sci. Technol.* **2014**, *30*, 307–310. [CrossRef]
78. Prekajski, M.; Mirković, M.; Todorović, B.; Matković, A.; Marinović-Cincović, M.; Luković, J.; Matović, B. Ouzo effect—New simple nanoemulsion method for synthesis of strontium hydroxyapatite nanospheres. *J. Eur. Ceram. Soc.* **2016**, *36*, 1293–1298. [CrossRef]
79. Jiang, B.; Ke, S.; Yang, B.; Chen, J.; Li, W.; Fang, M.; Huang, Z.; Sun, J.; Min, X.; Hu, X. Interfacial characteristics of bioglass diffusion zone between scaffold-like hydroxyapatite/wollastonite coating and dense zirconia substrate. *Ceram. Int.* **2023**, *49*, 37388–37395. [CrossRef]

Journal of *Functional Biomaterials*

MDPI

Article

Apatite-Forming Ability and Visible Light-Enhanced Antibacterial Activity of CuO-Supported TiO_2 Formed on Titanium by Chemical and Thermal Treatments

Po-Cheng Sung [1], Taishi Yokoi [2], Masaya Shimabukuro [2], Takayuki Mokudai [3,4] and Masakazu Kawashita [2,*]

1 Graduate School of Medical and Dental Sciences, Tokyo Medical and Dental University, Bunkyo-ku, Tokyo 113-8549, Japan; foster01506@gmail.com

2 Institute of Biomaterials and Bioengineering, Tokyo Medical and Dental University, 2-3-10 Kanda-Surugadai, Chiyoda-ku, Tokyo 101-0062, Japan; yokoi.taishi.bcr@tmd.ac.jp (T.Y.); shimabukuro.bcr@tmd.ac.jp (M.S.)

3 Joining and Welding Research Institute, Osaka University, 11-1 Mihogaoka, Ibaraki, Osaka 567-0047, Japan; tmokudai.jwri@osaka-u.ac.jp or tmokudai@tohoku.ac.jp

4 Institute of Materials Research, Tohoku University, 2-1-1 Katahira, Aoba-ku, Sendai, Miyagi 980-8577, Japan

* Correspondence: kawashita.bcr@tmd.ac.jp

Abstract: Titanium with apatite-forming ability as well as antibacterial activity is useful as a component of antibacterial dental implants. When Ti was subjected to hydrogen peroxide (H_2O_2), copper acetate ($Cu(OAc)_2$), and heat (H_2O_2-$Cu(OAc)_2$-heat) treatments, a network structure of anatase and rutile titanium dioxide (TiO_2) and fine copper oxide (CuO) particles was formed on the Ti surface. The resulting samples accumulated a dense and uniform apatite layer on the surface when incubated in simulated body fluid and showed enhanced antibacterial activity against *Escherichia coli* and *Staphylococcus aureus* under visible-light irradiation. Electron spin resonance spectra of H_2O_2-$Cu(OAc)_2$-heat-treated samples showed that hydroxyl radicals (·OH) were generated from the samples, and the concentration of ·OH increased with increasing Cu concentration of the $Cu(OAc)_2$ solution. The enhanced antibacterial activity of these samples under visible-light irradiation may be attributable to the generation of ·OH from samples. These results suggest that Ti implants obtained using H_2O_2-$Cu(OAc)_2$-heat treatments and subjected to regular or on-demand visible-light irradiation may provide a decreased risk of peri-implantitis.

Keywords: titanium; CuO; TiO_2; apatite; antibacterial activity; photocatalytic activity

Citation: Sung, P.-C.; Yokoi, T.; Shimabukuro, M.; Mokudai, T.; Kawashita, M. Apatite-Forming Ability and Visible Light-Enhanced Antibacterial Activity of CuO-Supported TiO_2 Formed on Titanium by Chemical and Thermal Treatments. *J. Funct. Biomater.* **2024**, *15*, 114. https://doi.org/10.3390/jfb15050114

Academic Editor: Christie Ying Kei Lung

Received: 19 March 2024
Revised: 19 April 2024
Accepted: 21 April 2024
Published: 24 April 2024

1. Introduction

With increases in the number of dental implants installed, peri-implant diseases have become more common. Such diseases can be broadly classified into peri-implant mucositis and peri-implantitis. Peri-implant mucositis, considered a precursor to peri-implantitis, is defined as a reversible inflammation that is localized to the surrounding mucosal tissue and does not involve bone resorption. On the other hand, peri-implantitis is defined as an irreversible inflammation that involves soft tissue inflammation and bone resorption around the implant [1]; prolonged peri-implantitis leads to implant failure. As assessed at greater than 5 years after dental implant placement, the prevalence of peri-implantitis is estimated to be 10% to 37% at the patient level and 7% to 27% at the implant level [2–7]. Classically, surgical treatment has been the main therapy for peri-implantitis; however, in recent years, chemotherapy using antibiotics and anti-inflammatory drugs has been employed to treat invasive and severe peri-implantitis [8,9]. However, surgical treatment is itself highly invasive, and there are concerns about the emergence of drug-resistant bacteria due to the use of antibiotics and anti-inflammatory drugs [10].

In recent years, there has been ongoing development of antimicrobial therapies using micromaterials and nanomaterials. Titanium-based metal organic frameworks (Ti-MOFs) have been investigated for use in therapeutic areas, including the treatment of bacterial infection, cancer, inflammation, and bone injury [11]; notably, MOF-loaded biohybrid magnetic microrobots have been shown to achieve good dispersion and targeted mobility with enhanced antibacterial efficiency [12]. Surface-modified nanomaterials also have been proposed to achieve passive or active targeted antibacterial effects [13,14]. Furthermore, materials that possess topographical features with high aspect ratios show bactericidal activity [15,16]. However, it remains unclear whether any of these materials are effective in treating peri-implantitis. The use of titanium dental implants with antibacterial activity as well as osteoconductivity has been shown to reduce the risk of peri-implantitis. To date, many surface treatments of Ti have been proposed to impart antibacterial activity as well as osteoconductivity to this metal [17–25]. Orthopedic implants coated with silver-containing hydroxyapatite [26] are employed clinically in Japan, but dental implants with both antibacterial activity and osteoconductivity are not yet in clinical use. If TiO_2 with visible-light-responsive photocatalytic antibacterial activity were generated on the abutment of Ti dental implants, the risk of peri-implantitis might be reduced by regular or on-demand visible-light exposure in the dental clinic. One candidate for the visible light source is a light-emitting diode (LED) light used in teeth whitening, but this LED light has been shown to exhibit difficulty in penetrating the gingiva and reaching the abutment of the dental implants. Therefore, the visible light-responsive antibacterial activity of the dental implants would be expected to be enhanced when the abutment is not completely covered by periodontal tissue. In previous work, our laboratory has attempted to generate copper-doped titanium dioxide (TiO_2) on Ti by chemical and thermal treatments [27,28]. This approach acknowledges that TiO_2 generated by such strategies has been shown to bond to living bone [29] and that Cu-doped TiO_2 demonstrates visible-light-induced photocatalytic antibacterial activity [30–32]. Some metals other than Cu, such as silver and zinc, also exhibit antibacterial effects, but the visible-light-induced antibacterial effects cannot be expected for Ag- or Zn-doped TiO_2. In this context, we reported that Cu-doped TiO_2 formed on Ti when Ti was subjected to (sequentially) immersion in a sodium hydroxide solution, immersion in a Cu-containing aqueous solution, and heating at 600 °C for 1 h. Apatite was observed to accumulate on the surface of the resulting Ti within 7 days of incubation in simulated body fluid (SBF) [27,28], and the resulting sample exhibited antibacterial activity against *Escherichia coli* under visible-light irradiation [28]. However, the apatite formed only on a part of the sample surface, indicating that the sample had limited apatite-forming ability. Furthermore, the antibacterial activity of the sample was not assessed against Gram-positive bacteria such as *Staphylococcus aureus*.

In the present study, we induced the formation of Cu-doped TiO_2 on Ti by subjecting the Ti to (sequentially) immersion in a hydrogen peroxide solution, immersion in a copper acetate ($Cu(OAc)_2$) solution, and thermal treatment. Previous work showed that a TiO_2 layer is formed on Ti by sequential treatment with an aqueous NaOH solution, hot water, and heat [29,33], but surface treatments involving immersion in a H_2O_2 solution also are known to generate a TiO_2 layer on Ti [34–38]. In that previous work, we found that the TiO_2 layer (which forms on Ti following such treatment) accumulates apatite when the sample is incubated in SBF [34,36–38]; this apatite permits bonding to living bone [35]. Therefore, from the perspective of simplifying surface treatment, treatment with H_2O_2 solution was employed to induce the formation of a TiO_2 layer on Ti in the present study. Furthermore, we used a $Cu(OAc)_2$ solution for Cu doping into TiO_2, a strategy that suppresses the undesired dissolution of the TiO_2 layer that otherwise results from immersion in acidic Cu-containing solutions such as copper nitrate ($Cu(NO_3)_2$) [27]. We additionally evaluated the resulting samples for surface structure, apatite-forming ability, and Cu release. Furthermore, we investigated the generation by the samples of hydroxyl radicals (·OH), a reactive oxygen species (ROS), and assessed the antibacterial activity of samples against *S. aureus* as well as *E. coli*, both with and without visible-light irradiation.

2. Materials and Methods

2.1. Sample Preparation

Commercially pure Ti (99.5%) (Test Materials Co., Ltd., Tokyo, Japan) was employed in this study. The Ti was provided in the form of 8-mm diameter cylindrical rods, which subsequently were sliced into 1.0-mm thick sections. Following slicing, the surfaces of the disks were subjected to mechanical polishing using a diamond pad (No. 400; Maruto Instrument Co., Ltd., Tokyo, Japan). The polished Ti disks were washed ultrasonically, for 10 min per cycle, once with acetone (99%; Nacalai Tesque, Inc., Kyoto, Japan) and then twice with ultrapure water. The washed Ti disks were then dried at room temperature and atmospheric pressure. Each dried Ti disk was immersed in 10 mL of a 30 vol% H_2O_2 aqueous solution (FUJIFILM Wako Pure Chemical Corp., Osaka, Japan) in an ECK-50ML-R polypropylene tube (AS-ONE Corp., Osaka, Japan), and the tube then was heated at 80 °C for 1 h using a BW101 water bath (Yamato Scientific Co., Ltd., Tokyo, Japan). The samples were removed from the H_2O_2 aqueous solution and immersed in $Cu(OAc)_2$ aqueous solutions at various concentrations (1, 10, or 100 mM) in a polypropylene tube at 80 °C for 24 h using the water bath. After the $Cu(OAc)_2$ treatment, the disk was removed from the solution, washed with ultrapure water, and then dried at room temperature and atmospheric pressure. Each Ti disk that had been treated sequentially with the H_2O_2 and $Cu(OAc)_2$ solutions was then heat-treated at 600 °C for 1 h using an FO101 electric furnace (Yamato Scientific Co., Ltd.). Surface treatments and the sample designations are listed in Table 1.

Table 1. Surface treatments and names of samples.

Surface Treatment	Sample Name
None	TI
H_2O_2 + heat	HP
H_2O_2 + 1 mM $Cu(OAc)_2$ + heat	Cu1
H_2O_2 + 10 mM $Cu(OAc)_2$ + heat	Cu10
H_2O_2 + 100 mM $Cu(OAc)_2$ + heat	Cu100

2.2. Surface Analysis of Samples

Many of the experimental protocols and conditions described in Sections 2.2–2.6 are essentially identical to those used in previous studies [28,39]. The surface morphology of each sample was observed using a JSM-7900F field emission scanning electron microscope (FE-SEM; JEOL, Tokyo, Japan). A mapping image of the sample was obtained using a JED-2300 energy dispersive X-ray spectrometer (JEOL) attached to the FE-SEM. The crystalline phase of the sample surface layer was identified using a MiniFlex600 X-ray diffractometer (XRD; Rigaku Corporation, Tokyo, Japan) with Cu Kα radiation.

The composition of the surface layer was evaluated using a JPS-9010MC X-ray photoelectron spectroscopy (XPS) instrument (JEOL). Monochromatic Mg Kα radiation (1253.6 eV) at 10 kV and 10 mA was used as the X-ray source. The C 1s photoelectron peak at 285.0 eV was used as a reference to calibrate the binding energy. The XPS peak was analyzed using CasaXPS (version 2.3.24; Casa Software, Ltd., Devon, UK). The chemical state of Cu on the samples was determined using a modified Auger parameter, which was calculated as described previously [22]. We conducted background subtraction using the CasaXPS software, following Shirley's method [40] to calculate the integrated intensity of peaks. The surface composition of the samples was calculated using CasaXPS while employing the following relative sensitivity factors: C 1s (1.00), O 1s (2.93), P 2p (1.192), Ca 2p (5.07), Ti 2p (7.81), and Cu $2p_{3/2}$ (16.73). The proportions of (Cu^0 + Cu^+) and Cu^{2+} at the sample surface were calculated using the following equations [41]:

$$\%\,(Cu^0 + Cu+) = (A - (A1_s/B_s) \times B)/(A + B) \times 100 \tag{1}$$

$$\% (Cu^{2+}) = B\,(1 + (A1_s/B_s))/(A + B) \times 100 \quad (2)$$

where A is the total area of the Cu $2p_{3/2}$ peak, B is the area of the shake-up peak, and $A1_s/B_s$ is the ratio of the Cu $2p_{3/2}$ peak/shake-up peak areas. In the present study, the $A1_s/B_s$ ratio was fixed at the value of 1.89 expected for CuO.

2.3. Evaluation of In Vitro Apatite-Forming Ability of Sample

The apatite-forming ability of the samples was evaluated using an SBF [42,43] that contained ions at concentrations (Na^+, 142.0 mM; K^+, 5.0 mM; Ca^{2+}, 2.5 mM; Mg^{2+}, 1.5 mM; Cl^-, 147.8 mM; HCO_3^-, 4.2 mM; HPO_4^{2-}: 1.0 mM; SO_4^{2-}, 0.5 mM) nearly identical to those found in human blood plasma. We prepared the SBF according to the ISO 23317:2014 [43] protocol. All chemicals used for SBF formulation were purchased from Nacalai Tesque, Inc. (Kyoto, Japan). A volume of 30 mL of the prepared SBF was dispensed into an ECK-50ML-R centrifuge tube. The samples were immersed in the SBF at 36.5 °C for 7 days. The samples were then removed from the SBF, gently rinsed with ultrapure water, and dried at approximately 25 °C and atmospheric pressure. The lower surface of each sample was observed using a VE8800 scanning electron microscope (SEM; Keyence Corp., Osaka, Japan) and XRD.

2.4. Measurement of Cu Ion-Release Behavior of Sample

To investigate the Cu ion-release behavior of each sample, an aliquot (10 mL) of phosphate-buffered saline (PBS; Catalog 166-23555; FUJIFILM Wako Pure Chemical Corp., Osaka, Japan) was dispensed into an ECK-50ML-R centrifuge tube. The sample (n = 3) was immersed in PBS and the tube was maintained at 36.5 °C. The PBS was refreshed at appropriate periods. The amounts of Cu ions (per time point and cumulative) released from the samples were calculated at 1, 3, 7, 14, and 28 days based on the Cu concentrations in the PBS, which were measured using an ICP-7000 inductively coupled plasma atomic emission spectroscopy (ICP-AES) instrument (Shimadzu Corp., Kyoto, Japan). Figure 1 shows a schematic diagram of the Cu ion-release assay.

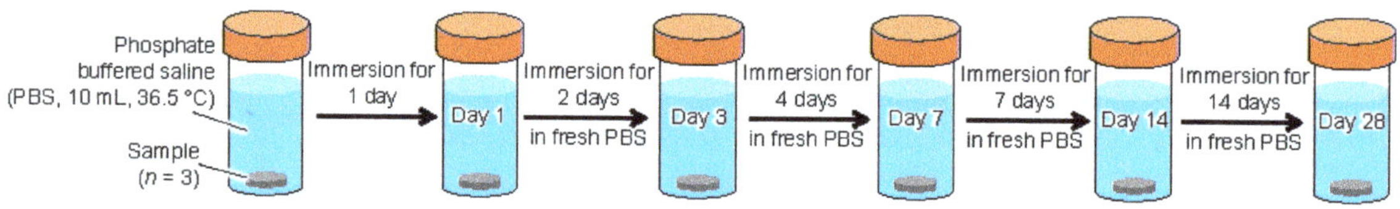

Figure 1. Schematic diagram of Cu ion-release assay. PBS, phosphate-buffered saline.

2.5. Evaluation of Antimicrobial Activity of Sample

The antibacterial activity of each sample was evaluated using strains of *S. aureus* (NBRC1221355; obtained from the NITE Biological Resource Center, Chiba, Japan) and *E. coli* (NBRC39725; also obtained from the NITE Biological Resource Center) according to the method described in JIS R 1752:2020 [44]. This experiment was approved by the Pathogenic Organisms Safety Management Committee of the Tokyo Medical and Dental University (Approval No. M22022-002; Approval date: 25 April 2022). *S. aureus* and *E. coli* were cultivated in soybean-casein digest (SCD) broth (Nissui Pharmaceutical Co., Ltd., Tokyo, Japan) or Luria–Bertani (LB) broth (MP Biomedicals, CA, USA), respectively, at 37 °C for 24 h. The resulting *S. aureus* cells were pelleted and resuspended in PBS at a density of 4.3×10^7 to 1.0×10^8 colony-forming units (CFU)/mL. Similarly, the resulting *E. coli* cells were pelleted and resuspended in PBS at a density of 3.2×10^7 to 1.0×10^8 CFU/mL. An antibacterial activity test was performed in quadruplicate (n = 4) for each sample, as described below.

Briefly, the sample (a Ti disc) was placed on a Falcon® bacteriological petri dish (Catalog No. 351029; Corning, Inc., Corning, NY, USA) with the sample surface facing upward; an aliquot (10 μL) of the test bacterial suspension then was dispensed onto the sample. Subsequently, the sample surface was covered with plastic film (Catalog No. JPG533090; A.R. Medicom Inc. (Asia), Ltd., Kobe, Japan) to ensure close contact. The petri dish then was placed in a GB-2.0L gas barrier box (AS ONE Corp., Osaka, Japan). To reduce the effects on the bacteria of increasing temperature and drying during visible-light irradiation, pure water was dispensed into the bottom of the gas barrier box to maintain a relative humidity exceeding 90%. A 460-nm LED lamp (Model SPA-10SW; Hayashi Clock Industry Co., Ltd., Tokyo, Japan) was used as the light source. The distance from the lower part of the lens to the sample surface was 10 cm; the irradiance and the irradiation period were 250 W/m^2 and 30 min, respectively. These irradiation parameters were defined based on the distance that might be expected between a visible-light source and the abutment of dental implants during dental treatment in the clinic. Following the visible-light irradiation, bacteria adhering to the sample and plastic film were recovered by washing with 1 mL of PBS. The resulting test bacterial suspension was subjected to 10-fold serial dilution in PBS, and aliquots (100 μL) of the 10-fold and 100-fold dilutions were plated on SCD agar (*S. aureus*) or LB agar (*E. coli*). After incubation at 37 °C for 24 h, the number of bacterial colonies formed on the plate surface was counted. The number of viable bacteria was calculated using the following formula:

$$N = Z \times E \times 10 \quad (3)$$

where N, Z, and E are the viable bacterial count, arithmetic mean number of colonies, and dilution factor, respectively.

2.6. Measurement of ROS Generated by Samples

It is difficult to directly measure ROS and free radicals at room temperature (approximately 25 °C). Instead, we employed electron spin resonance (ESR) using a spin-trapping method to measure the ROS generated from samples irradiated with visible light. ESR was conducted using a JES-FA-100 instrument (JEOL, Ltd.). The spin-trapping agent consisted of 5,5-dimethyl-1-pyrroline-*N*-oxide (DMPO; Labotech Co., Tokyo, Japan). The ESR measurement was conducted under the following conditions: microwave power, 4.0 mW; microwave frequency, 9428.954 MHz; magnetic width, 0.1 mT; field sweep width, ±5 mT; field modulation frequency, 100 kHz; modulation width, 0.1 mT; time constant, 0.03 s; and sweep time, 0.1 min. Each sample was placed in a well of a 24-well plate, and 300 mM DMPO was dispensed at 500 μL/well. The samples immersed in the DMPO solution were irradiated with visible light, using the same 460-nm LED lamp for 30 min under the same conditions as those in the antibacterial property test. Subsequently, 200 μL of the DMPO solution (in which the sample had been immersed) was removed from each well and assessed for ROS using the ESR spectrometer. Hydroxyl radicals were quantified using 4-hydroxy-2,2,6,6-tetramethylpiperidine-1-oxyl (TEMPOL; Sigma-Aldrich, St. Louis, MO, USA) as the standard. The areas of DMPO-OH spectra were compared with that of a 2 μM TEMPOL standard measured under identical settings to estimate the concentration of the DMPO-OH adduct.

2.7. Statistical Analysis

Statistical analyses were performed using Microsoft Excel version 2209 (Microsoft, Redmond, WA, USA). One-way analysis of variance followed by multiple-hypothesis tests (Tukey's honestly significant difference tests) was performed for comparisons between more than two groups. $p < 0.01$ was considered statistically significant.

3. Results

XRD peaks attributable to anatase- and rutile-type TiO_2 were observed for all samples other than Sample TI; peaks attributable to copper oxide (CuO) were observed for samples Cu1, Cu10, and Cu100 (Figure 2). As the $Cu(OAc)_2$ concentration increased, the intensity ratio of the XRD peak of rutile-type TiO_2 around 27.7° in 2θ to that of anatase-type TiO_2 around 25.5° in 2θ also increased.

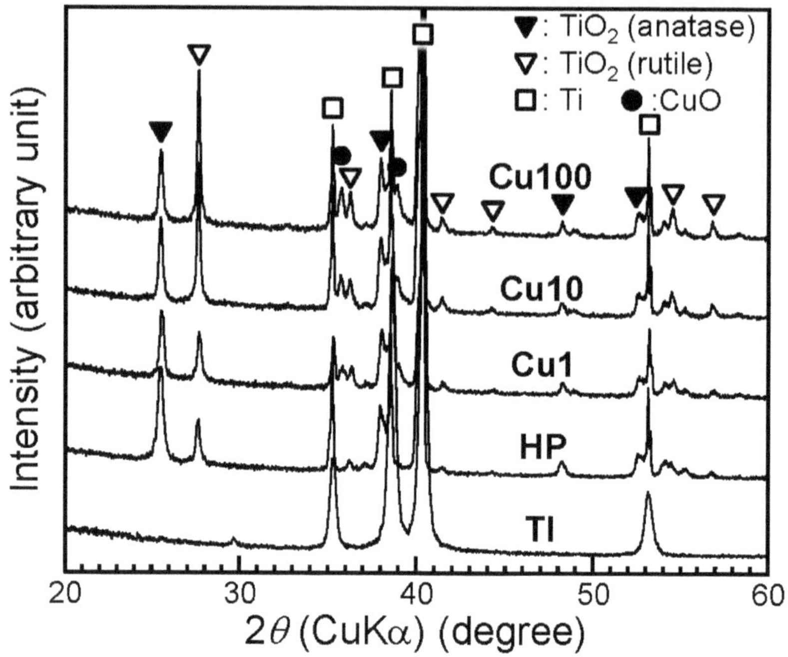

Figure 2. XRD patterns of samples. XRD: X-ray diffraction, TI: sample with no treatment, HP: sample with H_2O_2 treatment, Cu1: sample with H_2O_2, 1 mM $Cu(OAc)_2$, and heat treatment, Cu10: sample with H_2O_2, 10 mM $Cu(OAc)_2$, and heat treatment, Cu100: sample with H_2O_2, 100 mM $Cu(OAc)_2$, and heat treatment.

FE-SEM micrographs (Figure 3A) showed that a fine network structure formed on the surface of the HP, Cu1, Cu10, and Cu100 samples and that fine particles formed on the Cu1, Cu10, and Cu100 samples. At the macroscopic level, the HP sample exhibited a brown color, while the Cu1, Cu10, and Cu100 samples exhibited a black color (insets in Figure 3A). The EDS mapping images of the samples (Figure 3B) indicated that the fine particles that formed on the Cu1, Cu10, and Cu100 samples were composed primarily of Cu and O.

In the XPS spectra of the Cu1, Cu10, and Cu100 samples, XPS peaks located in the C 1s, Ti 2p, O 1s, and Cu 2p energy regions were detected on the surfaces of the samples. The binding energy of the Ti $2p_{3/2}$ peak for the Cu1, Cu10, and Cu100 samples was approximately 458.8 eV (Figure 4A), indicating that Ti existed primarily as TiO_2 in these samples [45,46]. The binding energies of the Cu $2p_{3/2}$ peaks for the Cu1, Cu10, and Cu100 samples were located at approximately 933.5, 933.7, and 933.7 eV, respectively (Figure 4B); the modified Auger parameters for the Cu1, Cu10, and Cu100 samples were measured at 1850.6, 1850.9, and 1850.7 eV, respectively. These results indicated that the Cu in these samples existed primarily as Cu^{2+} [22,41]. The concentrations of Cu on the surfaces of the Cu1, Cu10, and Cu100 samples were 6.8, 15.8, and 12.9 atomic%, respectively. For all three of these samples, the proportions of (Cu^0 + Cu^+) and Cu^{2+} ranged from 19.2 to 26.5% and 73.5–80.8%, respectively (Table 2).

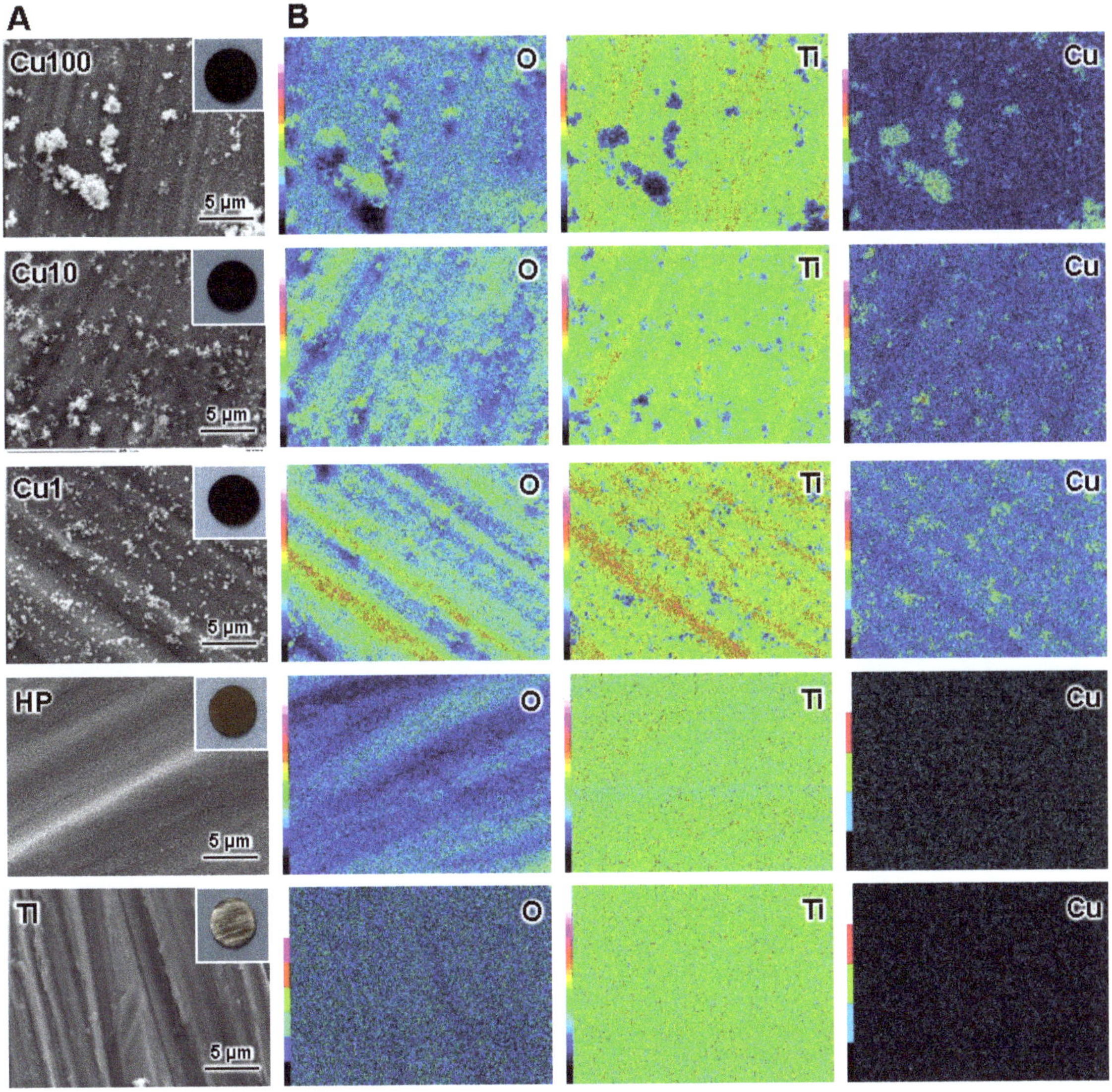

Figure 3. FE-SEM micrographs with insets showing macroscopic appearance (**A**), and EDS elemental mapping images (**B**) of samples. FE-SEM: field emission scanning electron microscope, EDS: energy dispersive X-ray spectrometer, TI: sample with no treatment, HP: sample with H_2O_2 treatment, Cu1: sample with H_2O_2, 1 mM $Cu(OAc)_2$, and heat treatment, Cu10: sample with H_2O_2, 10 mM $Cu(OAc)_2$, and heat treatment, Cu100: sample with H_2O_2, 100 mM $Cu(OAc)_2$, and heat treatment. The color bar represents the intensity of EDS signal (black: low intensity, red or pink: high intensity).

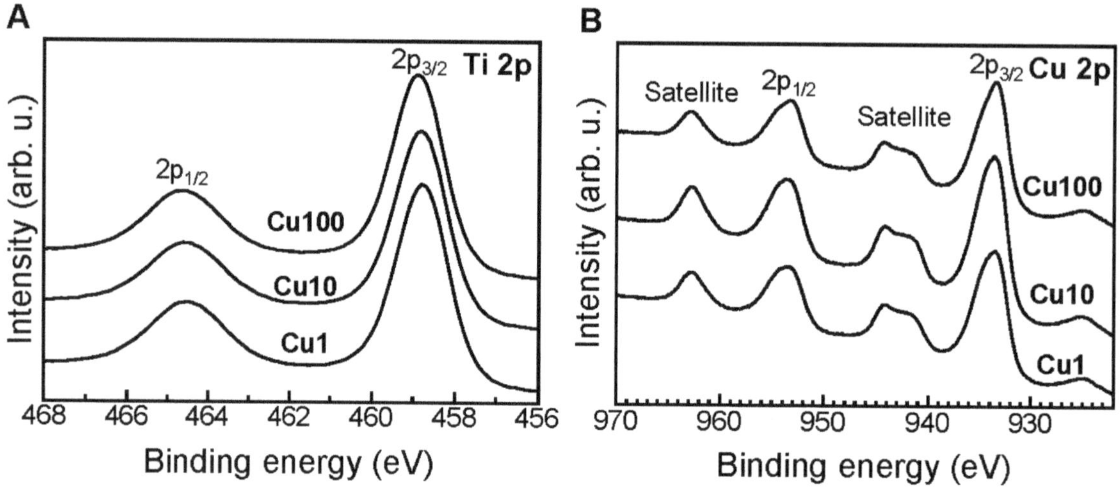

Figure 4. XPS narrow spectra in the Ti 2p (**A**) and Cu 2p (**B**) energy regions for the Cu1, Cu10, and Cu100 samples. XPS: X-ray photoelectron spectroscopy, arb. u.: arbitrary unit, Cu1: sample with H_2O_2, 1 mM $Cu(OAc)_2$, and heat treatment, Cu10: sample with H_2O_2, 10 mM $Cu(OAc)_2$, and heat treatment, Cu100: sample with H_2O_2, 100 mM $Cu(OAc)_2$, and heat treatment.

Table 2. Concentrations of Cu and proportions of Cu species at the surfaces of the Cu1, Cu10, and Cu100 samples. at.%: atomic percent, Cu1: sample with H_2O_2, 1 mM $Cu(OAc)_2$, and heat treatment, Cu10: sample with H_2O_2, 10 mM $Cu(OAc)_2$, and heat treatment, Cu100: sample with H_2O_2, 100 mM $Cu(OAc)_2$, and heat treatment.

Sample	Concentration of Cu (at.%)	Proportion of Cu Species (%)	
		$Cu^0 + Cu^+$	Cu^{2+}
Cu1	6.8	20.1	79.9
Cu10	15.8	19.2	80.8
Cu100	12.9	26.5	73.5

SEM photographs and XRD patterns of samples following 7 days of incubation in SBF are shown in Figure 5. The HP, Cu1, Cu10, and Cu100 samples accumulated dense and uniform layers after a week of immersion in SBF, and XRD peaks attributable to apatite were observed for these samples.

When samples were immersed in PBS for periods of up to 28 days, the concentration of Cu eluted from the samples increased gradually (Figure 6). However, no statistically significant differences were observed in the eluted Cu ion concentration when comparing the samples at a given immersion time point.

With or without visible-light irradiation, the viable counts of *S. aureus* and *E. coli* did not differ significantly when comparing between the HP and TI samples; in contrast, viable counts for both bacterial species were decreased significantly, with and without irradiation, in the Cu1, Cu10, and Cu100 samples (compared to the TI sample) (Figure 7). Notably, among samples irradiated with visible light, the viable counts of *E. coli* in the Cu1, Cu10, and Cu100 samples were decreased by more than two logs compared to similarly treated TI and HP samples.

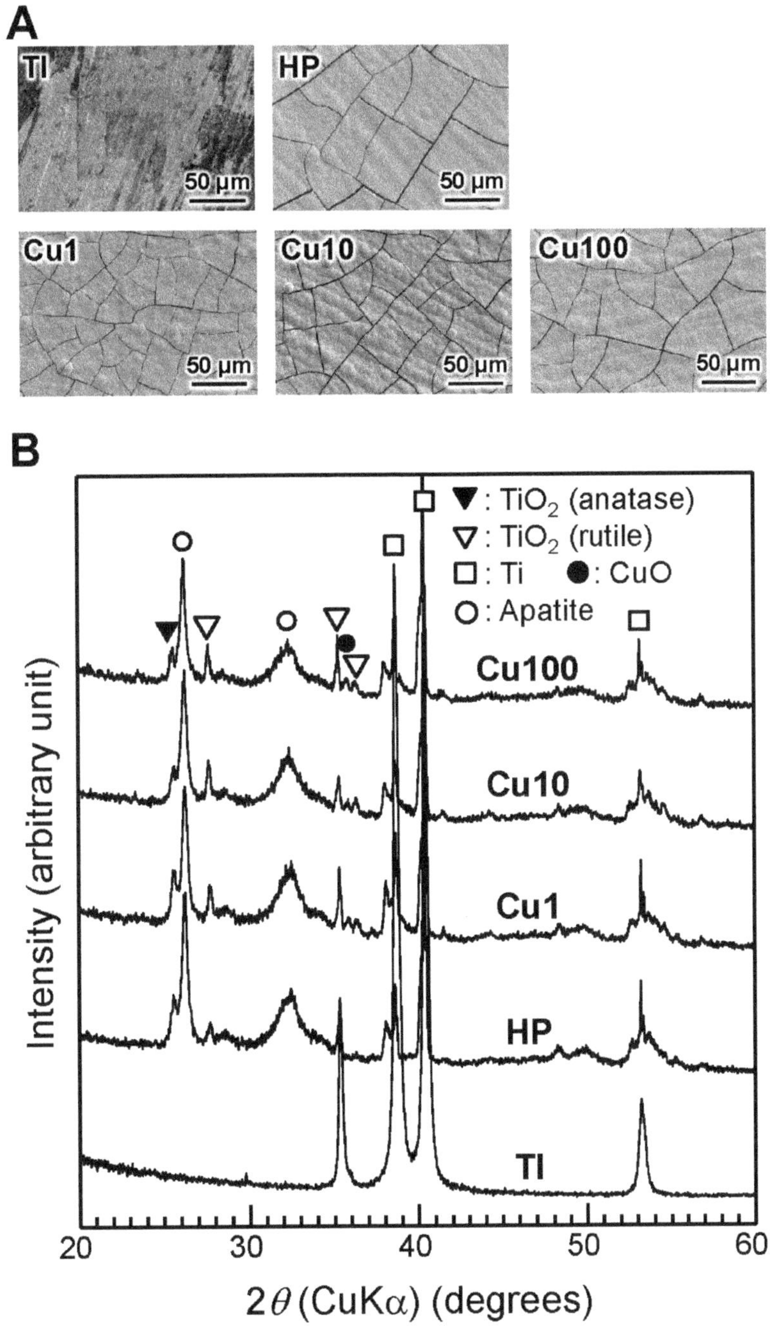

Figure 5. SEM photographs (**A**) and XRD patterns (**B**) of samples after 7 days of incubation in SBF. SEM: scanning electron microscope, TI: sample with no treatment, HP: sample with H_2O_2 treatment, Cu1: sample with H_2O_2, 1 mM $Cu(OAc)_2$, and heat treatment, Cu10: sample with H_2O_2, 10 mM $Cu(OAc)_2$, and heat treatment, Cu100: sample with H_2O_2, 100 mM $Cu(OAc)_2$, and heat treatment, SBF: simulated body fluid.

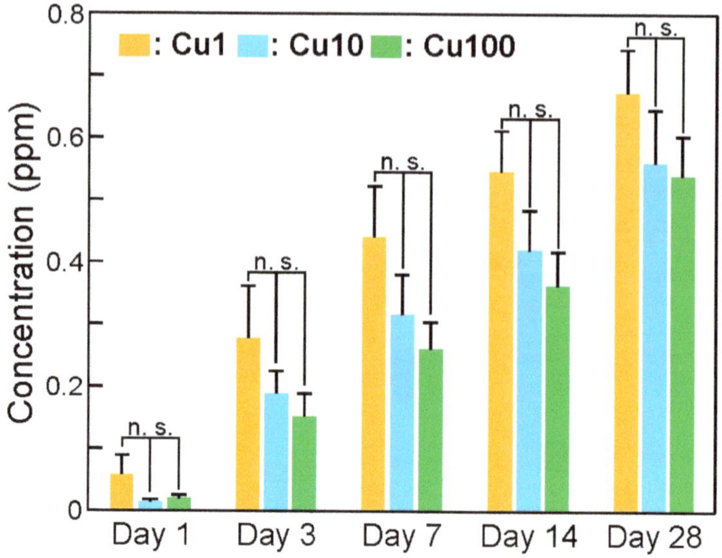

Figure 6. Concentration of Cu eluted from samples into PBS when immersed in PBS for periods of up to 28 days. Values are presented as arithmetic mean ± SD (n = 3). n.s. $p \geq 0.01$ (one-way analysis of variance followed by multiple-hypothesis tests (Tukey's honestly significant difference tests)). Cu1: sample with H_2O_2, 1 mM $Cu(OAc)_2$, and heat treatment, Cu10: sample with H_2O_2, 10 mM $Cu(OAc)_2$, and heat treatment, Cu100: sample with H_2O_2, 100 mM $Cu(OAc)_2$, and heat treatment, PBS: phosphate-buffered saline, n.s.: not significant.

The ESR spectra of the Cu1, Cu10, and Cu100 samples, with or without visible-light irradiation, exhibited two classes of ESR signals (Figure 8). The first showed hyperfine coupling constants of a_N = 1.49 and a_H = 1.49 mT and was assigned to DMPO-OH, a spin adduct of DMPO and ·OH; the second was an unknown signal. Notably, the intensity of the DMPO-OH peaks was increased by irradiation with visible light. The concentration of hydroxyl radicals (·OH) generated from the samples was elevated by irradiation with visible light, and the ·OH concentrations were nominally higher in the Cu1, Cu10, and Cu100 samples compared to the TI and HP samples (Table 3).

Table 3. Concentrations of hydroxyl radical (·OH) generated by samples with and without visible-light irradiation. TI: sample with no treatment, HP: sample with H_2O_2 treatment, Cu1: sample with H_2O_2, 1 mM $Cu(OAc)_2$, and heat treatment, Cu10: sample with H_2O_2, 10 mM $Cu(OAc)_2$, and heat treatment, Cu100: sample with H_2O_2, 100 mM $Cu(OAc)_2$, and heat treatment.

Sample	·OH Concentration (μM)	
	Without Visible-Light Irradiation	With Visible-Light Irradiation
TI	0.09	0.19
HP	0.21	0.51
Cu1	1.08	3.03
Cu10	1.44	3.23
Cu100	1.50	4.36

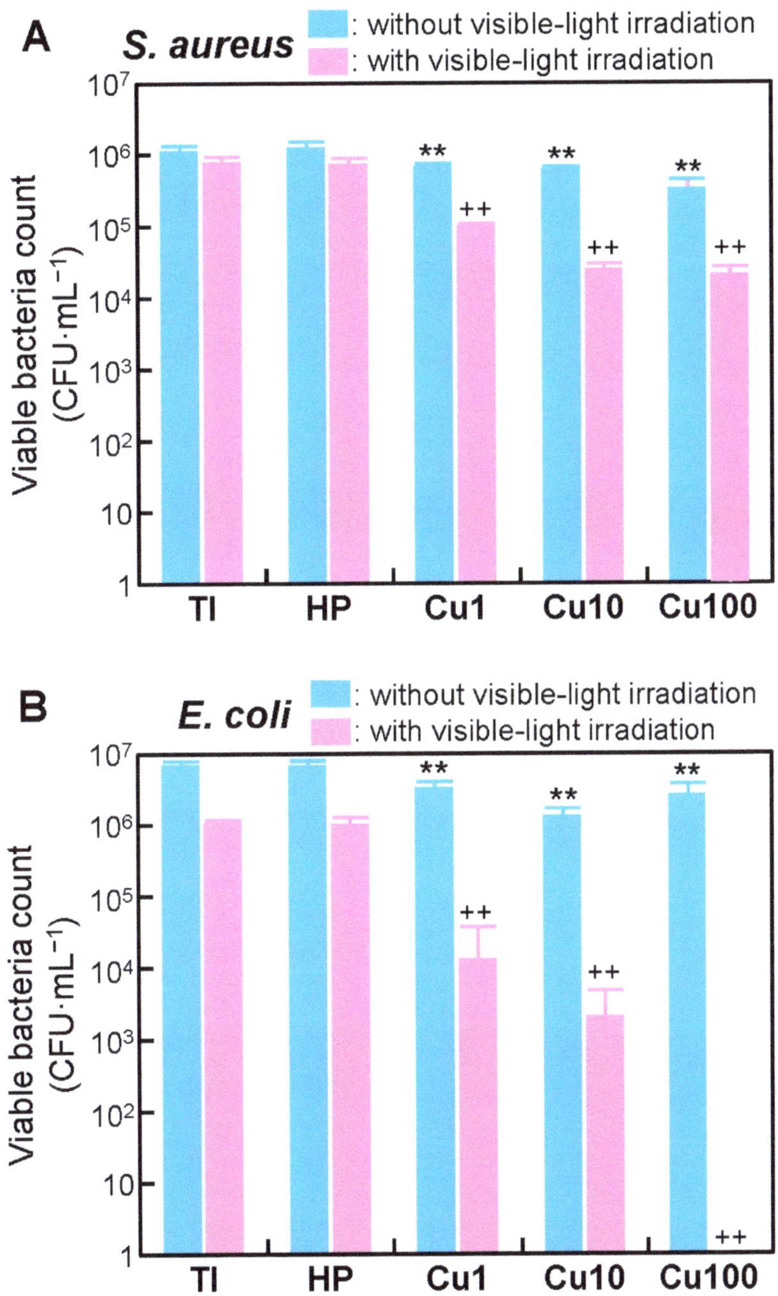

Figure 7. Viable counts of *S. aureus* (**A**) and *E. coli* (**B**) in samples without and with visible-light irradiation. Values are presented as arithmetic mean ± SD (n = 4). ** $p < 0.01$ vs. TI without visible-light irradiation, ++ $p < 0.01$ vs. TI with visible-light irradiation (one-way analysis of variance followed by multiple-hypothesis tests (Tukey's honestly significant difference tests)). CFU: colony-forming unit, TI: sample with no treatment, HP: sample with H_2O_2 treatment, Cu1: sample with H_2O_2, 1 mM $Cu(OAc)_2$, and heat treatment, Cu10: sample with H_2O_2, 10 mM $Cu(Oac)_2$, and heat treatment, Cu100: sample with H_2O_2, 100 mM $Cu(OAc)_2$, and heat treatment.

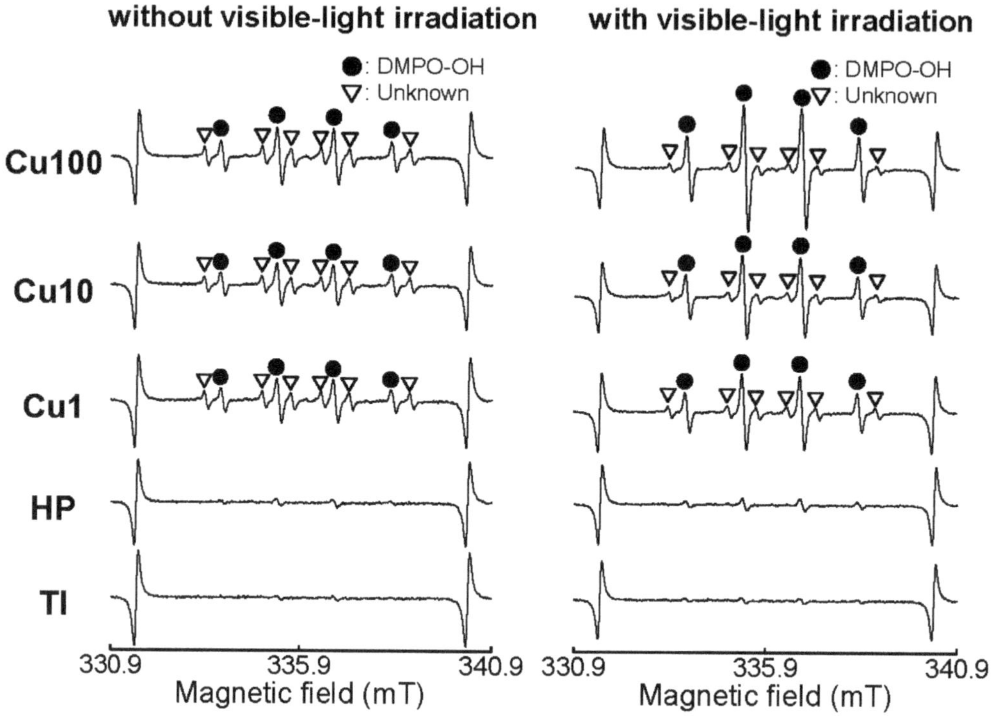

Figure 8. ESR spectra of samples without and with visible-light irradiation. ESR: electron spin resonance, DMPO: 5,5-dimethyl-1-pyrroline-*N*-oxide, TI: sample with no treatment, HP: sample with H_2O_2 treatment, Cu1: sample with H_2O_2, 1 mM $Cu(OAc)_2$, and heat treatment, Cu10: sample with H_2O_2, 10 mM $Cu(OAc)_2$, and heat treatment, Cu100: sample with H_2O_2, 100 mM $Cu(OAc)_2$, and heat treatment.

4. Discussion

Following sequential exposure to H_2O_2, $Cu(OAc)_2$, and heat (the procedure described in the present work), the surface of the HP sample exhibited a fine network structure consisting of anatase- and rutile-type TiO_2 (Figures 2 and 3A). H_2O_2 has been reported to react with Ti to form TiO_2 on Ti as follows [38].

$$Ti + 2H_2O_2 \rightarrow TiOOH + 1/2O_2 + H_3O+ \text{ (or } H^+\text{)} \tag{4}$$

$$TiOOH + 1/2O_2 + H_3O^+ \rightarrow Ti(OH)_4 \tag{5}$$

$$Ti(OH)_4 \rightarrow TiO_2 + 2H_2O \tag{6}$$

The TiO_2 fine network structure formed by immersion in H_2O_2 solution remained even after immersion in 100 mM $Cu(OAc)_2$ solution, indicating that the undesired dissolution of the TiO_2 layer seen upon exposure to strongly acidic Cu solutions was avoided by using a weakly acidic $Cu(OAc)_2$ solution for Cu doping of the TiO_2. We attribute the observed increase in the intensity ratio of the TiO_2 XRD peaks (ratio of rutile-type to anatase-type) in the Cu1, Cu10, and Cu100 samples (Figure 2) to the enhancement of anatase-to-rutile transformation by Cu doping of the TiO_2 [47–51].

Figures 2–4 and Table 2 indicated that the small particles formed on the Cu1, Cu10, and Cu100 samples are composed primarily of CuO. There are concerns about the stability and durability of small CuO particles [52], but we believe that the small CuO particles are stable and chemically durable, given that the concentration of Cu released from samples was less than 0.8 ppm, even after soaking in PBS for 28 days (Figure 6), and that the color of samples remained black even after the soaking in PBS. Additionally, the data in Table 2 indicated that the Cu concentration on the sample surface does not increase linearly with the $Cu(OAc)_2$ concentration, instead showing saturation at about 16 atomic%. In contrast, in our previous study on Ti subjected to NaOH-$Cu(NO_3)_2$-heat treatments [27], the Cu concentration on the sample surface peaked at approximately 12 atomic%. Although the mechanistic basis for this difference is unknown, the H_2O_2-$Cu(OAc)_2$-heat treatment proposed in the present work permits the doping of nominally greater levels of Cu onto the sample surface than does the previous NaOH-$Cu(NO_3)_2$-heat treatments.

All samples other than the TI sample accumulated a dense and uniform surface apatite layer during 7 days of incubation in SBF (Figure 4). Previous work [33] has shown that the anatase- and rutile-type TiO_2 on these sample surfaces (Figure 2) is responsible for the apatite formation during immersion in SBF. The apatite-forming ability of the Cu1, Cu10, and Cu100 samples appeared to be much higher than that seen previously in samples prepared by NaOH-$Cu(NO_3)_2$-heat treatments [28]. Therefore, the technique described here succeeded in forming Cu-doped TiO_2 with increased apatite-forming ability on Ti. Many factors affect the apatite-forming ability of TiO_2, including the acidity and basicity of Ti-OH groups [53–55], and the crystalline phase [33] and porosity [56] of TiO_2. In addition, the apatite-forming ability of metal-doped TiO_2 on Ti is affected by the chemical state of the doped metal [27]. To gain insight for the preparation of samples with higher apatite-forming ability, it is necessary to identify the factor(s) responsible for the higher apatite-forming ability of the present samples (compared to that of the previous samples). However, we are not able to define the relevant factor(s), given the many differences between the processes used in the present and previous work. Further study will be needed to identify the factor(s) responsible for the higher apatite-forming ability of the present samples. For clinical applications, samples prepared as in the present paper will need to be evaluated in vivo for their bone-binding ability. We conjecture that samples prepared as described here will form an apatite layer on their surface even in vivo, and that this apatite layer will facilitate direct bonding to living bone [42].

The Cu1, Cu10, and Cu100 samples also showed antibacterial activity against *S. aureus* and *E. coli*, even in the absence of visible-light irradiation (Figure 7). This activity may reflect the combined effects of copper ions released from the sample (Figure 6), contact between bacteria and small CuO particles present on the sample surface [57], and the small amount of ·OH generated from the samples (Table 3). The antibacterial activity of these samples was remarkably enhanced by visible-light irradiation. The generation of ·OH from these samples (Figure 8 and Table 3) may be responsible for the enhanced antibacterial activity of these samples following irradiation with visible light. Notably, the improvement in the antibacterial activity following visible-light irradiation appeared to be more pronounced against *E. coli* than against *S. aureus* (Figure 7B). This distinction may reflect the fact that the peptidoglycan layer in *E. coli*, a Gram-negative bacterium, is thinner than that that in *S. aureus*, a Gram-positive bacterium [58,59]. Given that ROS, including ·OH, can induce chronic inflammation, future in vivo studies will be needed to investigate any inflammatory responses induced by ROS generated by the samples. However, we conjecture that the inflammatory response(s) induced by ROS generated by the samples will be limited, since the concentration of ·OH was low, and increased only when the sample was irradiated with visible light (Table 3). The antibacterial activity of the samples against other bacteria that cause peri-implantitis, such as *Porphyromonas gingivalis*, should also be investigated in future studies.

In the ESR assessment, DMPO-OH peaks were observed for the Cu1, Cu10, and Cu100 samples, even in the absence of visible-light irradiation (Figure 8). This observation indicated that these samples generate ·OH even in the dark (Table 3). We hypothesize that ·OH is generated when CuO particles present on the sample surface react with the reagent or water used in this experiment [60,61]. The unknown peaks in the ESR spectrum of the sample may be attributable to carbon-centered radicals, such as methyl radicals [62] These carbon-centered radicals may originate from carbon in reagents decomposed by CuO particles present on the sample surface.

To the eye, the Cu1, Cu10, and Cu100 samples appeared black in color (inset photographs in Figure 3A). This black coloration presumably is attributable to the visible light absorption of the CuO [31,63–68] formed on these samples by heat treatment. For CuO-doped TiO_2, it has been reported that CuO doping extends the tailing of the absorption edge toward the visible region, resulting in band-gap narrowing [64–66]. In addition, the precipitation of fine particles of CuO on the TiO_2 surface (Figure 2) suggests that the visible-light-responsive photocatalytic effect may have been induced by interfacial charge transfer of electrons from TiO_2 to CuO [31,67,68]. Although no clear UV-visible spectrum (which would indicate narrowing of the band gap) was obtained for these samples, we speculate that the Cu-doped TiO_2 formed on Ti by this method exhibits visible-light-responsive photocatalytic activity due to the interface charge transfer between CuO and TiO_2; this activity would result in the observed enhancement of antibacterial activity by visible-light irradiation (Figure 7). Additionally, as shown in Figure 2, CuO small particles formed sparsely on Ti substrate, making it difficult to obtain clear UV-visible spectra of the CuO small particles themselves.

In the present study, we found that H_2O_2-$Cu(OAc)_2$-heat treatments induced the layering of TiO_2 with CuO fine particles; we further demonstrated that the TiO_2 layer showed apatite-forming ability during incubation in SBF and visible-light-enhanced antibacterial activity. The novelty of our work is that we have succeeded in preparing titanium substrates with CuO-supported TiO_2 surfaces; these substrates exhibited excellent apatite-forming ability and visible-light-enhanced antibacterial activity against both *E. coli* and *S. aureus*. Previous work has shown that apatite-coated TiO_2 powders show antibacterial activity against *E. coli* and *S. aureus* when irradiated with black light or visible light [69], and that TiO_2 layers formed on Ti substrates by Au-sputtering and thermal oxidation show antibacterial activity against *E. coli* when irradiated with visible light [70]. However, to the best of our knowledge, surface-modified titanium substrates with similar functions have not been reported to date. Future investigations will be needed to evaluate the toxicity and bone-bonding ability of samples prepared as described here. Nonetheless, we hypothesize that dental implants with Cu-doped TiO_2 on the abutments may decrease the risk of peri-implantitis when treated in the dental clinic by regular or on-demand irradiation with visible light. Furthermore, if employed for devices that are partially exposed outside the body (such as external fixators), materials with such surfaces are expected to exhibit antibacterial properties when exposed to visible light.

5. Conclusions

A network structure of anatase- and rutile-type TiO_2 with CuO fine particles was generated when Ti was subjected to H_2O_2-$Cu(OAc)_2$-heat treatments. Such surface-treated Ti accumulated a dense and uniform surface layer of apatite when incubated for 7 days in SBF, indicating enhanced apatite-forming ability compared to that previously reported for Ti subjected to NaOH-$Cu(NO_3)_2$-heat treatments. Furthermore, samples prepared as described here showed enhanced antibacterial activity against *E. coli* and *S. aureus* when subjected to visible-light irradiation. The increased antibacterial activity of these samples under visible-light irradiation may be attributable to the generation of ·OH by the samples. Together, these results suggest that Ti implants with reduced risk of peri-implantitis following on-demand visible-light irradiation can be obtained by the H_2O_2-$Cu(OAc)_2$-heat treatments described here.

Author Contributions: Conceptualization, M.K.; methodology, P.-C.S., M.S., T.Y. and T.M.; validation, P.-C.S., M.S., T.Y., T.M. and M.S.; formal analysis, P.-C.S., M.S., T.Y. and T.M.; investigation, P.-C.S., M.S. and T.M.; data curation, P.-C.S., M.S., T.Y. and T.M.; writing—original draft preparation, P.-C.S. and M.K.; writing—review and editing, P.-C.S., M.S., T.Y., T.M. and M.K.; supervision, M.K.; project administration, M.K.; funding acquisition, M.K. All authors have read and agreed to the published version of the manuscript.

Funding: This research was funded by the Japan Society for the Promotion of Science (JSPS) KAKENHI Grant No.: JP16H03177; the Kazuchika Okura Memorial Foundation; and a grant from the Ministry of Education, Culture, Sports, Science and Technology, Japan (MEXT) administered via the Institute of Biomaterials and Bioengineering, Tokyo Medical and Dental University (Project "Design and Engineering by Joint Inverse Innovation for Materials Architecture (DEJI^{2}MA)").

Data Availability Statement: The raw data supporting the conclusions of this article will be made available by the authors on request.

Conflicts of Interest: The authors declare no conflicts of interest.

References

1. Mombelli, A.; Lang, P. The diagnosis and treatment of peri-implantitis. *Periodontology 2000* **1998**, *17*, 63–76. [CrossRef] [PubMed]
2. Brägger, U.; Aeschlimann, S.; Bürgin, W.; Hämmerle, C.H.; Lang, N.P. Biological and technical complications and failures with fixed partial dentures (FPD) on implants and teeth after four to five years of function. *Clin. Oral Implant. Res.* **2001**, *12*, 26–34. [CrossRef] [PubMed]
3. Gruica, B.; Wang, H.Y.; Lang, N.P.; Buser, D. Impact of IL-1 genotype and smoking status on the prognosis of osseointegrated implants. *Clin. Oral Implants Res.* **2004**, *15*, 393–400. [CrossRef] [PubMed]
4. Fransson, C.; Lekholm, U.; Jemt, T.; Berglundh, T. Prevalence of subjects with progressive bone loss at implants. *Clin. Oral Implant. Res.* **2005**, *16*, 440–446. [CrossRef] [PubMed]
5. Roos-Jansåker, A.M.; Lindahl, C.; Renvert, H.; Renvert, S. Nine-to fourteen-year follow-up of implant treatment, Part I: Implant loss and associations to various factors. *J. Clin. Periodontol.* **2006**, *33*, 283–289. [CrossRef] [PubMed]
6. Dvorak, G.; Arnhart, C.; Heuberer, S.; Christian, D.; Huber, C.D.; Watzek, G.; Gruber, R. Peri-implantitis and late implant failures in postmenopausal women: A cross-sectional study. *J. Clin. Periodontol.* **2011**, *38*, 950–955. [CrossRef] [PubMed]
7. Lee, J.C.Y.; Mattheos, N.; Nixon, K.C.; Ivanovski, S. Residual periodontal pockets are a risk indicator for peri-implantitis in patients treated for periodontitis. *Clin. Oral Implant. Res.* **2012**, *23*, 325–333.
8. Renvert, S.; Lessem, J.; Dahlén, G.; Renvert, H.; Lindah, C. Mechanical and repeated antimicrobial therapy using a local drug delivery system in the treatment of peri-implantitis: A randomized clinical trial. *J. Periodontol.* **2008**, *79*, 836–844. [CrossRef] [PubMed]
9. Bassetti, M.; Schär, D.; Wicki, B.; Eick, S.; Ramseier, C.A.; Arweiler, N.B.; Sculean, A.; Salvi, G.E. Anti-infective therapy of peri-implantitis with adjunctive local drug delivery or photodynamic therapy: 12-month outcomes of a randomized controlled clinical trial. *Clin. Oral Implant. Res.* **2014**, *25*, 279–287. [CrossRef]
10. Dadgostar, P. Antimicrobial resistance: Implications and costs. *Infect. Drug Resist.* **2019**, *12*, 3903–3910. [CrossRef]
11. Chen, J.Y.; Cheng, F.; Luo, D.W.; Huang, J.F.; Ouyang, J.; Nezamzadeh-Ejhieh, A.; Khan, M.S.; Liu, J.Q.; Peng, Y.Q. Recent advances in Ti-based MOFs in biomedical applications. *Dalton Trans.* **2022**, *51*, 14817–14832. [CrossRef] [PubMed]
12. Gu, B.; Cai, J.; Peng, G.; Zhou, H.; Zhang, W.; Deyuan Zhang, D.; Gong, D. Metal organic framework-loaded biohybrid magnetic microrobots for enhanced antibacterial treatment. *Colloids Surf. A* **2024**, *685*, 133295. [CrossRef]
13. Veerapandian, M.; Yun, K. Functionalization of biomolecules on nanoparticles: Specialized for antibacterial applications. *Appl. Microbiol. Biotechnol.* **2011**, *90*, 1655–1667. [CrossRef]
14. Geng, Z.; Cao, Z.; Liu, J. Recent advances in targeted antibacterial therapy basing on nanomaterials. *Exploration* **2023**, *3*, 20210117. [CrossRef] [PubMed]
15. Ivanova, E.P.; Hasan, J.; Webb, H.K.; Gervinskas, G.; Juodkazis, S.; Truong, V.K.; Wu, A.H.F.; Lamb, R.N.; Baulin, V.A.; Watson, G.S.; et al. Bactericidal activity of black silicon. *Nat. Commun.* **2013**, *4*, 2838. [CrossRef] [PubMed]
16. Tarannum, T.; Ahmed, S. Recent development in antiviral surfaces: Impact of topography and environmental conditions. *Helyon* **2023**, *9*, e16698. [CrossRef] [PubMed]
17. Das, K.; Bose, S.; Bandyopadhyay, A.; Karandikar, B.; Gibbins, B.L. Surface coatings for improvement of bone cell materials and antimicrobial activities of Ti implants. *J. Biomed. Mater. Res. B Appl. Biomater.* **2008**, *87*, 455–460. [CrossRef] [PubMed]
18. Shimazaki, T.; Miyamoto, H.; Ando, Y.; Noda, I.; Yonekura, Y.; Kawano, S.; Miyazaki, M.; Mawatari, M.; Hotokebuchi, T. In vivo antibacterial and silver-releasing properties of novel thermal sprayed silver-containing hydroxyapatite coating. *J. Biomed. Mater. Res. B Appl. Biomater.* **2010**, *92*, 386–389. [CrossRef] [PubMed]
19. Shirai, T.; Shimizu, T.; Ohtani, K.; Zen, Y.; Takaya, M.; Tsuchiya, H. Antibacterial iodine-supported titanium implants. *Acta Biomater.* **2011**, *7*, 1928–1933. [CrossRef]
20. Kizuki, T.; Matsushita, T.; Kokubo, T. Antibacterial and bioactive calcium titanate layers formed on Ti metal and its alloys. *J. Mater. Sci. Mater. Med.* **2014**, *25*, 1737–1746. [CrossRef]
21. Shimabukuro, M.; Tsutsumi, Y.; Yamada, R.; Ashida, M.; Chen, P.; Doi, H.; Nozaki, K.; Nagai, A.; Hanawa, T. Investigation of realizing both antibacterial property and osteogenic cell compatibility on titanium surface by simple electrochemical treatment. *ACS Biomater. Sci. Eng.* **2019**, *5*, 5623–5630. [CrossRef]

22. Shimabukuro, M.; Tsutsumi, Y.; Nozaki, K.; Chen, P.; Yamada, R.; Ashida, M.; Doi, H.; Nagai, A.; Hanawa, T. Investigation of antibacterial effect of copper introduced titanium surface by electrochemical treatment against facultative anaerobic bacteria. *Dent. Mater. J.* **2020**, *39*, 639–647. [CrossRef] [PubMed]
23. Shimabukuro, M.; Manaka, T.; Tsutsumi, Y.; Nozaki, K.; Chen, P.; Ashida, M.; Nagai, A.; Hanawa, T. Corrosion behavior and bacterial viability on different surface states of copper. *Mater. Trans.* **2020**, *61*, 1143–1148. [CrossRef]
24. Shimabukuro, M. Antibacterial property and biocompatibility of silver, copper, and zinc in titanium dioxide layers incorporated by one-step micro-arc oxidation: A review. *Antibiotics* **2020**, *9*, 716. [CrossRef]
25. Aoki, S.; Shimabukuro, M.; Kishida, R.; Kyuno, K.; Noda, K.; Yokoi, T.; Kawashita, M. Electrochemical deposition of copper on bioactive porous titanium dioxide layer: Antibacterial and pro-osteogenic activities. *ACS Appl. Bio Mater.* **2023**, *6*, 5759–5767. [CrossRef] [PubMed]
26. Morimoto, T.; Hirata, H.; Eto, S.; Hashimoto, A.; Kii, S.; Kobayashi, T.; Tsukamoto, M.; Yoshihara, T.; Toda, Y.; Mawatari, M. Development of silver-containing hydroxyapatite-coated antimicrobial implants for orthopaedic and spinal surgery. *Medicina* **2022**, *58*, 519. [CrossRef]
27. Kawashita, M.; Iwabuchi, Y.; Suzuki, K.; Furuya, M.; Yokota, K.K.; Kanetaka, H. Surface structure and in vitro apatite-forming ability of titanium doped with various metals. *Colloids Surf. A* **2018**, *555*, 558–564. [CrossRef]
28. Suzuki, K.; Yokoi, T.; Iwatsu, M.; Mokudai, T.; Kanetaka, H.; Kawashita, M. Antibacterial properties of Cu-doped TiO_2 prepared by chemical and heat treatment of Ti metal. *J. Asian Ceram. Soc.* **2021**, *9*, 1448–1456. [CrossRef]
29. Fujibayashi, S.; Nakamura, T.; Nishiguchi, S.; Tamura, J.; Uchida, M.; Kim, H.M.; Kokubo, T. Bioactive titanium: Effect of sodium removal on the bone-bonding ability of bioactive titanium prepared by alkali and heat treatment. *J. Biomed. Mater. Res.* **2001**, *56*, 562–570. [CrossRef]
30. Yadav, H.M.; Otari, S.V.; Koli, V.B.; Mali, S.S.; Hong, C.K.; Pawar, S.H.; Delekar, S.D. Preparation and characterization of copper-doped anatase TiO_2 nanoparticles with visible light photocatalytic antibacterial activity. *J. Photochem. Photobiol. A* **2014**, *280*, 32–38. [CrossRef]
31. Liu, M.; Sunada, K.; Hashimoto, K.; Miyauchi, M. Visible-light sensitive Cu(II)-TiO_2 with sustained anti-viral activity for efficient indoor environmental remediation. *J. Mater. Chem. A* **2015**, *3*, 17312–17319. [CrossRef]
32. Mathew, S.; Ganguly, P.; Rhatigan, S.; Kumaravel, V.; Byrne, C.; Hinder, S.J.; Bartlett, J.; Nolan, M.; Pillai, S.C. Cu-doped TiO_2: Visible light assisted photocatalytic antimicrobial activity. *Appl. Sci.* **2018**, *8*, 2067. [CrossRef]
33. Uchida, M.; Kim, H.M.; Kokubo, T.; Fujibayashi, S.; Nakamura, T. Structural dependence of apatite formation on titania gels in a simulated body fluid. *J. Biomed. Mater. Res. Part A* **2003**, *64*, 164–170. [CrossRef] [PubMed]
34. Ohtsuki, C.; Iida, H.; Hayakawa, S.; Osaka, A. Bioactivity of titanium treated with hydrogen peroxide solutions containing metal chlorides. *J. Biomed. Mater. Res.* **1997**, *35*, 39–47. [CrossRef]
35. Kaneko, S.; Tsuru, K.; Hayakawa, S.; Takemoto, S.; Ohtsuki, C.; Ozaki, T.; Inoue, H.; Osaka, A. In vivo evaluation of bone-bonding of titanium metal chemically treated with a hydrogen peroxide solution containing tantalum chloride. *Biomaterials* **2001**, *22*, 875–881. [CrossRef] [PubMed]
36. Wu, J.M.; Hayakawa, S.; Tsuru, K.; Osaka, A. Crystallization of anatase from amorphous titania in hot water and in vitro biomineralization. *J. Ceram. Soc. Jpn.* **2002**, *110*, 78–80. [CrossRef]
37. Wang, X.X.; Hayakawa, S.; Tsuru, K.; Osaka, A. Bioactive titania gel layers formed by chemical treatment of Ti substrate with a H_2O_2/HCl solution. *Biomaterials* **2002**, *23*, 1353–1357. [CrossRef] [PubMed]
38. Osaka, A.; Tsuru, K.; Hayakawa, S. Titania derived from combined chemical and thermal treatments of titanium: In vitro apatite forming ability. *Phosphorus Res. Bull.* **2004**, *17*, 130–141. [CrossRef] [PubMed]
39. Suzuki, K.; Iwatsu, M.; Mokudai, T.; Furuya, M.; Yokota, K.; Kanetaka, H.; Shimabukuro, M.; Yokoi, T.; Kawashita, M. Visible-light-enhanced antibacterial activity of silver and copper co-doped titania formed on titanium via chemical and thermal treatments. *Molecules* **2023**, *28*, 650. [CrossRef]
40. Shirley, D.A. High-resolution X-ray photoemission spectrum of the valence bands of gold. *Phys. Rev. B* **1972**, *5*, 4709–4714. [CrossRef]
41. Biesinger, M.C. Advanced analysis of copper X-ray photoelectron spectra. *Surf. Interf. Anal.* **2017**, *49*, 1325–1334. [CrossRef]
42. Kokubo, T.; Takadama, H. How useful is SBF in predicting in vivo bone bioactivity? *Biomaterials* **2006**, *27*, 2907–2915. [CrossRef]
43. *ISO 23317:2014*; Implants for Surgery—In Vitro Evaluation for Apatite-Forming Ability of Implant Materials. International Organization for Standardization: Geneva, Switzerland, 2014.
44. *JIS R 1752:2020*; Fine Ceramics (Advanced Ceramics, Advanced Technical Ceramics)—Test Method for Antibacterial Activity of Photocatalytic Materials and Efficacy under Indoor Lighting Environment. Japanese Standards Association: Tokyo, Japan, 2020.
45. Georgiadou, I.; Spanos, N.; Papadopoulou, C.; Matralis, H.; Kordulis, C.; Lycourghiotis, A. Preparation and characterization of various titanias (anatase) used as supports for vanadia-supported catalysts. *Colloids Surf. A* **1995**, *98*, 155–165. [CrossRef]
46. Shimabukuro, M.; Kobayashi, M.; Kawashita, M. Metallic substrate influences on the osteogenic cell compatibility and antibacterial activity of silver-incorporated porous oxide layers formed by micro-arc oxidation. *ACS Appl. Eng. Mater.* **2023**, *1*, 2288–2294. [CrossRef]
47. Iida, Y.; Ozaki, S. Grain growth and phase transformation of titanium oxide during calcination. *J. Am. Ceram. Soc.* **1961**, *44*, 120–127. [CrossRef]
48. Shannon, R.D.; Pask, J.A. Kinetics of the anatase-rutile transformation. *J. Am. Ceram. Soc.* **1965**, *48*, 391–398. [CrossRef]

49. MacKenzie, K.J.D. Calcination of titania V. Kinetics and mechanism of the anatase-rutile transformation in the presence of additives. *Trans. J. Br. Ceram. Soc.* **1975**, *74*, 77–84.
50. Kim, D.W.; Kim, T.G.; Hong, K.S. Low-firing of CuO-doped anatase. *Mater. Res. Bull.* **1999**, *34*, 771–781. [CrossRef]
51. Hanaor, D.A.H.; Sorrell, C.C. Review of the anatase to rutile phase transformation. *J. Mater. Sci.* **2011**, *46*, 855–874. [CrossRef]
52. Sajjad, H.; Sajjad, A.; Haya, R.T.; Khan, M.M.; Zia, M. Copper oxide nanoparticles: In vitro and in vivo toxicity, mechanisms of action and factors influencing their toxicology. *Comp. Biochem. Physiol. C Toxicol. Pharmacol.* **2023**, *271*, 109682. [CrossRef]
53. Kasuga, T.; Kondo, H.; Nogami, M. Apatite formation on TiO_2 in simulated body fluid. *J. Crystal Growth* **2002**, *235*, 235–240. [CrossRef]
54. Kokubo, T.; Yamaguchi, S. Novel bioactive materials developed by simulated body fluid evaluation: Surface-modified Ti metal and its alloys. *Acta Biomater.* **2016**, *44*, 16–30. [CrossRef] [PubMed]
55. Ferraris, S.; Yamaguchi, S.; Barbani, N.; Cazzola, M.; Cristallini, C.; Miola, M.; Vernè, E.; Spriano, S. Bioactive materials: In vitro investigation of different mechanisms of hydroxyapatite precipitation. *Acta Biomater.* **2020**, *102*, 468–480. [CrossRef] [PubMed]
56. Li, Z.; Miyazaki, T.; Kawashita, M. Preparation and in vitro apatite-forming ability of porous and non-porous titania microspheres. *J. Ceram. Soc. Jpn.* **2013**, *121*, 782–787. [CrossRef]
57. Hans, M.; Erbe, A.; Mathews, S.; Chen, Y.; Solioz, M.; Mücklich, F. Role of copper oxides in contact killing of bacteria. *Langmuir* **2013**, *29*, 16160–16166. [CrossRef] [PubMed]
58. Louis, P.; Trüper, H.G.; Galinski, E.A. Survival of *Escherichia coli* during drying and storage in the presence of compatible solutes. *Appl. Microbiol. Biotechnol.* **1994**, *41*, 684–688. [CrossRef]
59. Sunada, K.; Watanabe, T.; Hashimoto, K. Studies on photokilling of bacteria on TiO_2 thin film. *J. Photochem. Photobiol. A* **2003**, *156*, 227–233. [CrossRef]
60. Hricovíni, M.; Mazúr, M.; Sîrbu, A.; Palamarciuc, O.; Arion, V.B.; Brezová, V. Copper (II) thiosemicarbazone complexes and their proligands upon UVA irradiation: An EPR and spectrophotometric steady-state study. *Molecules* **2018**, *23*, 721. [CrossRef] [PubMed]
61. Zhao, S.; Miao, D.; Zhu, K.; Tao, K.; Wang, C.; Sharma, V.K.; Jia, H. Interaction of benzo [a] pyrene with Cu (II)-montmorillonite: Generation and toxicity of environmentally persistent free radicals and reactive oxygen species. *Environ. Int.* **2019**, *129*, 154–163. [CrossRef] [PubMed]
62. Kuppusamy, P.; Zweier, J.L. Characterization of free radical generation by xanthine oxidase: Evidence for hydroxyl radical generation. *J. Biol. Chem.* **1989**, *264*, 9880–9884. [CrossRef]
63. Colón, G.; Maicu, M.; Hidalgo, M.C.; Navío, J.A. Cu-doped TiO_2 systems with improved photocatalytic activity. *Appl. Catal. B* **2006**, *67*, 41–51. [CrossRef]
64. Ni, Y.; Zhu, Y.; Ma, X. A simple solution combustion route for the preparation of metal-doped TiO_2 nanoparticles and their photocatalytic degradation properties. *Dalton Trans.* **2011**, *14*, 3689–3694. [CrossRef] [PubMed]
65. Ravishankar, T.N.; Vaz, M.d.O.; Teixeira, S.R. The effects of surfactant in the sol-gel synthesis of CuO/TiO_2 nanocomposites on its photocatalytic activities under UV-visible and visible light illuminations. *New J. Chem.* **2020**, *44*, 1888–1904. [CrossRef]
66. Hernández-Gordillo, A.; Arriaga, S. Mesoporous TiO_2 monoliths impregnated with CdS and CuO nanoparticles for airborne bacteria inactivation under visible light. *Catal. Lett.* **2022**, *152*, 629–640. [CrossRef] [PubMed]
67. Irie, H.; Kamiya, K.; Shibanuma, T.; Miura, S.; Tryk, D.A.; Yokoyama, T.; Hashimoto, K. Visible light-sensitive Cu(II)-grafted TiO_2 photocatalysts: Activities and X-ray absorption fine structure analyses. *J. Phys. Chem. C* **2009**, *113*, 10761–10766. [CrossRef]
68. Moniz, S.J.A.; Tang, J. Charge transfer and photocatalytic activity in CuO/TiO_2 nanoparticle heterojunctions synthesised through a rapid, one-pot, microwave solvothermal route. *ChemCatChem* **2015**, *7*, 1659–1667. [CrossRef]
69. Kangwansupamonkon, W.; Lauruengtana, V.; Surassmo, S.; Ruktanonchai, U. Antibacterial effect of apatite-coated titanium dioxide for textiles applications. *Nanomed. Nanotechnol. Biol. Med.* **2009**, *5*, 240–249. [CrossRef]
70. Ueda, T.; Koizumi, R.; Ueda, K.; Ito, K.; Ogasawara, K.; Kanetaka, H.; Narushima, T. Antibacterial properties of TiO_2 layers formed by Au-sputtering and thermal oxidation of titanium under visible light. *Mater. Trans.* **2023**, *64*, 155–164. [CrossRef]

Journal of
Functional
Biomaterials

MDPI

Article

Yeast Particle Encapsulation of Azole Fungicides for Enhanced Treatment of Azole-Resistant *Candida albicans*

Ernesto R. Soto, Florentina Rus and Gary R. Ostroff *

Program in Molecular Medicine, University of Massachusetts Medical School, Worcester, MA 01605, USA; ernesto.soto-villatoro@umassmed.edu (E.R.S.); florentina.rus@umassmed.edu (F.R.)
* Correspondence: gary.ostroff@umassmed.edu

Abstract: Addressing the growing problem of antifungal resistance in medicine and agriculture requires the development of new drugs and strategies to preserve the efficacy of existing fungicides. One approach is to utilize delivery technologies. Yeast particles (YPs) are 3–5 μm porous, hollow microspheres, a byproduct of food-grade *Saccharomyces cerevisiae* yeast extract manufacturing processes and an efficient and flexible drug delivery platform. Here, we report the use of YPs for encapsulation of tetraconazole (TET) and prothioconazole (PRO) with high payload capacity and stability. The YP PRO samples were active against both sensitive and azole-resistant strains of *Candida albicans*. The higher efficacy of YP PRO versus free PRO is due to interactions between PRO and saponifiable lipids in the YPs. Encapsulation of PRO in glucan lipid particles (GLPs), a highly purified form of YPs that do not contain saponifiable lipids, did not result in enhanced PRO activity. We evaluated the co-encapsulation of PRO with a mixture of the terpenes: geraniol, eugenol, and thymol. Samples co-encapsulating PRO and terpenes in YPs or GLPs were active on both sensitive and azole-resistant *C. albicans*. These approaches could lead to the development of more effective drug combinations co-encapsulated in YPs for agricultural or GLPs for pharmaceutical applications.

Keywords: yeast particles; glucan lipid particles; azole fungicides; drug resistance; terpenes; microencapsulation; biocide; controlled delivery

Citation: Soto, E.R.; Rus, F.; Ostroff, G.R. Yeast Particle Encapsulation of Azole Fungicides for Enhanced Treatment of Azole-Resistant *Candida albicans*. *J. Funct. Biomater.* **2024**, *15*, 203. https://doi.org/10.3390/jfb15080203

Academic Editor: Pankaj Vadgama

Received: 9 June 2024
Revised: 17 July 2024
Accepted: 18 July 2024
Published: 23 July 2024

1. Introduction

Antifungal drug resistance poses great challenges to agriculture, as well as veterinary and public health [1–7]. Species belonging to *Candida*, *Aspergillus*, *Pneumocystis* and *Cryptococcus* are among the most common fungal pathogens exhibiting drug resistance [4,5,8]. There are four classes of antifungals used to manage and prevent fungal infections: (1) polyenes, which bind ergosterol and produce membrane pores, (2) echinocandins that inhibit the synthesis of β-1,3-glucan, (3) pyrimidine analogs, which inhibit synthesis of DNA and RNA, and (4) azoles that function by inhibiting sterol 14 demethylase and reducing ergosterol levels in the plasma membrane [9]. Of these groups, azoles are the primary class of fungicides for the prevention and effective treatment of fungal diseases in plants, animals, and humans, and they are also widely used in cosmetics and material preservation [3,10]. The extensive use of azole fungicides across different fields is due to the broad-spectrum range of fungal targets and relatively low cost. Unfortunately, the widespread use of azole fungicides and the similarities in chemical structure and mechanism of action of agricultural and medical azoles increases the emergence of resistant pathogenic strains with potential crossover of resistant strains from environment to clinical settings [2,3].

Addressing the challenges of antifungal drug resistance requires efforts to optimize sustainable agricultural practices that minimize the application of antifungals, mitigation of antifungal leakage to the environment, development of novel antifungal agents, and strategies of drug repurposing and combination therapies [2,5,11]. Given the prominent role of the azole class of fungicides for treatment of fungal diseases in many fields, one main research focus should be to preserve this class of antifungals for continued use in all areas [2,5].

The strategies to reverse resistance to azole fungicides can include combination therapies with compounds that either elicit antifungal effects or increase the pathogen susceptibility to the original antifungal drug. In this context, naturally occurring compounds are a primary source of bioactive materials with the potential to be used as chemosensitizers in combination therapies with existing antifungals [12]. For example, natural lignin was shown to reverse resistance to fluconazole in *C. albicans* though alteration of the glycolytic pathway [13], and terpenoids, the naturally occurring compounds that constitute the primary component of essential oils, can enhance the activity of fluconazole and other antifungal drugs against *Candida* strains [11,14]. The possibility of reversing antifungal resistance can provide the opportunity to resurrect the efficacy of existing, cost-effective drugs.

Yeast particles (YPs) are 3–5 μm porous, hollow particles derived from Baker's yeast (*Saccharomyces cerevisiae*). We have developed the use of YPs as an efficient and flexible platform for the encapsulation of a wide range of small molecules and macromolecules within the hollow cavity of the particles for drug delivery and agricultural applications [15–26]. In this article, we present the results of the encapsulation of the agricultural azole fungicides tetraconazole (TET) and prothioconazole (PRO). These compounds were encapsulated in YPs with high efficiency (>90%) and high payload loading capacity (1:1 *w*/*w*). The YP-encapsulated PRO samples (YP PRO) were active against both sensitive and azole-resistant strains of *C. albicans*. The higher efficacy of YP PRO compared to unencapsulated PRO is due to the synergy between PRO and saponifiable lipids co-encapsulated in the YPs. These YPs are suitable for agricultural applications but cannot be used in most parenteral pharmaceutical applications due to their potential immunogenic effects. A highly purified form of YPs, glucan lipid particles (GLPs), were used to encapsulate PRO. The GLPs lack the saponifiable lipids of YPs and samples of GLP PRO are not active against azole-resistant *C. albicans*. The synergy of co-encapsulating a mixture of terpenes (geraniol, eugenol, thymol, GET 2:1:2 weight ratio) and PRO in both GLP and YP were evaluated, resulting in both GLP and YP samples being active against sensitive and azole-resistant *C. albicans*. The development of these yeast-particle-encapsulated fungicides has potential applications to target antifungal resistance in agriculture (YP fungicides) and pharmaceutical (GLP fungicides) applications.

2. Materials and Methods

Yeast particles (YPs) were purchased from Biorigin (Louisville, KY, USA). Technical-grade prothioconazole and tetraconazole were supplied from Sipcam Oxon (Lodi, Italy) and Eden Research plc (Oxfordshire, England), respectively. Terpenes (eugenol, thymol, and geraniol) were procured from Penta Manufacturing (Livingston, NJ, USA). All other reagents and solvents were obtained from Fisher Scientific (Waltham, MA, USA) or Sigma Aldrich (St. Louis, MO, USA). Yeast peptone dextrose (YPD) was prepared from Difco™ yeast extract, Difco™ Bacto peptone and dextrose (all materials obtained from Fisher Scientific) at a composition of 1% yeast extract, 2% peptone and 2% dextrose *w*/*v*.

2.1. Synthesis of Glucan Lipid Particles (GLPs)

The glucan lipid particles were prepared from YPs following the method previously reported for glucan particles (GPs) [15] but without the washing steps with organic solvents to avoid removal of non-saponifiable lipids. Yeast particles (100 g) are suspended in 1 L of 1M NaOH and heated to 85 °C. The cell suspension is stirred vigorously for 1 h at this temperature. The insoluble material containing the yeast cell walls is recovered by centrifugation (14,000 rpm, 30′). The pellet is then suspended in 1 M NaOH, heated, and stirred vigorously for 1 h at 85 °C. The suspension is allowed to cool to room temperature (20–25 °C) and the extraction is continued for a further 16 h. The insoluble residue is recovered by centrifugation, suspended in 500 mL of water and the pH of the particle suspension is brought to pH 4.5 with HCl. The insoluble residue is recovered by centrifugation and washed three times with water. The resulting slurry is placed in glass trays and dried under reduced pressure to produce a fine light-yellow powder.

2.2. YP Loading of Tetraconazole (YP TET)

Dry YPs were mixed with water (1 µL water/mg YP) for 12–18 h at 4 °C to obtain a uniform hydrated YP sample. Then, TET was added at weight ratios of 1:1 or 3:1 TET:YP and incubated at room temperature (20–25 °C) for a minimum of 24 h to allow for complete TET loading.

2.3. YP or GLP Loading of Prothioconazole (YP PRO or GLP PRO)

Dry YPs or GLPs were mixed with water (0.5 µL water/mg YP) for 12–18 h at 4 °C to obtain a uniform hydrated YP or GLP sample. Then, PRO was absorbed into YPs or GLPs by swelling the particles with a solution of 400 mg PRO/mL in acetone (2.5 µL acetone solution/mg YP or GLP) and incubated at room temperature (20–25 °C) for a minimum of 24 h to allow for complete PRO loading. The sample was dried under reduced pressure to remove solvent. To maximize loading of PRO into the hollow cavity of the particles, the hydration and absorption steps were repeated with 0.5 µL water/mg YP and 2 µL acetone/mg YP.

2.4. YP or GLP Co-Encapsulation of Terpenes and Prothioconazole

Samples co-encapsulating a mixture of terpenes—geraniol, eugenol and thymol (2:1:2 geraniol:eugenol:thymol, GET212)—and PRO were prepared following the procedure described above for YP PRO using the GET212 mixture as solvent for PRO (0.5 µL water/mg YP, 1.1 mg GET212/mg YP, 0.055 mg PRO/mg GLP or YP).

2.5. Characterization of Payload Loading Efficiency

Samples of YP TET and YP PRO were stained with Nile red to qualitatively assess loading inside the hollow cavity of the particles by fluorescence microscopy of the encapsulated fluorescent fungicide-Nile red complex. Microscopy images were collected with an Olympus BX60 upright compound fluorescent microscope (ex 550 nm/em 570 nm, Olympus, Tokyo, Japan). Samples of YP- or GLP-encapsulated payloads (5–10 mg YP or GLP) were weighed, suspended in 1 mL of water, centrifuged at low speed (3000 rpm for 5 min) and the supernatant was collected. The pellets were extracted with a mixture of 10% water and 90% methanol (PRO and GET) or 10% water and 90% acetone (TET samples). The azole fungicides and terpenes were quantified in both supernatant (unencapsulated payload) and extracted pellets (encapsulated) by HPLC operated with 32 Karat™ software version 7.0 (Beckman Coulter, Inc., Brea, CA, USA), using a Waters Symmetry® C18 column (3.5 µm, 4.6 × 150 mm), flow rate of 0.8 mL/min, injection volume of 10 µL, detection by absorbance at 210 nm. The isocratic mobile phase conditions were acetonitrile:water 70:30 for quantification of samples containing only TET or PRO, and acetonitrile:water 60:30 for mixtures containing PRO and terpenes. The quantification of each payload was performed by measuring the peak area and interpolating the concentration using a calibration curve.

2.6. Payload Release from YPs or GLPs

Samples were suspended in water at a concentration of 10 mg PRO or TET/mL, and a set of samples were prepared by 10-fold serial dilutions down to a concentration of 0.01 mg PRO or TET/mL. The samples were incubated at 23 °C and aliquots were collected at predetermined times, centrifuged and the supernatant collected to measure payload released from the particles by HPLC.

2.7. Antifungal Activity Assays against Sensitive and Azole-Resistant Candida albicans Strains

The antifungal activity of YP TET, YP PRO and GLP PRO samples was evaluated using an azole-sensitive *C. albicans* strain (wild-type, WT SC5134), and the two azole-resistant strains ATCC 11651 and White 1. The strain White 1 is a fluconazole-resistant clinical isolate from Professor Theodore White's laboratory (University of Missouri Kansas City) [27]. The antifungal activity was evaluated using a modified published microplate assay procedure previously described for the antimicrobial activity of YP-encapsulated terpenes [18,28]. The minimum inhibitory concentration (MIC) was determined as the concentration of fungicide that inhibits fungal growth, as measured by absorbance (650 nm), by more than 75%.

2.8. Checkerboard Microplate Assays

To characterize and quantify the antifungal activity of PRO, GET and the drug combinations over a range of concentrations, these compounds (unencapsulated or encapsulated in different YPs) were tested in a checkerboard 96-well plate format. To perform the assay, 100 μL of YPD was added to each well, followed by 100 μL of PRO or YP PRO to Rows A-H of column 1. Next, 1:1 serial dilution were performed from column 1 down to column 11, finally removing 100 μL from column 11. The second drug (GET or YP GET) was added to Row A of columns 1–12, and 1:1 serial dilution were performed from Row A down to Row G, finally removing 100 μL from Row G. Diluted *C. albicans* cells (100 μL, 10^6 cells/mL) were added to all wells of the plate except H12. Initial (t = 0) and final (t = 18 h, 37 °C) absorbance readings were taken at 650 nm. The minimum inhibitory concentration was determined as the concentration that inhibits fungal growth as measured by absorbance by more than 75%. Heat maps of growth inhibition were generated using GraphPrism v 9.0. Contour plots (2D) and synergy calculations using zero interaction potency (ZIP) synergy scoring were generated using the SynergyFinder R online tool (https://synergyfinder.org, accessed on 14 March 2024) [29,30].

2.9. Statistical Analysis

All experiments were conducted with a minimum of three replicates and the reported data correspond to average values with standard deviation. For all two-group comparisons, a two-tailed Student's test was used.

3. Results

3.1. Yeast Particle Encapsulation of Azoles

Yeast particles (YPs) are hollow, porous microparticles (3–5 μm) derived from Baker's yeast [31]. The porous cell wall structure makes these particles excellent absorbent materials. We have previously demonstrated the high loading capacity of terpene encapsulation inside the hydrophobic cavity of YPs prepared by the passive diffusion of terpene through the porous cell walls of YPs [18,19]. Here, two loading methods were used for the encapsulation of azole fungicides, and the selection of the method for each payload was based on the physical state of the payload at room temperature (see Table 1 for chemical properties relevant to the selection of loading method in YPs). The first method (Figure 1A) is the solvent-free method previously used for encapsulation of terpenes [18,19]. This approach requires only a minimum volume of water to hydrate the pores of YPs, and then a payload with a high octanol/water partition coefficient (log P) that is a liquid or oil at room temperature (e.g., TET) is loaded by passive diffusion into the hydrophobic cavity of the particles. This is a very efficient method that does not require the use of organic solvents. The second method (Figure 1B) requires hydration of the YPs in a minimum volume of water followed by absorption of a highly concentrated solution of the payload in an organic solvent (e.g., PRO in acetone at 400 mg/mL). A minor disadvantage of this second method is the need to include steps for solvent removal.

Table 1. Chemical structures of tetraconazole (TET) and prothioconazole (PRO) and their chemical properties relevant for the selection of loading method in YPs: TET can be loaded following a solvent-free method, and PRO can be loaded in YPs using acetone as the loading solvent.

	Tetraconazole (TET)	Prothioconazole (PRO)
Chemical structure	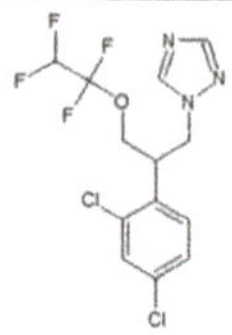	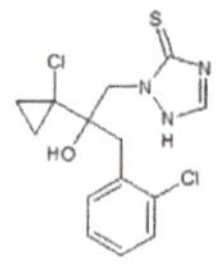
Log P	3.56	4.05
Melting point (°C)	6	139–144
Solubility in water	150 mg/L	22 mg/L
Solubility in acetone	-	440 mg/mL [1]

[1] All property values are from PubChem [32], except for the solubility of PRO in acetone (solubility value was determined in our laboratory).

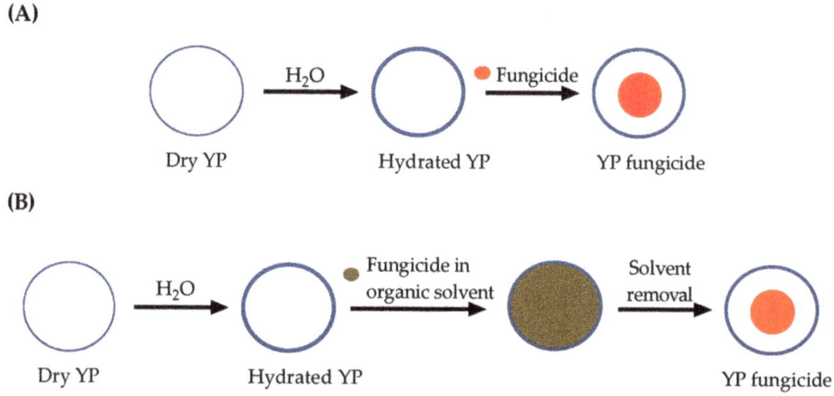

Figure 1. Schematics of two methods for loading of fungicides in yeast particles: (**A**) solvent-free loading method and (**B**) loading of payload in an organic solvent.

Encapsulation of both fungicides was achieved with high efficiency, as determined quantitatively by HPLC analysis and qualitatively by microscopic Nile red staining, which allows for visualization of the payload loaded inside the hollow cavity of YPs (Figure 2). The solvent-free loading method was also used to prepare hyper-loaded TET samples similar to our previously reported hyper-loaded YP terpenes [18]. Samples with a 3:1 TET:YP ratio were produced with >90% TET encapsulation efficiency (Supplementary Materials).

The process of sustained fungicide release from YPs is based on diffusion of the payload of the particles and is a function of payload solubility in water. The encapsulation stability of the samples was evaluated by incubation of YP TET or YP PRO samples diluted at different fungicide concentrations in water. The results in Figure 3 show that YP TET remains stably encapsulated at 1 mg TET/mL (>80% TET inside YPs after 24, no burst release of TET in an emulsified form), but it is rapidly released upon dilution to 0.1 mg TET/mL (~20% TET inside YPs after <1 h). PRO is less water soluble than TET, and PRO remains stably encapsulated in YPs at 1 and 0.1 mg PRO/mL and releases the payload upon dilution at 0.01 mg PRO/mL.

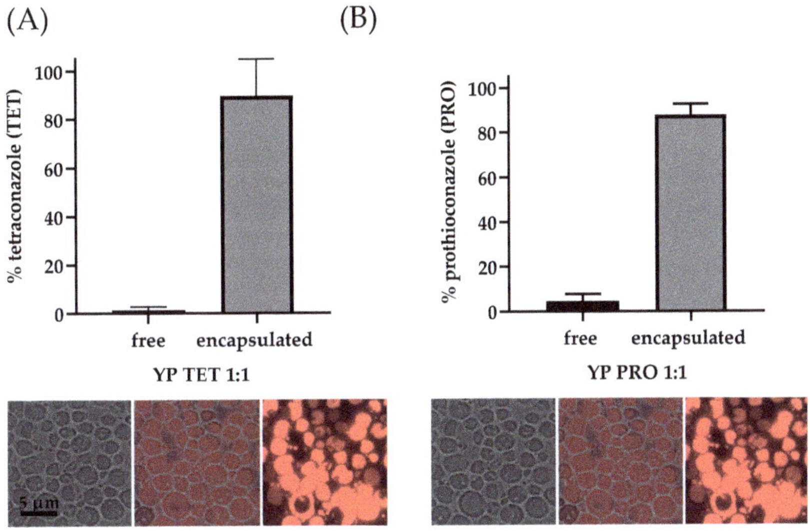

Figure 2. Encapsulation efficiency and microscopy images showing Nile-red-stained fungicides inside the cavity of YPs: (**A**) tetraconazole and (**B**) prothioconazole. Both compounds were loaded at a target weight ratio of 1:1 fungicide:YP.

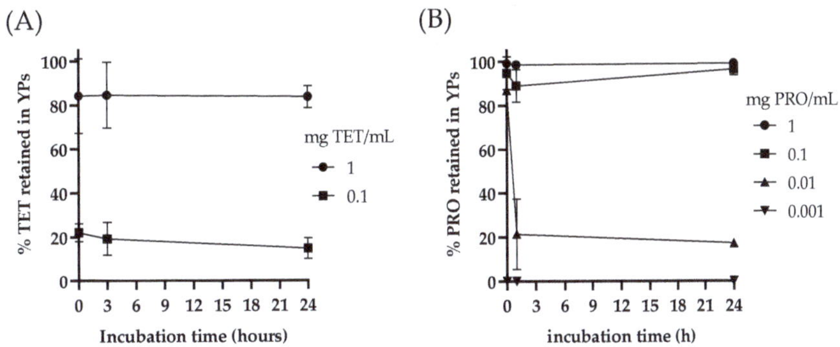

Figure 3. Kinetics of free and YP-encapsulated TET (Tetraconazole (**A**)) and PRO (prothioconazole (**B**)) released over time following dilution of YP fungicide samples in water at different fungicide concentrations and incubation at 23 °C.

3.2. Antifungal Activity of YP Azoles

The antifungal activities of YP TET and YP PRO were evaluated in vitro against an azole-sensitive (wild-type) and two azole-resistant (White 1 and ATCC 11651) *C. albicans* strains. The results in Table 2 show that YP TET and YP PRO are more effective than the corresponding unencapsulated fungicide controls on both sensitive and azole-resistant strains. The ratio of active concentrations (MIC 75%) between the unencapsulated and encapsulated samples plotted in Figure 4 illustrates the enhanced effect of delivering the fungicides in YPs. Samples of YP TET were ~2.5 to 8 times more effective than unencapsulated TET (i.e., the MIC 75% of YP TET is ~2.5 to 8 times lower than MIC 75% of the unencapsulated TET) and YP PRO samples were six times more effective than unencapsulated PRO on sensitive *C. albicans* strain and 20–25 times more effective on azole-resistant strains.

Table 2. In vitro antifungal activity of YP PRO 1:1, unencapsulated (free) PRO suspension, unencapsulated PRO mixed with empty YPs and negative empty YP control on a susceptible (wild-type) and two azole-resistant *Candida albicans* strains. The MIC results represent the average of six replicate experiments. Statistically significant results were obtained between the paired samples indicated with letter superscripts (^{A} $p < 0.005$; $^{B–E}$ $p < 0.01$; $^{F–J}$ $p < 0.05$).

Sample	Minimum Inhibitory Concentration (MIC 75%) in μg PRO/mL on Wild-Type and Azole-Resistant *Candida albicans* Strains		
	Wild-Type	**White 1**	**ATCC 11651**
YP TET 1:1	10 ± 6^{B}	18 ± 11	11 ± 6^{F}
Unencapsulated TET	44 ± 17^{B}	44 ± 26	60 ± 21^{F}
YP PRO 1:1	$12 \pm 4^{C,D}$	$19 \pm 7^{A,I,J}$	12 ± 5^{G}
Unencapsulated PRO	$75 \pm 28^{C,H}$	$375 \pm 161^{A,E,J}$	302 ± 108^{G}
Unencapsulated PRO + empty YP	$87 \pm 34^{D,E}$	$400 \pm 137^{A,E,J}$	302 ± 118
Empty YP	Not active	Not active	Not active

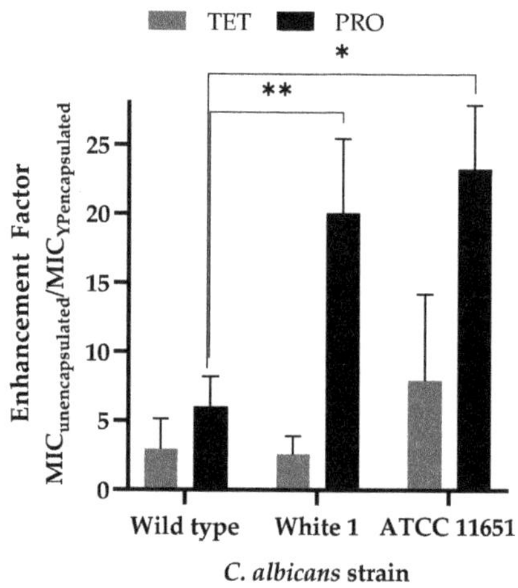

Figure 4. Effect of YP encapsulation on antifungal activity of azole fungicides reported as the ratio of the MIC 75% of unencapsulated and YP-encapsulated samples (statistically significant results were obtained between the paired samples, * $p < 0.1$, ** $p < 0.05$).

The improvement of YP fungicides over unencapsulated fungicides on the azole-sensitive strain is likely due to the delivery of a more homogeneous YP fungicide suspension than the unencapsulated PRO powder suspension or TET oil mixture in YPD. This effect was previously observed for YP-encapsulated terpenes, with the encapsulated terpenes exhibiting a four-fold enhancement of antimicrobial activity compared to the unencapsulated compounds [18].

The effect of YP encapsulation on azole-resistant strains is significantly higher for PRO than for TET due to the strains not being highly resistant to unencapsulated TET. Further efforts of this investigation were focused on the YP PRO samples. Empty YPs were evaluated at the same YP concentrations as YP-PRO samples to confirm that the empty particles lack fungicidal activity. Empty particles admixed with unencapsulated PRO were also evaluated to determine if the enhanced effect on azole-strains can be obtained without the need for PRO encapsulation in YPs. The admixed samples have similar MIC 75% values as unencapsulated PRO for wild-type and White 1 strains and are only 2.5 times more active against the ATCC 11651 strain.

We hypothesized that the enhanced activity of YP PRO on azole-resistant strains was due to the presence of saponifiable lipids in YPs that could improve solubilization of PRO and uptake into the target pathogen. Improvement of physicochemical properties, such as solubility of fungicides to increase interaction with pathogens is critical to the development of novel fungicide formulations [33]. To confirm this hypothesis, PRO was encapsulated in glucan lipid particles (GLPs). GLPs are a highly purified form of YPs that retain non-saponifiable lipids necessary for the loading of hydrophobic payloads into the hollow cavity of the particles but lack the saponifiable lipids of YPs that could improve solubilization and release of hydrophobic payloads. Samples of GLP PRO were prepared following the same procedure as with YPs, and encapsulation at a target 1:1 PRO:GLP ratio was achieved with >90% efficiency (Supplementary Materials).

The *C. albicans* growth response curves (Figure 5) with unencapsulated PRO and PRO encapsulated in YPs or GLPs show that encapsulation of PRO in GLPs inhibits PRO activity on the wild-type strain compared to unencapsulated PRO or YP PRO, and that GLP PRO has similar activity as unencapsulated PRO on azole-resistant/strains. These results support the hypothesis that the saponifiable lipid fraction present in YPs is largely responsible for enhancing the activity of PRO on azole-resistant strains.

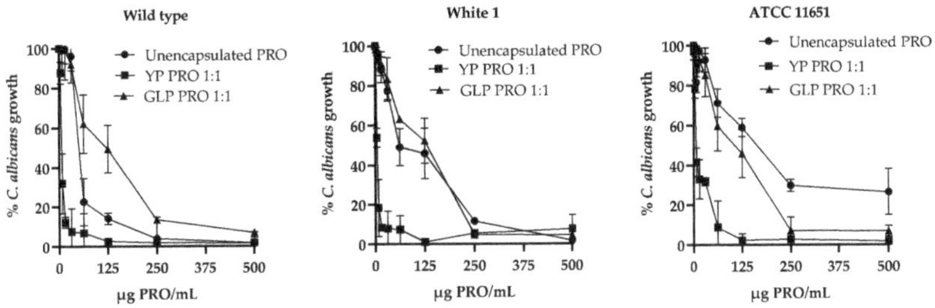

Figure 5. *Candida albicans* growth response curves with unencapsulated PRO and PRO encapsulated in YPs or GLPs.

The effect of YP PRO on azole-resistant strains requires further investigation on agricultural fungal pathogens. The use of YPs is suitable for agricultural applications but is not acceptable for parenteral pharmaceutical applications. The GLPs are suitable for pharmaceutical development, but it will be necessary to improve the activity of GLP fungicide samples by co-encapsulation with a sustained release agent to maximize formulation effectiveness.

We next focused on evaluating the possible synergy on azole-resistant strains by co-encapsulation of PRO and terpenes in YPs and GLPs.

3.3. Evaluation of Synergy of Prothioconazole and Terpenes Encapsulated in Yeast Particles and Glucan Lipid Particles against Azole-Resistant C. albicans

As previously published, the encapsulation of mixtures of the terpenes geraniol, eugenol and thymol (GET 2:1:2 weigh ratio) in YPs has been successfully implemented to develop and commercialize a fungicide product and mixtures of the terpenes geraniol and thymol as a nematicide product for agricultural applications [19,34]. A potential benefit of terpenes is that, given their generalized and multiple mechanisms of activity that evolved over millennia as plant defense molecules, terpenes are extremely difficult to make resistant to target pathogens. The combination of terpenes with antifungals (e.g., fluconazole) has been shown to act synergistically to overcome antifungal resistance [11,13,14].

The synergistic effects of PRO and GET212 were evaluated by conducting checkerboard MIC assays with unencapsulated and YP-encapsulated samples on the *C. albicans* strains. The heat maps in Figure 6 illustrate the inhibition activity of the different PRO and GET212 concentrations evaluated in the checkerboard assay. The results indicate that the YP PRO and YP GET212 samples are more potent (i.e., high % inhibition at low doses) than unencapsulated samples on the three *Candida* strains.

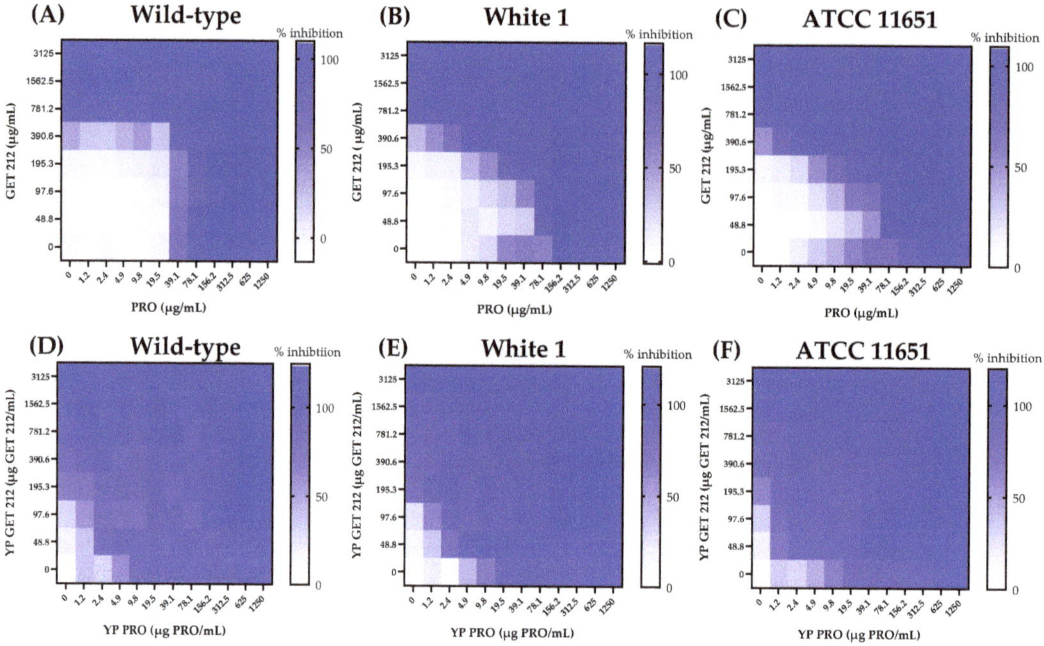

Figure 6. Heat maps showing effect of unencapsulated GET 212 and unencapsulated PRO (heat maps (**A–C**)), and YP GET 212 and YP PRO (heat maps (**D–F**)) on growth inhibition of sensitive and azole-resistant *C. albicans* strains.

To assess if there is a synergistic effect between PRO and GET, the inhibition response from checkerboard assays was analyzed by ZIP synergy scoring generated using the SynergyFinder R online tool. The ZIP model captures interactions between two drugs by comparing the change in the potency of the inhibition response curves between the individual drugs and their combinations. The 2D contour plots in Figure 7 show the ZIP scoring values for the different free drug and YP-encapsulated drug checkerboard combinations on the three *C. albicans* strains. The interaction between the drugs is antagonistic for ZIP scores < −10, additive for score values between −10 and +10, and synergistic for ZIP scores > +10. The analysis of these 2D contour plots indicates that the unencapsulated PRO and GET212 primarily have an additive effect on wild-type *C. albicans*, and there is a concentration range that shows synergy in both azole-resistant strains (2.4 to 39.9 µg PRO/mL and 195.3 to 500 µg GET212/mL). The synergistic concentration range is at lower concentrations of PRO (1.2 to 9.8 µg/mL) and GET212 (24.4 to 146.5 µg/mL) when encapsulated in YPs, and the encapsulated samples show synergy on all three *Candida* strains. The improved synergy effect of YP-encapsulated samples compared to unencapsulated compounds is due to the production of a more homogenous YP drug suspension. The effect of YP encapsulation was also observed in the fungicidal results on wild-type *C. albicans* (Table 2) and previous work on the development of hyper-loaded YP terpenes [18]. The optimization of unencapsulated fungicide and terpenes will require a completely different formulation approach to identify proper ratios, solvent and surfactant conditions to produce a stable emulsion-based delivery system.

Next, we proceeded to prepare samples co-encapsulating PRO and GET212 at a ratio of 20:1 GET212:PRO in both YP and GLPs. The selected ratio corresponds to the ratio of the maximum YP PRO and YP GET212 concentrations showing synergy in the 2D contour analysis shown in Figure 7.

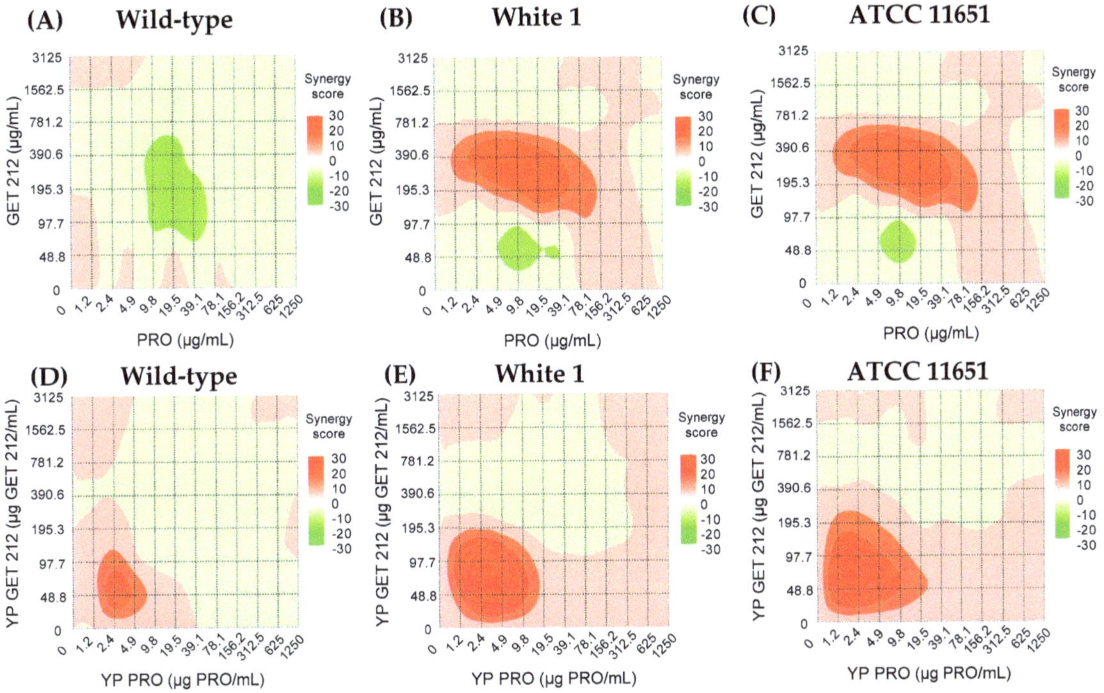

Figure 7. Two-dimensional contour plots showing synergy scores of unencapsulated GET212 and PRO (top (**A–C**)), and YP GET212 and YP PRO (bottom (**D–F**)). ZIP synergy scoring and 2D contour plots were generated using SynergyFinder R online tool (https://synergyfinder.org, accessed on 14 March 2024).

YP and GLP samples co-encapsulating PRO and GET212 were prepared using GET212 as a loading solvent for PRO (Figure 8A) at a target weight ratio of 0.055:1.1:1 PRO:GET212:YP or GLP. The ratio of GET212:YP or GLP is the same that we have previously used to prepare YP GET212 for the development of a fungicide product for agricultural applications [18,19]. The ratio of PRO:GET212 of 1:20 is based on the synergy calculations results described above. Both payloads were efficiently encapsulated in GLPs and YPs (Figure 8B). YP and GLP samples encapsulating PRO alone or in combination with GET212 were evaluated for kinetics of PRO release. Unexpectedly, the results in Figure 8C show that the addition of GET212 slows the release of PRO from both YP and GLP, even at a concentration 2× below the maximum solubility of PRO in water.

The samples co-encapsulating PRO and GET212 were evaluated for antifungal activity in microdilution assays. The results in Table 3 show that YP PRO and YP PRO GET212 have similar MIC 75% for PRO on sensitive and azole-resistant strains, but the activity of YP PRO GET212 is achieved with samples encapsulating <5% *w*/*w* PRO compared to the 50% *w*/*w* in YP PRO samples. Further, GLP PRO samples (1:1 PRO:GLP ratio) require high PRO concentrations on all strains, but samples co-encapsulating PRO with GET212 show improvement of PRO activity on sensitive and azole-resistant strains with MIC 75% values, similar to those of YP PRO and YP PRO GET. The MIC 75% ratio of unencapsulated and encapsulated PRO plotted in Figure 9 clearly illustrates a 20–40-fold improvement of YP encapsulation over unencapsulated PRO on azole-resistant strains and YP or GLP encapsulation in combination with terpenes (50–85-fold improvement). Although the results in Figure 7C indicate that GET212 does not improve solubilization of PRO, it is possible that GET212 released from the particles permeabilizes the *C. albicans* cell membrane or wall, allowing for improved uptake of PRO.

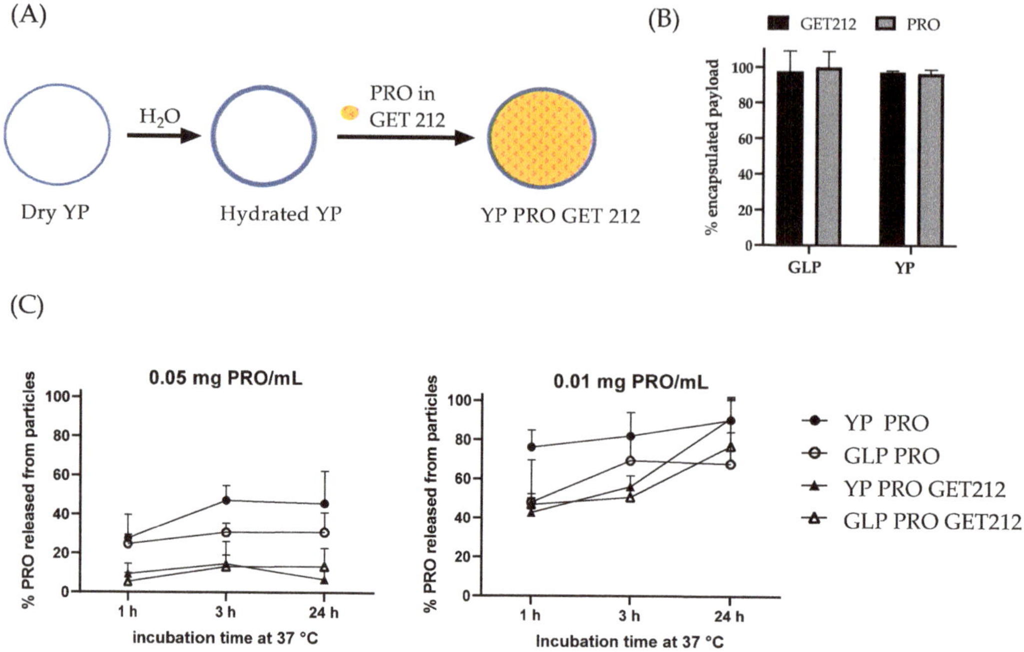

Figure 8. (**A**) Schematics of GET 212 and PRO co-encapsulation in YPs and GLPs, (**B**) GET212 and PRO encapsulation efficiency in YPs and GLPs in samples prepared at 0.055:1.1:1 PRO:GET 212:YP or GLP weight ratio, and (**C**) PRO release in water at 37 °C from control YP and GLP particles containing only PRO and samples co-encapsulating PRO and GET; samples were diluted at 0.05 mg PRO/mL (~2.3× higher concentration than maximum solubility of PRO in water) and at 0.01 mg PRO/mL (~2.2× lower concentration than maximum solubility of PRO in water).

Table 3. In vitro antifungal activity of unencapsulated (free) PRO suspension, particles encapsulating GET or PRO, and particles co-encapsulating both payloads at 0.055:1.1:1 PRO:GET212:particle weight ratio on a susceptible (wild-type) and two azole-resistant *C. albicans* strains. The MIC results represent the average of six replicate experiments. Statistically significant results were obtained between the paired samples indicated with letter superscripts ($^{A–J}$ $p < 0.005$, ^{K} $p < 0.01$, $^{L–P}$ $p < 0.05$).

Sample	Dry Weight Composition			Minimum Inhibitory Concentration (MIC 75%) in µg/mL on Wild-Type and Azole-Resistant *Candida albicans* Strains					
	%PRO	%GET	%YP	Wild-Type		White 1		ATCC 11651	
				PRO	GET	PRO	GET	PRO	GET
Free PRO	100	-	-	$78 \pm 38^{L,M,N}$	-	$500 \pm 2^{B,C,D,E}$	-	417 ± 129^{G}	-
YP GET	-	52.4	47.6	-	330 ± 16	-	327 ± 17	-	327 ± 17
YP PRO 1:1	50	-	50	$11 \pm 4^{A,M}$		$16 \pm 1^{A,D,K}$	-	10 ± 5^{H}	-
YP GET PRO	2.6	51	47.4	3.9 ± 0.1^{L}	82 ± 4	6 ± 2^{B}	121 ± 41	$6 \pm 2^{G,I,P}$	121 ± 39
GLP PRO	50	-	50	$200 \pm 68^{A,N,O}$	-	$175 \pm 68^{E,F,K}$	-	$333 \pm 12^{H,I,J,P}$	-
GLP PRO GET	2.6	51	47.4	8 ± 1^{O}	164 ± 8	$6 \pm 2^{C,F}$	121 ± 39	8 ± 6^{J}	173 ± 112

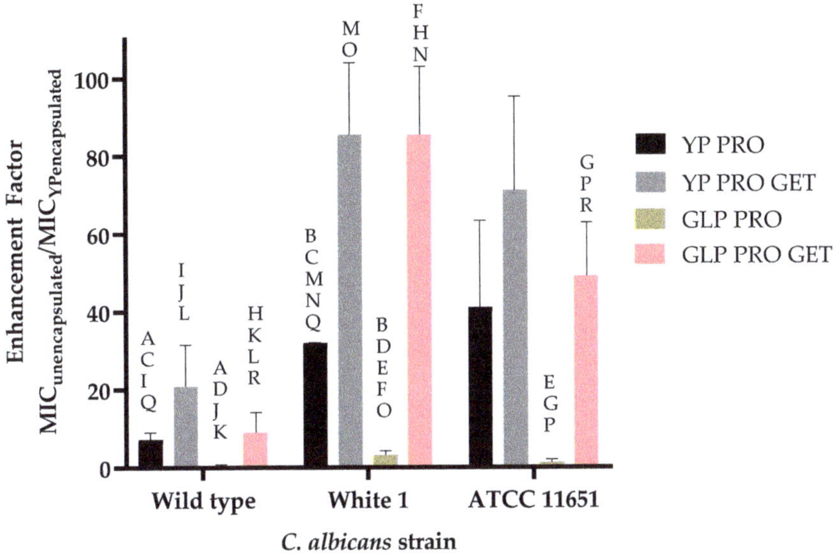

Figure 9. Effect of YP or GLP encapsulation on activity of prothioconazole, reported as the ratio of the MIC 75% of unencapsulated PRO and YP- or GLP-encapsulated PRO samples. Statistically significant results were obtained between the paired samples indicated with letter superscripts (A–E $p < 0.005$, F–H $p < 0.01$, I–R $p < 0.05$).

4. Discussion

Yeast particles offer several advantages as delivery vehicles, such as high payload loading capacity in the hollow cavity of the particles, payload protection from environmental stresses, biocompatibility, biodegradability, and possibility of controlled payload release. Additionally, highly purified forms of YPs developed for pharmaceutical applications provide for receptor-mediated targeted delivery to macrophages and dendritic via interaction of the of β-1,3-d glucan on the particle surface with cellular glucan receptors.

The encapsulation of hydrophobic payloads in YPs occurs by passive diffusion through the YP shell into its hydrophobic interior. This has been previously demonstrated for the efficient encapsulation of terpenes. Here, we show results of high payload loading capacity following the same solvent-free loading method as terpenes for the fungicide tetraconazole and the loading of prothioconazole using a suitable organic solvent for loading in YPs by a diffusion method. The YP TET and YP PRO samples prepared at 1:1 PRO or TET:YP weight ratios can be used to generate stable YP suspensions at 150 g YP/L of YP fungicide (15% *w*/*v*). These loading levels are higher than typical concentrations of TET in commercial products (e.g., 11.6% TET in Eminent 125 from Isagro [35]). Commercial PRO products like 4LSelect™ from Albaugh or Proline from Bayer Crop contain 41% TET *w*/*v* [36]. Although the 15% PRO in a formulation at 150 g YP/L is lower than commercial products, the preparation of YP PRO formulations has the advantages of eliminating the use of surfactants, the potential targeting of azole-resistant strains and reduction in PRO dosage by co-encapsulation with compounds like terpenes that exhibit a synergistic effect with PRO.

The release of fungicide from YP is based on diffusion out of the particles and is a function of the payload solubility in water. Both TET and PRO remain stably encapsulated in YP fungicide aqueous suspensions above the maximum solubility of the fungicide in water and release upon dilution below the maximum solubility of the fungicide in water (Figure 3). This encapsulation stability and diffusion-based release process is similar to the previously developed and commercialized YP terpene formulations for agricultural applications and is the basis for sustained active release.

The encapsulated TET and PRO samples exhibit slightly better activity than unencapsulated TET or PRO on azole-sensitive *C. albicans* and 5–20-fold better activity on azole-resistant strains (Table 2, Figure 4). The reduction in MIC for both YP TET and YP PRO on the azole-sensitive strain is likely due to delivering a more homogeneous fungicide suspension in YPs than the unencapsulated samples. The effect on azole-resistant strains was significantly higher (>10-fold enhancement) for PRO samples than TET, as none of the *C. albicans* strains used in this study showed strong resistance to unencapsulated TET, and further work evaluating the effect of YP encapsulation was focused on PRO samples. The enhanced activity of PRO in YP PRO samples is likely due to the interaction of PRO with saponifiable lipids in YPs that could improve PRO release from YPs and uptake into the *C. albicans* strains. The effect of the saponifiable lipids in YPs was confirmed by encapsulation of PRO in GLPs, a highly purified form of YPs that do not contain these saponifiable lipids. The growth inhibition dose response curves (Figure 5) show reduced activity of GLP PRO on both sensitive and azole-resistant strains compared to unencapsulated PRO or YP PRO samples.

The use of GLPs is critical for encapsulation of hydrophobic payloads for pharmaceutical applications. To trigger PRO release from GLPs and potentially target PRO to azole-resistant strains, we evaluated the combination of terpenes (GET212) with PRO. There are several advantages of using terpenes as chemosensitizers for antifungal application, such as (1) terpenes being natural compounds, (2) possessing antifungal activity but not at low enough concentrations like antifungal drugs, and (3) the main functions of terpenes being to permeabilize the cell wall and disrupt the fungal stress response, which can result in an additive or synergistic interaction with the primary antifungal drug [12]. Checkerboard assays of unencapsulated PRO and GET212 or YP-encapsulated drugs indicate that there is a synergistic effect between PRO and the GET212 terpene mixture at weight ratios from 10:1 to 20:1 GET212:PRO. Samples co-encapsulating GET212 and PRO at a 20:1 ratio in YPs or GLPs are active on sensitive and azole-resistant strains. The combination of GET212 and PRO offers the advantages of (1) ~10-fold reduction in the PRO weight content required for encapsulation (50% in YP PRO, 5.5% in YP or GLP PRO GET) and (2) overcomes the limitation of GLP PRO not being active due to the lack of saponifiable lipids, as the GET212 mixture improves PRO activity. The release kinetics results (Figure 8C) show that GET212 inhibited PRO release from YP or GLP PRO GET212 compared to YP or GLP PRO. These results do not correlate with the improved activity of YP and GLP PRO GET on *C. albicans*. It was expected that GET would improve PRO release from the particles and uptake by the target strains. However, it is possible that both payloads do not have to release from particles simultaneously to show synergy—a two-step mechanism could be taking place with (1) GET releasing faster from particles and disrupting *C. albicans* cell wall membrane and (2) PRO releasing slower and getting taken up by *C. albicans* due to a leaky membrane.

The methods developed to encapsulate PRO and TET in YPs or the combination of PRO and GET212 in YP and GLPs could be used for other synergistic antimicrobial hydrophobic drug combinations. The development of these materials is significant for the use of antimicrobial YP and GLP compositions for agricultural and pharmaceutical applications to combat the growing problem of drug-resistance.

5. Conclusions

Yeast particles can be used for the encapsulation of fungicides with high payload loading capacity, encapsulation efficiency and encapsulation stability. The presence of saponifiable lipids in YPs enhanced the activity of encapsulated PRO on azole-resistant *C. albicans* strains (YP PRO is >10-fold more active than unencapsulated PRO). The combination of PRO and terpenes (GET212) at a ratio of 20:1 GET212:PRO encapsulated in YPs or GLPs shows synergistic activity on both sensitive and azole-resistant *C. albicans* strains. These YP and GLP PRO GET212 combinations possess properties of interest for the development of YP and GLP materials for the targeting of fungal drug resistance in agricultural and pharmaceutical applications. The methods presented in this paper can be further investigated for other antimicrobial applications targeting drug resistance.

6. Patents

Hyperloaded Yeast Cell Wall Particle and Uses Thereof, G.R. Ostroff, E.R. Soto and F. Rus. US Patent App. 63/346,012. 26 May 2022.

Supplementary Materials: The following supporting information can be downloaded at: https://www.mdpi.com/article/10.3390/jfb15080203/s1, Figure S1: En-capsulation efficiency of tetraconazole (TET) encapsulated in YPs at a target weight ratio of 3:1 TET:YP; Figure S2: Encapsulation efficiency of prothioconazole (PRO) encapsulated in GLPs at a target weight ratio of 1:1 PRO:YP.

Author Contributions: Conceptualization, E.R.S. and G.R.O.; methodology, E.R.S. and F.R.; resources, G.R.O.; writing—original draft preparation, E.R.S.; writing—review and editing, G.R.O.; funding acquisition, G.R.O. All authors have read and agreed to the published version of the manuscript.

Funding: Portions of this research were funded by a Sponsored Research Agreement from Eden Research plc. Eden-UMMS SRA 03012015.

Data Availability Statement: The original contributions presented in the study are included in the article/Supplementary Materials, further inquiries can be directed to the corresponding authors.

Conflicts of Interest: All authors are inventors of US Patent App. 63/346,012, Hyperloaded Yeast Cell Wall Particle and Uses Thereof assigned to University of Massachusetts Medical School. The patent is managed by the Bridge Office of Technology Management and the Conflicts of Interest policies of the University of Massachusetts.

References

1. Perfect, J.R.; Ghannoum, M. Emerging Issues in Antifungal Resistance. *Infect. Dis. Clin.* **2020**, *34*, 921–943. [CrossRef] [PubMed]
2. Fisher, M.C.; Alastruey-Izquierdo, A.; Berman, J.; Bicanic, T.; Bignell, E.M.; Bowyer, P.; Bromley, M.; Brüggemann, R.; Garber, G.; Cornely, O.A.; et al. Tackling the emerging threat of antifungal resistance to human health. *Nat. Rev. Microbiol.* **2022**, *20*, 557–571. [CrossRef] [PubMed]
3. Verweij, P.E.; Lucas, J.A.; Arendrup, M.C.; Bowyer, P.; Brinkmann, A.J.; Denning, D.W.; Dyer, P.S.; Fisher, M.C.; Geenen, P.L.; Gisi, U.; et al. The one health problem of azole resistance in Aspergillus fumigatus: Current insights and future research agenda. *Fungal Biol. Rev.* **2020**, *34*, 202–214. [CrossRef]
4. Arastehfar, A.; Gabaldón, T.; Garcia-Rubio, R.; Jenks, J.D.; Hoenigl, M.; Salzer, H.J.F.; Ilkit, M.; Lass-Flörl, C.; Perlin, D.S. Drug-resistant fungi: An emerging challenge threatening our limited antifungal armamentarium. *Antibiotics* **2020**, *9*, 877. [CrossRef] [PubMed]
5. Woods, M.; McAlister, J.A.; Geddes-McAlister, J. A One Health approach to overcoming fungal disease and antifungal resistance. *WIREs Mech. Dis.* **2023**, *15*, 1–21. [CrossRef] [PubMed]
6. Vitiello, A.; Ferrara, F.; Boccellino, M.; Ponzo, A.; Cimmino, C.; Comberiati, E.; Zovi, A.; Clemente, S.; Sabbatucci, M. Antifungal Drug Resistance: An Emergent Health Threat. *Biomedicines* **2023**, *4*, 1063. [CrossRef]
7. Stevenson, E.M.; Gaze, W.H.; Gow, N.A.R.; Hart, A.; Schmidt, W.; Usher, J.; Warris, A.; Wilkinson, H.; Murray, A.K. Antifungal Exposure and Resistance Development: Defining Minimal Selective Antifungal Concentrations and Testing Methodologies. *Front. Fungal Biol.* **2002**, *3*, 1–15. [CrossRef] [PubMed]
8. Bhattacharya, S.; Sae-Tia, S.; Fries, B.C. Candidiasis and mechanisms of antifungal resistance. *Antibiotics* **2020**, *9*, 312. [CrossRef] [PubMed]
9. De Oliveira, H.C.; Bezerra, B.T.; Rodrigues, M.L. Antifungal Development and the Urgency of Minimizing the Impact of Fungal Diseases on Public Health. *ACS Bio Med. Chem. Au* **2023**, *3*, 137–146. [CrossRef] [PubMed]
10. Assress, H.A.; Selvarajan, R.; Nyoni, H.; Mamba, B.B.; Msagati, T.A.M. Antifungal azoles and azole resistance in the environment: Current status and future perspectives—A review. *Rev. Environ. Sci. Biotechnol.* **2021**, *20*, 1011–1041. [CrossRef]
11. de Lima Silva, M.G.; Ferreira de Lima, L.; Alencar Fonseca, V.J.; Santos da Silva, L.Y.; Donelardy, A.C.C.; de Almeida, R.S.; Oliveira-Tintino, C.D.d.M.; Martins, A.O.B.P.B.; Ribeiro-Filho, J.; Morais-Braga, M.F.B.; et al. Enhancing the Antifungal Efficacy of Fluconazole with a Diterpene: Abietic Acid as a Promising Adjuvant to Combat Antifungal Resistance in Candida spp. *Antibiotics* **2023**, *12*, 1565. [CrossRef] [PubMed]
12. Campbell, B.C.; Chan, K.L.; Kim, J.H. Chemosensitization as a means to augment commercial antifungal agents. *Front. Microbiol.* **2012**, *3*, 20983. [CrossRef] [PubMed]
13. Li, X.N.; Zhang, L.M.; Wang, Y.Y.; Zhang, Y.; Jin, Z.H.; Li, J.; Wang, R.R.; Xiao, W.L. SWL-1 Reverses Fluconazole Resistance in Candida albicans by Regulating the Glycolytic Pathway. *Front. Microbiol.* **2020**, *11*, 572608. [CrossRef] [PubMed]

14. Gil, F.; Laiolo, J.; Bayona-Pacheco, B.; Cannon, R.D.; Ferreira-Pereira, A.; Carpinella, M.C. Extracts from Argentinian native plants reverse fluconazole resistance in Candida species by inhibiting the efflux transporters Mdr1 and Cdr1. *BMC Complement. Med. Ther.* **2022**, *22*, 1–12. [CrossRef]
15. Soto, E.R.; Ostroff, G.R. Characterization of Multilayered Nanoparticles Encapsulated in Yeast Cell Wall Particles for DNA. *Bioconjug. Chem.* **2008**, *19*, 840–848. [CrossRef] [PubMed]
16. Mirza, Z.; Soto, E.R.; Dikengil, F.; Levitz, S.M.; Ostroff, G.R. Beta-Glucan Particles as Vaccine Adjuvant Carriers. In *Vaccines for Invasive Fungal Infections*; Humana Press: New York, NY, USA, 2017; pp. 143–157. [CrossRef]
17. Soto, E.R.; Rus, F.; Ostroff, G.R. Yeast Particles Hyper-Loaded with Terpenes for Biocide Applications. *Molecules* **2022**, *27*, 3580. [CrossRef] [PubMed]
18. Soto, E.R.; Rus, F.; Mirza, Z.; Ostroff, G.R. Yeast Particles for Encapsulation of Terpenes and Essential Oils. *Molecules* **2023**, *28*, 2273. [CrossRef]
19. Bastos, R.; Oliveira, P.G.; Gaspar, V.M.; Mano, J.F.; Coimbra, M.A.; Coelho, E. Brewer's yeast polysaccharides—A review of their exquisite structural features and biomedical applications. *Carbohydr. Polym.* **2022**, *277*, 118826. [CrossRef] [PubMed]
20. Tan, Y.; Chen, L.; Li, K.; Lou, B.; Liu, Y.; Liu, Z. Yeast as carrier for drug delivery and vaccine construction. *J. Control. Release* **2022**, *346*, 358–379. [CrossRef]
21. Wu, Y.; Li, P.; Jiang, Z.; Sun, X.; He, H.; Yan, P.; Xu, Y.; Liu, Y. Bioinspired yeast-based Β-glucan system for oral drug delivery. *Carbohydr. Polym.* **2023**, *319*, 121163. [CrossRef] [PubMed]
22. Yang, F.; Shang, S.; Qi, M.; Xiang, Y.; Wang, L.; Wang, X.; Lin, T.; Hao, D.; Chen, J.; Liu, J.; et al. Yeast glucan particles: An express train for oral targeted drug delivery. *Int. J. Biol. Macromol.* **2023**, *253*, 127131. [CrossRef] [PubMed]
23. de Macedo, L.S.; Sousa de Pinho, S.; Duarte Silva, A.J.; de Moura, I.A.; Flores Espinoza, B.C.; Viana Invencao, M.C.; Silva Novis, P.V.; Turiah Machado da Gama, M.A.; do Nascimento Carvalho, M.; Sales Leal, L.R.; et al. Understanding yeast shells: Structure, properties and applications. *ADMET DMPK* **2024**, *12*, 299–317. [CrossRef] [PubMed]
24. He, L.; Zhu, Z.; Qi, C. β-glucan—A promising immunocyte-targeting drug delivery vehicle: Superiority, applications and future prospects. *Carbohydr. Polym.* **2024**, *339*, 122252. [CrossRef] [PubMed]
25. Lin, H.; Han, R.; Wu, W. Glucans and applications in drug delivery. *Carbohydr. Polym.* **2024**, *322*, 121904. [CrossRef] [PubMed]
26. Tan, C.; Huang, M.; McClements, D.J.; Sun, B.; Wang, J. Yeast cell-derived delivery systems for bioactives. *Trends Food Sci. Technol.* **2021**, *118*, 362–373. [CrossRef]
27. White, T.C.; Holleman, S.; Dy, F.; Mirels, L.F.; Stevens, D.A. Resistance mechanisms in clinical isolates of Candida albicans. *Antimicrob. Agents Chemother.* **2002**, *46*, 1704–1713. [CrossRef] [PubMed]
28. Sultanbawa, Y.; Cusack, A.; Currie, M.; Davis, C. An Innovative Microplate Assay to Facilitate the Detection of Antimicrobial Activity in Plant Extracts. *J. Rapid Meth. Aut. Mic.* **2009**, *17*, 519–534. [CrossRef]
29. Yadav, B.; Wennerberg, K.; Aittokallio, T.; Tang, J. Searching for Drug Synergy in Complex Dose-Response Landscapes Using an Interaction Potency Model. *Comput. Struct. Biotechnol. J.* **2015**, *13*, 504–513. [CrossRef] [PubMed]
30. Zheng, S.; Wang, W.; Aldadooh, J.; Malyutina, A.; Shadbahr, T.; Tanoli, Z.; Pessia, A.; Tang, J. SynergyFinder Plus: Toward Better Interpretation and Annotation of Drug Combination Screening Datasets. *Genom. Proteom. Bioinform.* **2022**, *20*, 587–596. [CrossRef] [PubMed]
31. Garcia, A.A.; Oden, P.; Knipping, U.; Ostroff, G.; Druyour, R. Characterization of a Β-glucan particle using the scanning tunneling and atomic force microscopes. In *Synthetic Microstructures in Biological Research*; Schnur, J.M., Peckerer, M., Eds.; Plenum Press: New York, NY, USA, 1992; pp. 131–141.
32. PubChem. Available online: https://pubchem.ncbi.nlm.nih.gov/ (accessed on 15 April 2024).
33. Tleuova, A.B.; Wielogorska, E.; Talluri, V.S.S.L.P.; Stepanek, F.; Elliot, C.T.; Grigoriev, D.O. Recent advances and remaining barriers to producing novel formulations of fungicides for safe and sustainable agriculture. *J. Control. Release* **2020**, *326*, 468–481. [CrossRef]
34. Franklin, L.; Ostroff, G. Nematicidal Compositions and Methods of Using Them. U.S. Patent No 10,004,229, 26 June 2018.
35. Available online: https://www.isagro.com/static/upload/emi/eminent-125.pdf (accessed on 20 May 2024).
36. Available online: https://www.cropscience.bayer.us/d/proline-480-sc-fungicide (accessed on 20 May 2024).

Journal of *Functional Biomaterials*

MDPI

Review

Applications of Bioactive Strontium Compounds in Dentistry

Mohamed Mahmoud Abdalla [1,2], Osama Sayed [3], Christie Ying Kei Lung [4], Vidhyashree Rajasekar [1] and Cynthia Kar Yung Yiu [1,*]

1 Paediatric Dentistry, Faculty of Dentistry, The University of Hong Kong, Hong Kong, China; mohamabd@hku.hk (M.M.A.); vidhya@connect.hku.hk (V.R.)
2 Dental Biomaterials, Faculty of Dental Medicine, Al-Azhar University, Cairo 11651, Egypt
3 Faculty of Dentistry, Fayoum University, Faiyum 63514, Egypt; osamamedhat@outlook.com
4 Restorative Dental Sciences, Faculty of Dentistry, The University of Hong Kong, Hong Kong, China; cyklung@hku.hk
* Correspondence: ckyyiu@hku.hk

Abstract: Divalent cations have captured the interest of researchers in biomedical and dental fields due to their beneficial effects on bone formation. These metallic elements are similar to trace elements found in human bone. Strontium is a divalent cation commonly found in various biomaterials. Since strontium has a radius similar to calcium, it has been used to replace calcium in many calcium-containing biomaterials. Strontium has the ability to inhibit bone resorption and increase bone deposition, making it useful in the treatment of osteoporosis. Strontium has also been used as a radiopacifier in dentistry and has been incorporated into a variety of dental materials to improve their radiopacity. Furthermore, strontium has been shown to improve the antimicrobial and mechanical properties of dental materials, promote enamel remineralization, alleviate dentin hypersensitivity, and enhance dentin regeneration. The objective of this review is to provide a comprehensive review of the applications of strontium in dentistry.

Keywords: strontium; dentistry; radiopacity; regeneration; hypersensitivity; antibacterial

Citation: Abdalla, M.M.; Sayed, O.; Lung, C.Y.K.; Rajasekar, V.; Yiu, C.K.Y. Applications of Bioactive Strontium Compounds in Dentistry. *J. Funct. Biomater.* **2024**, *15*, 216. https://doi.org/10.3390/jfb15080216

Academic Editor: Valeria Ambrogi

Received: 8 July 2024
Revised: 26 July 2024
Accepted: 30 July 2024
Published: 31 July 2024

1. Introduction

Strontium (Sr) belongs to group 2 of the periodic table and has two electrons in its valence shell. It loses two electrons to become Sr^{2+}, a divalent cation, after the formation of strontium compounds. These compounds have been used in dental materials for a long time. Strontium was discovered in Strontian, a village in Scotland, in 1790, and successfully isolated in 1808 by Davy [1,2]. With an atomic number of 38 and atomic mass of 87, Sr is classified as an alkaline earth metal in the periodic table, belonging to the same group as calcium (Ca) [3]. It is widely available and constitutes approximately 0.02–0.03% of the Earth's crust. Due to strontium's chemical similarity to Ca, fruits and vegetables absorb it from the soil [4]. As a result, the Sr content in our diet reflects the level found in the soil, albeit at a relatively low concentration compared to Ca. For every 1 mg of Sr, there is 125 mg of Ca [5]. The ratio of Ca to Sr in bodily tissues and fluids mirrors that of the dietary intake [6]. Although Sr is not considered an essential element from a biological standpoint [4], it comprises about 320 mg of the human body and is found in trace amounts in teeth [7]. Our daily intake of Sr ranges from 2.1 to 2.4 mg, primarily from food and water sources [8].

In 1870, Papillon made a significant discovery regarding the biological role of Sr through an in vitro study. After feeding Sr to a pigeon, the analysis demonstrated that it can naturally incorporate into bone [9] through ion exchange with Ca [10,11]. The bone-seeking property of Sr led to further investigations into its systemic effects, revealing that, similarly to Ca, Sr has the ability to influence cardiac contractility, modulate parathyroid secretions, trigger uterine contractions, and incorporate into tooth structure [12,13]. Despite the isomorphism of Sr hydroxyapatite and Ca hydroxyapatite, hydroxyapatite crystals show a preference for Ca ions over Sr ions due to the Sr ion being bigger (1.13 Å) than Ca (0.99 Å) [14]. Due to its similarity with Ca, Sr has been used as a substitute in many Ca-containing compounds [14–16].

Interestingly, the benefits of Sr have expanded its applications in dentistry. Strontium is commonly used as a radiopacifier in restorative materials [17], an antimicrobial agent in combination with fluoride [18], and for alleviating dentin hypersensitivity when incorporated in some dentifrices [19]. Furthermore, Sr has been found to stimulate the osteo/odontogenic differentiation of mesenchymal stems cell and human dental pulp stem cells [20,21]. In dental materials, Sr is incorporated in various formulations such as Sr oxide, Sr carbonate, Sr chloride, Sr acetate, and Sr fluoride. It is also used as a substitute for Ca in numerous biomaterials.

Considering the significant benefits of Sr in different dental and biomedical applications and the lack of literature reviews that discuss all the applications of strontium in dentistry (Figure 1), this review aimed to provide a comprehensive overview of the uses and benefits of strontium, the mechanisms of its biological interaction, and its applications in dentistry.

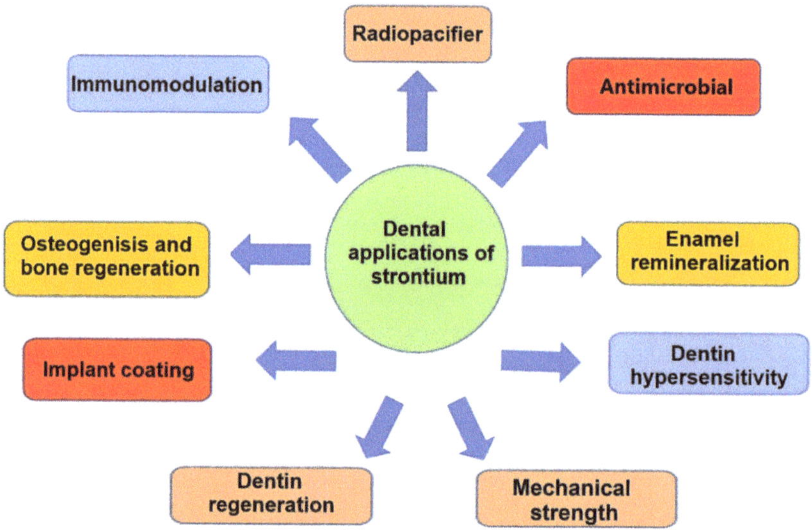

Figure 1. Applications of strontium in dentistry.

2. Radiopacifying Properties of Strontium Compounds

Strontium compounds have been widely used to improve the radiopacity of various materials. Ideally, dental materials should have sufficient radiopacity to be detected by X-ray during dental inspection and distinguished from adjacent carious tooth structure [22]. According to ISO standards and ADA recommendations, the minimal radiopacity of a 1 mm material should be equivalent to 1 mm of aluminum [22,23]. Strontium, a heavy metal with an atomic mass of 87.62 [3], has a higher density (Figure 2) and consequently higher radiopacity when detected by X-ray [24].

Strontium also has a low systemic toxic effect [25], making it an appropriate element for enhancing the radiopacity of many dental materials. Due to its similarity with Ca, Sr has the capability to replace Ca in dental materials [25]. In the 1980s, a patent was filed defining Sr as a radiopacifying agent in glass [26], which is utilized as a filler in dental resin composite restorative materials. In addition, Sr was added to glass ionomer cement to replace some of the Ca and act as a radiopacifying element. It is noteworthy that the incorporation of Sr did not adversely affect the visual opacity of glass ionomer cement [22].

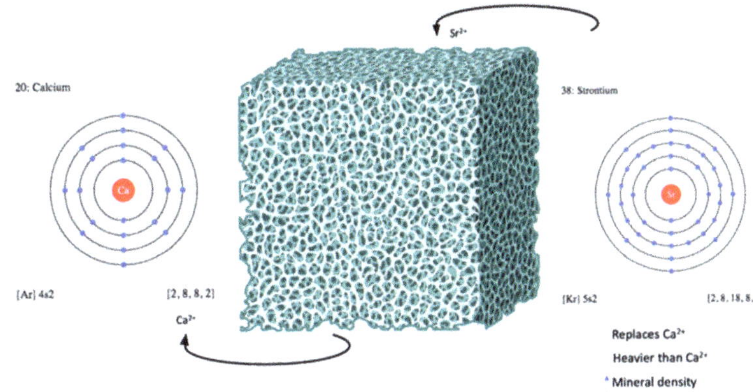

Figure 2. Strontium properties and calcium replacement.

Similarly, incorporating Sr fluoride as a radiopacifying agent in MTA endodontic cement and calcium silicate cement has been shown to significantly enhance radiopacity [24,27]. In a study investigating the impact of incorporating Sr into Biodentine, it was observed that higher radiopacity and compressive strength were achieved with a high fluoride + Sr bioglass modification, although this modification also resulted in an extended setting time [28]. In the context of dental adhesive systems, the inadequate opacity can make it challenging to identify secondary caries radiographically. One study addressed this issue by adding Sr to adhesive resin, which increased the resin's opacity without significantly affecting the degree of conversion or Young's modulus [29]. Another study found that the radiopacity of glass ceramics was enhanced with the addition of Sr, attributed to the precipitation of Sr fluorapatite crystals in the structure, instead of calcium fluorapatite, due to the larger ion and higher atomic mass of Sr2+ compared to Ca2+ in the apatite crystal [30]. Additionally, incorporating Sr carbonate into calcium phosphate cement has been shown to significantly improve radiopacity compared to using calcium phosphate alone [31].

3. Antimicrobial Effects of Strontium Compounds

Strontium compounds have demonstrated varying antibacterial effects against different bacterial strains, making them suitable for enhancing the antimicrobial properties of medical devices. These compounds can inhibit bacterial growth, impede the permeability of the cytoplasmic membrane, and slow the replication of bacterial chromosomes and cell metabolism [32,33]. In recent years, Sr has been incorporated into dental and orthopedic biomaterials to minimize the risk of secondary caries in dental restorations [34]. It has also been used in injectable bone cement for minimally invasive delivery of antimicrobials in vertebral compression fractures, aiming to inhibit microbial contamination and prevent implant-related infections [35]. Furthermore, Sr-coated implants have been explored for the prevention of peri-implantitis [36].

In dental restorations, the antimicrobial function provided by the restorative material is beneficial in reducing pulpal damage and improving durability of the restoration. Sr-containing glass ionomers have been developed commercially, where Sr ions effectively replace Ca in the composition. In this type of restoration, Sr can be used as a substitute for Ca by incorporating SrO and SrF_2 instead of CaO and CaF_2 in the glass-forming mixture [22]. These components are usually insoluble in neutral environments, but in acidic conditions due to oral bacteria, Sr ions are released in greater quantities, exerting antimicrobial effects against bacterial pathogens [37].

Microbiological analysis of secondary caries biofilm has identified that *Streptococcus mutans* and *Actinomyces* are dominant microflora, followed by *porphyromonas gingivalis* [38]. Glass ionomer cements containing Sr exhibited significant antibacterial activity against *Streptococcus mutans* (ATCC 25175) and *Actinomyces viscosus* (ATCC 19246), with Sr contributing particularly to the antibacterial activity against A. viscosus [33]. However, several investigated cements had no detectable antibacterial action against this species. The antibacterial activity of Sr-substituted bioglass on *Porphyromonas gingivalis* and *A. actinomycetemcomitans* was also investigated, and it was found that the antimicrobial activity increased considerably as Sr substitution increased from 0% to 100% [39].

Furthermore, research on injectable Sr-releasing bone cements based on bioglass and polyacrylic acid demonstrated a significant bactericidal effect on *Staphylococcus aureus* (NCIMB6571) and *Streptococcus faecalis* (NCIMB775) [35]. However, Dabsie et al. found that Sr has no inherent antibacterial properties, suggesting that its synergistic reaction with fluoride may enhance the antibacterial activity of dental restorations [40]. The proposed mechanism of antibacterial action of Sr could be due to interference with membrane stability, cellular proteins, and enzymes, as well as the formation of reactive oxygen species that cause severe cell damage [36]. It is also predicted that Sr facilitates fluoride entry by disrupting microbial cell membranes, thereby boosting its effect. Furthermore, as a divalent cation, Sr can alter the composition and structure of dental biofilm by interacting with exopolysaccharides (EPS) [41]. In conclusion, there is no clear consensus on whether Sr has an inhibitory effect on bacteria. However, it is suggested that the Sr2+ ion has a synergistic effect with F- in inhibiting bacterial activity.

4. Enamel Remineralization with Strontium Compounds

The remineralization of enamel is a natural, spontaneous healing process primarily based on the deposition of Ca and phosphate ions present in saliva, forming a new layer on the demineralized enamel surface. Modern caries management strategies are designed to prevent and halt the progression of caries [42]. The role of Sr in managing dental caries has been investigated for decades. However, the published data have shown conflicting results regarding whether Sr alone has a remineralizing effect or if its action is synergistic when combined with fluoride [42–46]. The discrepancies in the findings may be attributed to variations in the concentrations and compounds used. Epidemiological research has linked Sr to a lower prevalence of caries [47].

A study conducted by Thuy et al. [48] showed that using a remineralization solution with 10 ppm Sr resulted in the highest rate of mineral gain compared to other Sr concentrations. However, this study did not compare the mineralizing solutions with a control solution without Sr. Similarly, Wang et al. [49] incorporated different Sr concentrations into an acidic solution containing enamel samples and assessed the effect on enamel demineralization. They found that Sr ions at a concentration of 10-2 M could reduce the erosive acidic effect on enamel, leading to less reduction in enamel hardness and phosphorus ion dissolvement compared to acidic solutions without the Sr ions. However, Yassen et al. [47] explored the remineralizing potential of low concentrations of Sr (10, 15 ppm), and found that Sr at a concentration of 10 ppm could remineralize enamel only when combined with fluoride. The same study also suggested that F might be the major cause of remineralization. The integration of fluoride and Sr was proposed to reduce apatite solubility and improve crystallinity in enamel [50].

Another possibility is that carrying fluoride in an uncharged complex with Sr increases the diffusion rate of fluoride through enamel [51]. Studies investigating the remineralizing potential of different concentrations of F alone and in combination with 10 ppm Sr revealed that the enamel remineralization was significantly enhanced with the incorporation of Sr at 0.1 and 0.05ppm F concentrations, suggesting a synergistic enhancement in remineralization due to the interaction between Sr and fluoride [52]. Furthermore, toothpaste containing Sr has been found to increase the Sr content in enamel and reduce enamel solubility, suggesting that the positively charged Sr forms a layer on enamel known as the stern layer, which can mitigate the effects of citric acid [53].

One study substituted Ca in nano-hydroxyapatite with Sr to address issues related to large crystals, acid reactivity, and material strength. This substitution resulted in increased crystallinity and reduced particle size, which facilitated diffusion through small carious lesions and white-spot areas, ultimately leading to improved enamel remineralization [54]. Similarly, Rajendran et al. [55] synthesized Sr-substituted nano-hydroxyapatite in a paste form for better handling and observed enhanced enamel remineralization. Another interesting study by Dai et al. [56] examined the effect of Sr-doped bioactive glass and fluoride on apatite crystal formation during mineralization. The study demonstrated that Sr could replace Ca in the hydroxyapatite crystal lattice, forming an Sr–hydroxyapatite that is bioactive and can bond easily to bony tissues. The formation of two phases, Sr–hydroxyapatite and Sr–fluorhydroxyapatite, was observed, which exhibited greater chemical stability and resistance to dissolution. The study also noted that the larger radius of Sr compared to Ca led to the formation of distorted crystals, but at the low Sr concentration used in their study, this distortion was not observed. The presence of F further enhanced the stability of the structure. The study by Dai et al. highlighted the potential formation and stability of Sr–hydroxyapatite. However, it should be noted that in vitro models do not fully mimic the oral environment. Future research is recommended to determine the exact effect of Sr on remineralization and whether F plays a key role in the process or if Sr can enhance remineralization on its own.

5. Managing Dentin Hypersensitivity with Strontium Compounds

Exposed dentinal tubules can cause patients to experience a sharp, short pain in response to external factors such as thermal, chemical, or tactile stimuli. This phenomenon is known as dentin hypersensitivity (DH) and affects nearly one-third of adults globally [57,58]. The use of Sr compounds for managing DH dates back to 1956, when Pawlowska et al. [59] reported the efficacy of a 25% Sr chloride solution in treating DH. They interpreted this result as an indication that Sr salt has a special influence on both the nervous system and the hard tissues, emphasizing its ability to "transform the soft surface" of carious teeth into a hard and smooth structure. Since then, several studies have examined the effectiveness and mechanism of Sr in the treatment of DH.

The hydrodynamic theory (shown in Figure 3) proposes a widely accepted explanation for the mechanism of the DH pain mechanism. When adequate stimuli are applied to exposed dentine, the inward or outward flow of dentinal fluids in the tubule network increases, stimulating the baroreceptor nerve fibers in and around the dentine–pulp interface [60]. Histological findings that correlate DH with the percentage of exposed dentinal tubules in sensitive teeth compared to sound teeth support this explanation [61]. To manage DH symptoms, two methodologies are often used. The first is to properly occlude the dentinal tubules to prevent intratubular fluid movement, which has the benefit of a quick onset [62]. The second approach acts on desensitizing the nerve ending by modifying the extracellular potassium concentration to reduce its excitability [63]. Two different approaches have been used to evaluate the effectiveness of Sr in managing DH. The first approach involves assessing the permeability and morphological changes of dentin in vitro. The second approach involves evaluating the pain-relieving effect of Sr in vivo.

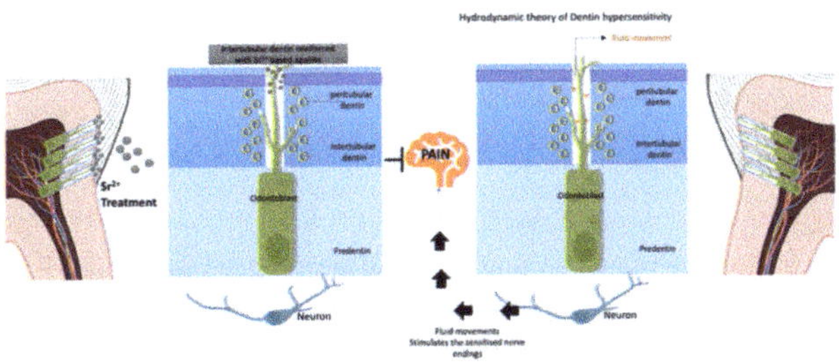

Figure 3. Hydrodynamic theory and mechanism of strontium's effect on occluding dentinal tubules.

Many studies have investigated how Sr works to occlude dentinal tubules and manage DH (Table 1). As an alkaline earth metal element, Sr is expected to have a strong natural absorptive ability for calcified tissues, particularly dentine, due to its high organic composition [64]. It induces precipitations into organic connective tissues, including the odontoblastic processes, to form a sealing film that can occlude open dentinal tubules. This inhibits dentinal fluid movement and helps alleviate DH [65]. Saeki et al. [66] confirmed that Sr can occlude dentinal tubules, as a thick layer containing Sr reaching 20 μm into the tubules was found. Abrasives were not used in the investigation, so the occlusion observed was solely attributable to the impact of Sr. However, this dentinal occlusion was weak and water-soluble, as the layer was undetectable after samples were immersed in DI water. This may explain why Oberg et al. [67] reported that a 10% Sr chloride gel did not show significant dentinal tubular occlusion compared to a negative control group. The analysis was performed after intense rinsing with running distilled water for five minutes.

On the other hand, combining Sr with abrasives in dentifrices has produced a more stable dentinal plug that resists acid and water abrasion, particularly in silica-based products. Olley et al. [68] reported that 8% Sr acetate-based dentifrice in silica base induced dentine tubular occlusion that was resistant to erosive dietary intake with narrower dentine tubules. In contrast, the control paste (1450 ppm sodium fluoride in a silica base) resulted in significantly more open dentinal tubules. Although many studies have found that the occluding materials in theses dentifrices is primarily silica with a few Sr surface deposits [69–72], the presence of Sr along with silica appears to be related to forming an acid-resistant dentinal occlusion. However, many dentifrices may contain silica-based abrasives and cannot induce dentinal occlusion. This may be due to the absence of sodium lauryl sulfate (SLS) in Sr-based dentifrices, as SLS may compete with silica for attachment to dentine [73].

It is interesting to note that Sr acetate has shown greater clinical effectiveness compared to Sr chloride, and it is also compatible with fluoride and potassium nitrate [74]. However, Dotta et al. [75] conducted a study using strontium carbonate and strontium-substituted calcium carbonate nanoparticles instead of strontium acetate and compared them to calcium carbonate. The hypothesis was that the use of Sr–carbonate nanoparticles would benefit from both the effects of strontium on dentin mineralization and the abrasive properties of carbonates. The study concluded that the performance of the strontium-containing nanoparticles, in terms of tubule obliteration and resistance to acid attack, was better than calcium carbonate and even better than the commercially available product Sensodyne® Rapid Relief. The authors suggested that the selective binding of strontium carbonate and strontium substituted calcium carbonate nanoparticles to the dentin surface is a better explanation for these improved characteristics, rather than simply blocking the exposed dentin tubules and desensitizing the pulp nerve.

Recently, Sr has been incorporated in bioactive glass formulations for the treatment of DH. Bioactive glass is known for its ability to induce the formation of an apatite layer in the presence of body fluid such as saliva [76,77]. It is primarily composed of specific proportions of SiO_2, Na_2O, CaO, and P_2O_5 [78]. Xia et al. [79] used a toothpaste containing 10 wt. % Sr-substituted Ca phosphate spheres (SCPSs) and found that after one day, the SCPSs had penetrated the dentinal tubules and apatite crystallites had formed on dentin surfaces. After 7 days, the dentin surfaces were completely covered by newly mineralized apatite, and the exposed tubules were fully covered. This freshly mineralized layer consisted of two layers: a porous layer with larger crystals on top and a thick layer beneath. While the porous layer was easily susceptible to mechanical or chemical attacks, the thick layer proved to be more stable [79]. In another study, Acevedo et al. [80] used fluoride varnish as a carrier to apply Sr-incorporated bioactive glass powder on open dentinal tubules. This approach aimed to achieve a synergistic effect in treating DH, resulting in a significant 90% reduction of dentinal permeability.

The efficacy of Sr-containing agents for pain relief in DH is a topic of debate. Martins et al. conducted a review of current clinical research and concluded that Sr is effective only for relieving tactile sensitivity associated with DH [81]. On the other hand, several other reviews (Table 2) have expressed doubts about the significance of Sr in treating

DH [82–87]. However, a recent randomized clinical trial suggested the combination of photobiomodulation (PBM) with Sr can be effective in treating post-bleaching hypersensitivity. Barros et al. [88] reported that over a one-week period, the combination of PBM and 8% Sr acetate effectively reduced dentin sensitivity compared to imitation PBM + toothpaste without an active ingredient, imitation PBM imitation + toothpaste with 8% Sr acetate, and PBM + toothpaste without the active ingredient [89]. This suggests that the combination of PBM and Sr may have potential in providing relief from dentin hypersensitivity, at least in specific scenarios such as post-bleaching sensitivity.

Table 1. In vitro evaluation of Sr efficacy in dentinal tubule occlusion.

Study	Intervention	Outcomes of SEM Analysis
Kodaka et al., 2001 [69]	Sr chloride-based desensitizing toothpaste	91.45% dentinal occlusion after 2 weeks. The occluding material contained artificial silica abrasive within the dentin sludge.
Arrais et al., 2003 [72]	Sr chloride-based desensitizing toothpaste	80.1% dentinal occlusion after 7 days. Deposition of crystal-like structures within the dentinal tubules consisting of Ca carbonate, the abrasive system of the dentifrice.
Banfield et al., 2004 [70]	Sr acetate-based desensitizing toothpaste Sr chloride-based desensitizing toothpaste	90% dentinal occlusion immediately. The occluding material was artificial silica abrasive. >70% dentinal occlusion immediately. The occluding material was artificial silica abrasive.
Oberg et al., 2009 [67]	10% Sr chloride gel	Open and partially obliterated dentin tubules like the no treatment group. Only traces of Sr were detected in the peritubular dentin deposits.
Saeki et al., 2016 [66]	Sr acetate-based desensitizing toothpaste	Clear thin layer of silicon covered the dentine surface and openings of dentine tubules.
	10% Sr acetate solution	Thick Sr-containing layer reaching 20 µm into dentinal tubules. After specimens were soaked in DI water, Sr-containing layer could not be detected.
	Sr chloride-based desensitizing toothpaste	50.54% dentinal occlusion after 7 days. The occlusion material was not reported.

Table 2. Systematic reviews evaluating the in vivo efficacy of Sr-containing agents in DH treatment.

Study	Efficacy of Sr-Containing Agents in DH Treatment
Martins et al., 2020 [81]	Sr was effective only for tactile stimulus relief.
Hu et al., 2019 [86]	Similar effects of Sr compared to fluoride, placebo, and potassium-containing toothpastes.
Cruz et al., 2019 [87]	Sr when not combined with potassium was no better than the negative control.
Hu et al., 2018 [85]	Sr when not combined with potassium had no desensitizing activity.
Bae et al., 2015 [84]	There was no statistically significant difference between Sr-containing toothpaste and placebo.
West et al., 2015 [83]	Sr acetate had equivocal pain-relieving effects when compared to arginine and was more effective than fluoride control. There is a lack of high-quality data supporting the use of Sr chloride salts for pain relief in dentine hypersensitivity; additional research is needed to determine whether this salt is useful.
Karim et al., 2013 [82]	There is inadequate evidence to state whether Sr salts per se are effective in reducing DH.

6. Dentin Regeneration with Strontium Compounds

The differentiation of human dental pulp stem cells (HDPSCs) into mature odontoblast-like cells is essential for the process of dentin regeneration. Studies have shown that Sr significantly promotes odontogenic differentiation of HDPSCs. However, there have been few investigations on the odontogenic potential of Sr on HDPSCs. Huang et al. [21] reported that Sr chloride elicited the odontogenic differentiation of HDPSCs by modulating the expression and secretion of dentin sialophosphoprotein (DSPP) and dentin matrix protein 1 (DMP-1) in vitro via the CaSR pathway, which shares similarities with osteoblast differentiation. Sr also enhances matrix production and mineralization. The mineralization activity of HDPSCs was further explored using Sr-modified bioglass, where it was found that the alkaline phosphatase activity increased in a dose-dependent manner compared to the control bioglass, while cellular proliferation was inhibited by the Sr-modified bioglass. However, the mechanism behind this inhibition remains unclear.

Abdalla et al. [20] reported that a calcium strontium silicate compound enhanced HDPSC differentiation, setting time, and cytocompatibility compared to calcium silicate alone. Another study conducted by Lee et al. [90] demonstrated an effective increase in the expression levels of genes related to odontogenic differentiation of HDPSCs (collagen type 1 alpha (COL1 A), DSPP, and DMP-1) when treated with a mesoporous bioglass doped with Sr ions (Sr–MBG) compared to non-Sr-containing MBG. Furthermore, Mandakhbayar et al. [91] reported that a Sr-containing nano-bioactive cement (Sr–NBC) enhanced odontogenic differentiation of HDPSCs in vitro and promoted better dentin regeneration in vivo compared to Sr-free NBC. Another study by Basheer et al. [92], which evaluated the effect of Sr-incorporated tetracalcium phosphate cement (STTCP) on HDPSC mineralization potential, discovered that STTCP resulted in a similar improvement in the mineralization and differentiation of HDPSCs to MTA. Considering the significant stimulatory effect of Sr on the odontogenic differentiation of HDPSCs, Sr should be considered in dental biomaterials for pulp regeneration. The enhanced odontogenic differentiation of HDPSCs by Sr is believed to occur through the regulation of Wnt/β-catenin signaling and TGF-β pathways, as well as the regulation of BMP pathway [21].

7. Osteogenic Effect and Bone Repair Potential of Strontium Compounds

Bone is one of the key organs in the human body that plays vital roles, such as providing a framework for the muscles, assisting with body movements, protecting the internal organs from physical injuries, and most importantly in the production of blood cells. However, several factors can affect the homeostasis of bone, such as orthopedic trauma caused by accidents, osteoporosis, and other infection-related bone loss, such as osteomyelitis and periodontal disease [93]. Periodontitis is a chronic inflammatory disease that progressively affects the alveolar bone, periodontal ligament, and root cementum. In addition to infection, osteoporosis has also been shown to intensify periodontal bone loss. Although the body has an inherent mechanism to overcome bone defects through endogenous regeneration, in some cases, the natural healing functions are not sufficient to retain the original vitality of the bone [94,95]. Bone tissue engineering is a rapidly growing field that aims to develop materials that can assist or enhance the osteoconductive, osteoinductive, and osteogenic capacity at the site of the bone defect. However, the mechanism of Sr upon osteogenic differentiation is not yet completely understood, and many studies have tried to explore the possible pathways and mechanisms involved in Sr's effects [96].

The major components of the bone extracellular matrix (ECM) are collagen types I, III, and V, as well as hydroxyapatite (HA). HA, also known as calcium apatite, contains Ca minerals within the matrix. Sr is an important trace element found in the body and shares similar chemical properties and structure to Ca. As a result, Sr can actively incorporate into the crystal lattice of HA [97,98]. Initially, Sr was suggested as an adjunct for osteoporotic treatment by stimulating bone formation through osteoblasts and suppressing bone resorption by inhibiting the osteoclasts. Although the mechanism by which Sr induced the osteoblasts is not fully understood, the similarities between Ca and Sr suggest that Sr also follows the Ca-sensing receptor (CaSR) pathway. When Sr binds to these receptors, it activates a cascade of pathways, by phosphorylating the PI3K–Akt, Ras–MAPK, and Wnt pathways to enhance the osteogenesis of the osteoblasts and mesenchymal stem cells (MSCs) [99]. Sr also activates the Cn–NFATc pathway, where Sr induces calcineurin to dephosphorylate NFATc (activated state). In the inactive state, this transcription factor remains in the cytoplasm, but upon activation, it is translocated inside the nucleus and promotes the expression of ALP and Runx2 genes to induce osteogenesis [100]. The proliferating osteoblasts then produces M-CSF, and RANKL ligands, which bind with RANK to activate the cascade of signaling pathways, including MAPK, NF-κB, and PI3K–Akt, to stimulate osteoclastogenesis, where osteoclast cells promote bone resorption (Figure 4). The osteoprotegerin (OPG) receptor, belonging to the TNF superfamily, acts as a decoy receptor for RANKL, reducing its interaction with RANK and consequently suppressing osteoclastogenesis. Sr that binds to CaSR enhances the expression of OPG, which in turn suppresses osteoclasts. Sr also inhibits bone resorption by enhancing the apoptosis of mature osteoclasts [101].

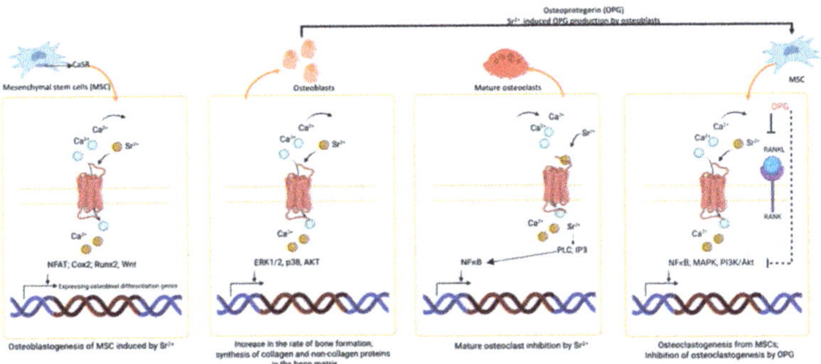

Figure 4. The role of Sr in osteogenesis.

Due to their common origin from MSCs, there is always a dilemma in determining the cell fate of endothelial cells and osteoblasts. In particular, aging disrupts the balance of differentiation by favoring the differentiation of MSCs into the adipogenic lineage rather than the osteogenic lineage. Increased levels of ROS accelerate the differentiation of MSCs into adipocytes. The signaling pathway tightly regulates this cascade, with CCAAT/enhancer-binding proteins (C/EBPs) and peroxisome proliferator-activated receptor γ (PPARγ) promoting adipogenesis, while TAZ and Runx2 pathways drive osteogenesis. Sr enhances the osteogenic differentiation of MSCs by exhibiting antioxidative properties, thereby increasing osteogenic differentiation and reducing adipogenic differentiation [94,95].

Considering the significant effect of Sr on bone regeneration, it has garnered increased attention in the field of dentistry. Periodontitis, a major cause of bone loss in the oral cavity, has led to increased interest in Sr-releasing biomaterials. These biomaterials have been proven to enhance the expression of ALP, bone matrix synthesis, and various osteogenic genes such as osteocalcin (OCN), Runx2, bone sialoprotein (BSP), and bone morphogenic protein (BMP) in pre-osteoblasts and osteoblasts [102]. Osteoblasts originate from bone marrow mesenchymal stem cells (BMSCs), and thus the commitment of BMSCs and MSCs located in the periodontal ligament to differentiate into osteoblasts plays a key role in bone regeneration. Novel tissue engineering scaffolds have incorporated Sr into mineralizing scaffolds to promote osteogenesis. This section discusses the Sr-incorporated biomaterials that have demonstrated enhanced bone regeneration in osteoblasts, BMSCs, and PDL stem cells.

The abundance of HA in the bone matrix has led to the widespread use of HA as the primary material in most bone tissue engineering scaffolds. This is because HA can mimic the extracellular matrix (ECM) of the bone. Derivatives of HA ceramics, including α-tricalcium phosphate, biphasic calcium phosphate (BCP), and β-tricalcium phosphate cement, have been shown to possess excellent biocompatibility and bioactivity [103]. However, these materials alone are not sufficient to stimulate an osteogenic effect on osteoblasts or mesenchymal stem cells (MSCs). To overcome the limitations of HA-based synthetic bioceramics, such as bioactive glass, Sr is incorporated into them to enhance the osteogenic activity. Metal ions like magnesium (Mg), zinc (Zn), and Sr are naturally present as trace elements in the bone matrix. Sr can actively substitute Ca in the crystal lattice of HA, thereby enhancing the solubility and degradability of the bioceramics. Sr-substituted bioceramics have been developed in various forms, such as powders, granules, fibers, and three-dimensional (3D) scaffolds.

Tsai et al. [104] developed a nanofiber using Sr-substituted HA. It is widely recognized that cell interaction with nanofibers is more efficient compared to compact powder-based fillings. The mesoporous structure and high surface area of the electrospun scaffold facilitate better cell absorption and attachment. The Sr-substituted hydroxyapatite nanofibers (Sr-HNAFs) exhibited excellent drug loading efficiency and sustained release of the antimicrobial drug (tetracycline). Additionally, the Sr-HNAFs also enhanced the osteogenic differentiation of MG63 osteoblast-like cells by increasing ALP activity. The alignment of the nanofibers also influenced cellular behavior, with the aligned nanofibers of Sr-HA showing higher osteoblast inductive efficacy compared to randomly substituted Sr-HNAFs [105,106].

While there are several biomaterials that incorporate Sr, the field of dentistry primarily focuses on utilizing Sr-coated dental implant surfaces. Titanium (Ti) implants are considered the ideal scaffold materials for dental and orthopedic implants due to their excellent mechanical properties and biocompatibility. By coating or incorporating Sr ions onto the implant surface, the osseointegration of localized precursor cells can be enhanced.

8. Strontium and Implant Coatings

Titanium, its alloys, and zirconia are widely used biomaterials for orthopedic and dental implants due to their excellent biocompatibility, mechanical properties, and corrosion resistance [107]. However, these materials are biologically inert, which can lead to a prolonged healing period after surgery. Achieving early osseointegration and implant fixation is essential for ensuring long-term implant stability and reducing the duration of clinical treatment.

Surface modification techniques have been developed to improve osseointegration and accelerate the healing time for titanium and zirconia implants [108]. Various methods, such as the sandblasted and acid-etched (SLA) process, sol–gel process, plasma spraying, and chemical vapor deposition, have been employed for dental implant coating [109,110]. These coating methods can induce morphological changes, such as increased surface roughness, and modify the chemical composition by facilitating the formation of hydroxyapatite formed on the implant surface.

Bioactive materials like hydroxyapatite and bioactive glass are frequently used to modify the surfaces of titanium and zirconia implants. These materials exhibit remarkable biocompatibility and bioactivity. Hydroxyapatite possesses a chemical composition and structure similar to human bones and teeth, enabling it to form a strong chemical bond with bone tissue cells [111–113]. The application of these coating materials can greatly enhance the performance of titanium and zirconia implants.

The substitution of Ca with Sr in hydroxyapatite and bioactive glass implant coatings has been found to enhance their physical, mechanical, and biological properties [114,115]. This substitution has been shown to improve the osseointegration and bone healing of these coated implants [116,117]. In vitro studies have demonstrated that Sr-substituted hydroxyapatite and bioactive glass coatings promote the proliferation and differentiation of osteoblast cells, while also inhibiting the production and proliferation of osteoclasts [118,119]. In vivo studies have shown that Sr-substituted hydroxyapatite and bioactive glass coatings enhance new bone formation, implant fixation, and bone-to-implant contact during the healing period [120–122] (Table 3).

Recently, bioactive Sr-functionalized coatings on titanium and zirconia implants have been developed using two different techniques: micro-arc oxidation and magnetron sputtering [123,124]. One example of such a coating is the nanoporous micro-arc oxidized Sr–titanium (MAO-Sr-Ti) coating on titanium, which is fabricated through electrochemical surface treatment [123]. In this process, titanium undergoes micro-arc oxidation, and then Sr ions are incorporated into the surface through electrochemical treatment using a Sr dichloride solution. The resulting nanoporous structure increases stem cell adhesion, while the MAO-Sr-Ti coating promotes the proliferation and differentiation of bone marrow-derived mesenchymal stem cells. Furthermore, the coating facilitates early osseointegration and the formation of new bone.

Another example of a Sr-functionalized coating is the development of a Sr–titanate nanocoating on zirconia implants using magnetron sputtering [124]. In this process, the zirconia surface is first treated by sandblasting and acid-etched with hydrofluoric acid (SA-Zr). Then, Sr titanate is deposited onto the SA-Zr surface using magnetron sputtering. The resulting nanocoating enhances the spreading and adhesion of pre-osteoblast cells. It also promotes the proliferation and differentiation of osteoblasts, leading to improved osseointegration compared to the surface treated with sandblasting and acid-etching alone.

Indeed, Sr has been proven to enhance bone growth and regeneration by increasing the proliferation and differentiation of osteoblasts [125] (Figure 5). Additionally, Sr has the ability to inhibit apoptosis in osteoblast cells. Furthermore, Sr has been found to inhibit the formation and differentiation of osteoclast cells and promote their apoptosis, leading to a reduction in bone resorption [126]. Therefore, the incorporation of Sr in functionalized coating can significantly improve the biological properties of titanium and zirconia implants.

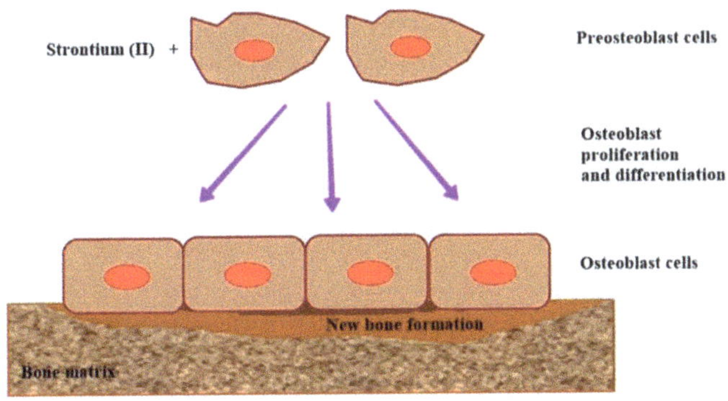

Figure 5. Schematic diagram of strontium's effect on bone formation.

Table 3. Sr-doped biomaterials and implants with enhanced osteogenic differentiation of BMSCs, PDL stem cells, and osteoblasts.

Materials	Material Characteristics	Biological Characteristics	Reference
Sr-incorporated HA (SrHA)	- SrHA bone cement was developed in an injectable form - Compressive strength was 73.4 MPa - Favorable injectability (100%) - Setting time (initial setting time 240 s and final (420 s) - Excellent radiopacity	- Excellent biocompatibility - Promoted osteogenic differentiation of PDL stem cells and dental pulp stem cells (DPSCs)	Dai et al., 2021 [127]
	- Micro–nano hybrid surfaces doped with different concentrations of Sr (Sr–mnHAP)	- Promoted attachment of BMSCs to scaffold and enhanced osteogenic differentiation through upregulation of ALP and OCN - Promoted bone regeneration in rats with calvaria defects and enhanced vascularization	Jiang et al., 2022 [98]

Table 3. *Cont.*

Materials	Material Characteristics	Biological Characteristics	Reference
SrHA with natural and synthetic polymers	- SrHA–SF biocomposite nanospheres - Mechanical strength was not calculated	- Enhanced osteogenic property of BMSCs	Wang et al., 2020 [97]
	- SrHA nanofibers synthesized with PVP, PCL, and PLLA	- Electrospun nanofibers showed extensive cellular homing for MG63 osteoblast-like cells - Enhanced osteogenic activity of MG63 cells	Tsai et al., 2018 [104]
Sr with bioactive glass	- Sr-incorporated mesoporous bioactive glass (Sr–BG)	- Periodontal defects were created through bilateral ovariectomy (OVX) - The filled periodontal defects with Sr–BG showed a 46.67% increase in bone formation when compared to BC scaffold alone - Increased expression of epigenetic regulator Setd2 in Sr–GB-induced bone formation	Jia et al., 2017 [94]; Zhang et al., 2014 [128]
Sr incorporation on dental implants	- Calcium–strontium–zinc phosphate coating on the Ti implant	- Enhanced osteogenic-related factors - Polarized the macrophages from M1 to M2 phase - Effects on PDL stem cells have not been explored	Zhao et al., 2021 [129]
	- Sr-doped titanium via sandblasted, large-grit, and acid-etching Sr-SLA	- Improved osseointegration and differentiation of BMSCs - Reduced intracellular ROS level, favorable for osteogenesis - Inhibited adipogenic differentiation - Enhanced anti-inflammatory properties	Zhou et al., 2019 [130]; Choi et al., 2018 [131]
	- Ti implant releasing both Sr and silver (Ag) ions	- Incorporation of Ag did not affect the effect of Sr on osteogenic property - Ag exhibited antibacterial property	Okuzu et al., 2021 [132]
	- Strontium-functionalized grade 4 Ti implants - Sustained release of strontium	- Assessed on rabbit femur model, showed increased bone-to-impact contact - No difference in % bone formation	Offermanns et al., 2018 [133]
	- Polyetheretherketone implant decorated with Sr and adiponectin	- Increased osteogenic activity of MC3T3-E1 cells - Adiponectin increased osteogenic properties - Sr and adiponectin had a synergistic effect on osteogenic property	Wang et al., 2019 [134]

Table 3. *Cont.*

Materials	Material Characteristics	Biological Characteristics	Reference
	- Polybutylcyanoacrylate (PBCA) loaded with Sr, synthesized by emulsion polymerization methods.	- Higher expression of bone matrix deposition when treated on porcine mandibular bone block - Increased osteogenic differentiation of MSCs	Chang et al., 2021 [135]
Sr with hormone	- Parathyroid hormone (PTH) and Sr-containing poloxamer implant tablets	- PTH enhanced bone and antiresorptive molecule; Sr enhanced bone formation - PTH alone promoted bone formation in 4 weeks and declined after 3 weeks - Sr alone had no positive effects on bone formation - PTH + Sr showed significantly higher bone formation	Goker et al., 2018 [136]

9. The Influence of Strontium on Mechanical Properties

The mechanical properties of dental materials are crucial for ensuring durability, functionality, and biocompatibility in oral environments. They directly impact the long-term success of dental treatments and patient comfort. The incorporation of strontium into dental materials significantly enhances their mechanical properties through various mechanisms, including grain refinement and structural modification [137,138]. Research has shown that the inclusion of strontium in dental composites and adhesives can enhance their mechanical properties, such as flexural strength and wear resistance [139]. This improvement is attributed to the ability of strontium to reinforce the composite matrix and promote better bonding between the filler and the resin matrix. Arepalli et al. [137] investigated the effects of strontium substitution in bioactive glasses. Their study demonstrated that strontium substitution led to enhanced bioactivity, biocompatibility, and mechanical behavior of bioactive glasses. The incorporation of strontium improved the compressive strength and fracture toughness, making the material more suitable for dental applications. Another study investigated the mechanical properties of strontium ion-doped mesoporous bioactive glass. Their findings revealed that the doping of strontium ions significantly improved the compressive strength and elastic modulus of the mesoporous bioactive glass [140]. The Sr^{2+} in the glass phase serves as a network modifier, where it enhances the degree of cross-linking compared to unmodified cements. This suggests that the inclusion of the larger Sr^{2+} ion in the system amplifies the disturbance within the glass network. Consequently, a decrease in interatomic spacing causes the densification of Sr-1. This then leads to greater compactness within the glass structure, potentially contributing to an increase in elastic moduli [137]. On the other hand, the incorporation of Sr into calcium silicate-based materials did not favor the enhancement of the mechanical properties [141]. With the addition of Sr and gradual increase in concentration, the calcium silicate network might experience further loosening. This happens as the larger Sr^{2+} ions progressively replace the smaller calcium ions, causing a linear increase in lattice constants. This leads to considerable disorganization and a weakening of network connectivity [142]. The incorporation of strontium can have diverse effects on the mechanical properties of dental materials, depending on its interaction with the material structure. Hence, while strontium plays a crucial role in influencing the properties of dental materials, it is essential to carefully evaluate the mechanical properties and assess the potentially positive or negative impact of strontium on these properties.

10. Immunomodulatory Potential of Strontium

The host immune response plays a crucial role in the process of bone formation and homeostasis. This response involves a large number of cytokines, signaling molecules, and specific proteins that have a substantial role in immunomodulation and osteogenesis [143,144]. Many immune cells modulate and influence inflammatory responses. Macrophages are among the most effective cells in the immune system, as they play significant roles in the inflammatory response to external biomaterials [145]. There are two phenotypes of macrophages: M1 and M2. M1 macrophages have a pro-inflammatory effect, while M2 macrophages have anti-inflammatory potential and promote cellular proliferation and tissue repair [146]. Sr has been found to regulate the response of macrophages and suppress interleukin 6 (IL-6) [147].

Lee et al. [147] incorporated Ca and Sr in nanostructured titanium (Ti) surfaces and concluded that the chemically modified Ti disks with divalent cations regulated the cell shape of adherent macrophages and markedly upregulated the expression of the M2 macrophage phenotype when combined with the nanostructured Ti surface. The effect of Sr and Ca ions favored early wound healing by modulating the early M1 macrophage cells (after implantation of the nanostructured Ti implants). In another study, the immunoregulation potential of Sr-modified Ti surfaces using a novel phase transited lysozyme (PTL) was evaluated [148]. The Sr-modified group exhibited greater osteogenesis and controlled inflammation at the implant–tissue interface. Moreover, the expression levels of pro-inflammatory genes of macrophages on Sr-modified Ti implants were lower than on Ti implants without Sr [148]. When macrophages are activated into the M1 subtype, it increases the expression levels of pro-inflammatory factors, contributing to inflammation and foreign body reaction with subsequent fibrosis. On the other hand, when macrophages are activated towards the M2 subtype, they produce anti-inflammatory cytokines that promote healing and create favorable conditions for new tissue formation [149].

In a study conducted by Zhang et al. [150], the effect of Sr-substituted submicrometer bioactive glass (Sr–SBG) on macrophages was investigated. The interaction between Sr-SBG and macrophages both in vitro and in vivo was examined using histological assessment. Zhang et al. found that Sr–SBG enhanced osteogenesis and suppressed osteoclastogenesis. They attributed this effect to the stimulation of macrophage polarization from M1 to M2 by Sr–SBG, creating more suitable conditions for osteogenesis compared to SBG alone. Furthermore, the macrophages conditioned with Sr–SBG exhibited a significant decrease in IL-6 cytokine levels compared to the SBG group. This decrease may have contributed to the inhibition of osteoclastic activity. The inhibitory effect of Sr may be attributed to the downregulation of tumor necrosis factor α (TNF α) and the suppressed effect of NFkB, which hinders the differentiation of pre-osteoclasts.

Li et al. [151] developed a Sr–Cu borosilicate glass bone cement that could regulate bone healing by modulating inflammation, enhancing vascularization, and improving osteogenic differentiation of bone marrow-derived stem cells. They reported that the sustained controlled release of Sr and Cu ions upregulated anti-inflammatory genes (IL-1Ra and TGF-β1) while downregulating the expression of pro-inflammatory genes (IL-1β and IL-6) in macrophages. Several other studies incorporated Sr in different biomaterials, scaffolds, and coatings [129,131,152–160], and reported the same outcome, i.e., that Sr is capable of modulating macrophage polarization, which significantly promotes osteogenesis and suppresses osteoclastic activity. Interestingly, it can be theorized that the ability of Sr to induce bone formation and inhibit bone resorption may arise from its effect on modulating the macrophage inflammatory phenotypes.

11. Future Perspectives

Future research should focus on the regenerative potential of Sr and its role in alleviating reversible and irreversible pulp inflammation (pulpitis) by modulation of macrophages to improve the success of vital pulp therapies and dentin regeneration. Furthermore, more evidence is needed to validate the antimicrobial potential of strontium and its role in arresting dental caries.

The development of new bioactive strontium compounds such as strontium-based mixed-oxide ceramics can enhance biological performance, such as antimicrobial and bone regeneration properties [161,162].

12. Conclusions

The incorporation of strontium into dental biomaterials presents a significant advancement in the field of dentistry. Strontium enhances dental material visibility in radiographic imaging, supporting accurate diagnosis and treatment. It also exhibits antimicrobial properties that reduce secondary caries risk and improve restoration longevity. Its ability to promote enamel remineralization offers potential for managing dental caries and strengthening tooth structure. Though mixed results exist, strontium compounds have demonstrated effectiveness in addressing dentin hypersensitivity and potential for dentin regeneration, highlighting their potential in vital pulp therapies and regenerative dentistry. Strontium's osteogenic effects make it promising for bone tissue engineering and dental implant biomaterials, with studies showing improved osseointegration and accelerated healing. Additionally, its immunomodulatory properties, particularly in macrophage polarization, contribute to favorable bone regeneration and healing environments. Strontium's multifaceted benefits underscore its potential in advancing biomaterials that support tissue regeneration and modulate host immune responses for improved treatment outcomes.

Author Contributions: M.M.A.: conceptualization, interpretation, data curation, and original draft preparation. O.S.: data curation, writing, and review. C.Y.K.L.: writing and review. V.R.: writing and review. C.K.Y.Y.: critical revision and editing all the drafts of the manuscript. All authors have read and agreed to the published version of the manuscript.

Funding: The research project was supported by the Research Grants Council of Hong Kong (17110422).

Conflicts of Interest: The authors declare no conflicts of interest.

References

1. Davy, H. XXIII. Electro-chemical researches, on the decomposition of the earths; with observations on the metals obtained from the alkaline earths, and on the amalgam procured from ammonia. *Philos. Trans. R. Soc. Lond.* **1808**, *98*, 333–370.
2. Partington, J. Early history of strontium. In *Handbook of Stable Strontium*; Springer: New York, NY, USA, 1981; pp. 1–9.
3. Rana, R.; Höcker, M.; Myers, E.G. Atomic masses of strontium and ytterbium. *Phys. Rev. A* **2012**, *86*, 050502. [CrossRef]
4. Amata, R.; Diamond, G.L.; Dorsey, A.; Fransen, M.E. *Toxicological Profile for Strontium*; U.S. Department of Health and Human Services, Public Health Service, Agency for Toxic Substances and Disease Registry: Atlanta, GA, USA, 2004.
5. Fleischer, M. *Recent Estimates of the Abundances of the Elements in the Earth's Crust*; US Department of the Interior, Geological Survey: Reston, VA, USA, 1953.
6. Bronner, F. *Dynamics and Function of Calcium in Mineral Metabolism*; Comar, C.L., Bronner, F., Eds.; Academic Press: New York, NY, USA, 1964.
7. Pan, H.-B.; Li, Z.-Y.; Wang, T.; Lam, W.; Wong, C.; Darvell, B.; Luk, K.; Hu, Y.; Lu, W. Nucleation of strontium-substituted apatite. *Cryst. Growth Des.* **2009**, *9*, 3342–3345. [CrossRef]
8. Heurich, E.; Beyer, M.; Jandt, K.D.; Reichert, J.; Herold, V.; Schnabelrauch, M.; Sigusch, B.W. Quantification of dental erosion—A comparison of stylus profilometry and confocal laser scanning microscopy (CLSM). *Dent. Mater.* **2010**, *26*, 326–336. [CrossRef]
9. Papillon, M. Recherches experimentales sur les modifications de la composition immediate des os. *C. R. Acad. Sci.* **1870**, *71*, 372–374.
10. Schoenberg, H.P. Extent of strontium substitution for calcium in hydroxyapatite. *Biochim. Biophys. Acta* **1963**, *75*, 96–103. [CrossRef]
11. Harrison, G.; Lumsden, E.; Raymond, W.; Sutton, A.; Boyd, J.; Neuman, W.; Hodge, H. On the mechanisms of skeletal fixation of strontium. Parts I and II. *Arch. Biochem. Biophys.* **1959**, *80*, 97–113. [CrossRef]
12. Oliveira, J.P.; Querido, W.; Caldas, R.J.; Campos, A.P.; Abraçado, L.G.; Farina, M. Strontium is incorporated in different levels into bones and teeth of rats treated with strontium ranelate. *Calcif. Tissue Int.* **2012**, *91*, 186–195. [CrossRef] [PubMed]
13. Dow, E.C.; Stanbury, J.B. Strontium and calcium metabolism in metabolic bone diseases. *J. Clin. Investig.* **1960**, *39*, 885–903. [CrossRef]
14. D'Onofrio, A.; Kent, N.W.; Shahdad, S.A.; Hill, R.G. Development of novel strontium containing bioactive glass based calcium phosphate cement. *Dent. Mater.* **2016**, *32*, 703–712. [CrossRef]

15. Wang, J.; Zhang, L.; Sun, X.; Chen, X.; Xie, K.; Lin, M.; Yang, G.; Xu, S.; Xia, W.; Gou, Z. Preparation and in vitro evaluation of strontium-doped calcium silicate/gypsum bioactive bone cement. *Biomed. Mater.* **2014**, *9*, 045002. [CrossRef] [PubMed]
16. Huang, T.-H.; Kao, C.-T.; Shen, Y.-F.; Lin, Y.-T.; Liu, Y.-T.; Yen, S.-Y.; Ho, C.-C. Substitutions of strontium in bioactive calcium silicate bone cements stimulate osteogenic differentiation in human mesenchymal stem cells. *J. Mater. Sci. Mater. Med.* **2019**, *30*, 68. [CrossRef] [PubMed]
17. Xuereb, M.; Sorrentino, F.; Damidot, D.; Camilleri, J. Development of novel tricalcium silicate-based endodontic cements with sintered radiopacifier phase. *Clin. Oral Investig.* **2016**, *20*, 967–982. [CrossRef] [PubMed]
18. Lippert, F.; Hara, A.T. Strontium and caries: A long and complicated relationship. *Caries Res.* **2013**, *47*, 34–49. [CrossRef] [PubMed]
19. Femiano, F.; Femiano, R.; Femiano, L.; Nucci, L.; Minervini, G.; Antonelli, A.; Bennardo, F.; Barone, S.; Scotti, N.; Sorice, V.; et al. A New Combined Protocol to Treat the Dentin Hypersensitivity Associated with Non-Carious Cervical Lesions: A Randomized Controlled Trial. *Appl. Sci.* **2021**, *11*, 187. [CrossRef]
20. Abdalla, M.M.; Lung, C.Y.K.; Bijle, M.N.; Yiu, C.K.Y. Physicochemical Properties and Inductive Effect of Calcium Strontium Silicate on the Differentiation of Human Dental Pulp Stem Cells for Vital Pulp Therapies: An In Vitro Study. *Materials* **2022**, *15*, 5854. [CrossRef] [PubMed]
21. Huang, M.; Hill, R.G.; Rawlinson, S.C. Strontium (Sr) elicits odontogenic differentiation of human dental pulp stem cells (hDPSCs): A therapeutic role for Sr in dentine repair? *Acta Biomater.* **2016**, *38*, 201–211. [CrossRef]
22. Shahid, S.; Hassan, U.; Billington, R.; Hill, R.; Anderson, P. Glass ionomer cements: Effect of strontium substitution on esthetics, radiopacity and fluoride release. *Dent. Mater.* **2014**, *30*, 308–313. [CrossRef]
23. *ISO 9917-1*; Dentistry-Water-Based Cements—Part 1: Powder/Liquid Acid–Base Cements. International Organization for Standardization: Geneva, Switzerland, 2007.
24. You, J.; Yoo, J.-S.; Kum, K.-Y.; Hong, S.-H. Hydration behavior and radiopacity of strontium substituted Ca_3SiO_5 cement. *J. Korean Ceram. Soc.* **2021**, *58*, 330–336. [CrossRef]
25. Pelepenko, L.E.; Marciano, M.A.; Francati, T.M.; Bombarda, G.; Antunes, T.B.M.; Sorrentino, F.; Martin, R.A.; Boanini, E.; Cooper, P.R.; Shelton, R.M. Can strontium replace calcium in bioactive materials for dental applications? *J. Biomed. Mater. Res. Part A* **2022**, *110*, 1892–1911. [CrossRef]
26. Billington, R. Glass Fiber Sizing Agent. U.S. Patent US4067835A, 10 January 1978.
27. Wu, C.; Ramaswamy, Y.; Kwik, D.; Zreiqat, H. The effect of strontium incorporation into $CaSiO_3$ ceramics on their physical and biological properties. *Biomaterials* **2007**, *28*, 3171–3181. [CrossRef] [PubMed]
28. Karpukhina, N.; Bushby, A. Incorporation of Strontium and Fluoride Containing Bioglass into Biodentine: Impact on Bioactivity, Radiopacity, Compressive Strength and Setting Time. Master's Thesis, University of London, London, UK, 2012.
29. Carvalho, E.; De Paula, D.; Neto, D.A.; Costa, L.; Dias, D.; Feitosa, V.; Fechine, P. Radiopacity and mechanical properties of dental adhesives with strontium hydroxyapatite nanofillers. *J. Mech. Behav. Biomed. Mater.* **2020**, *101*, 103447. [CrossRef] [PubMed]
30. Höland, W.; Schweiger, M.; Dittmer, M.; Ritzberger, C. Radiopaque strontium fluoroapatite glass-ceramics. *Front. Bioeng. Biotechnol.* **2015**, *3*, 149. [CrossRef] [PubMed]
31. Romieu, G.; Garric, X.; Munier, S.; Vert, M.; Boudeville, P. Calcium–strontium mixed phosphate as novel injectable and radio-opaque hydraulic cement. *Acta Biomater.* **2010**, *6*, 3208–3215. [CrossRef] [PubMed]
32. Alkhraisat, M.H.; Rueda, C.; Cabrejos-Azama, J.; Lucas-Aparicio, J.; Mariño, F.T.; García-Denche, J.T.; Jerez, L.B.; Gbureck, U.; Cabarcos, E.L. Loading and release of doxycycline hyclate from strontium-substituted calcium phosphate cement. *Acta Biomater.* **2010**, *6*, 1522–1528. [CrossRef] [PubMed]
33. Guida, A.; Towler, M.; Wall, J.; Hill, R.; Eramo, S. Preliminary work on the antibacterial effect of strontium in glass ionomer cements. *J. Mater. Sci. Lett.* **2003**, *22*, 1401–1403. [CrossRef]
34. Jayasree, R.; Kumar, T.; Mahalaxmi, S.; Abburi, S.; Rubaiya, Y.; Doble, M. Dentin remineralizing ability and enhanced antibacterial activity of strontium and hydroxyl ion co-releasing radiopaque hydroxyapatite cement. *J. Mater. Sci. Mater. Med.* **2017**, *28*, 95. [CrossRef] [PubMed]
35. Brauer, D.S.; Karpukhina, N.; Kedia, G.; Bhat, A.; Law, R.V.; Radecka, I.; Hill, R.G. Bactericidal strontium-releasing injectable bone cements based on bioactive glasses. *J. R. Soc. Interface* **2013**, *10*, 20120647. [CrossRef] [PubMed]
36. Alshammari, H.; Neilands, J.; Svensäter, G.; Stavropoulos, A. Antimicrobial potential of strontium hydroxide on bacteria associated with peri-implantitis. *Antibiotics* **2021**, *10*, 150. [CrossRef]
37. Forss, H. Release of fluoride and other elements from light-cured glass lonomers in neutral and acidic conditions. *J. Dent. Res.* **1993**, *72*, 1257–1262. [CrossRef]
38. Mo, S.-S.; Bao, W.; Lai, G.-Y.; Wang, J.; Li, M.-Y. The microfloral analysis of secondary caries biofilm around Class I and Class II composite and amalgam fillings. *BMC Infect. Dis.* **2010**, *10*, 241. [CrossRef]
39. Liu, J.; Rawlinson, S.C.; Hill, R.G.; Fortune, F. Strontium-substituted bioactive glasses in vitro osteogenic and antibacterial effects. *Dent. Mater.* **2016**, *32*, 412–422. [CrossRef]
40. Dabsie, F.; Gregoire, G.; Sixou, M.; Sharrock, P. Does strontium play a role in the cariostatic activity of glass ionomer? Strontium diffusion and antibacterial activity. *J. Dent.* **2009**, *37*, 554–559. [CrossRef]
41. Yan, P.; Xia, J.S.; Chen, Y.P.; Liu, Z.P.; Guo, J.S.; Shen, Y.; Zhang, C.C.; Wang, J. Thermodynamics of binding interactions between extracellular polymeric substances and heavy metals by isothermal titration microcalorimetry. *Bioresour. Technol.* **2017**, *232*, 354–363. [CrossRef]

42. Dai, L.L.; Mei, M.L.; Chu, C.H.; Lo, E.C.M. Remineralizing effect of a new strontium-doped bioactive glass and fluoride on demineralized enamel and dentine. *J. Dent.* **2021**, *108*, 103633. [CrossRef] [PubMed]
43. Spets-Happonen, S.; Luoma, H.; Seppä, L.; Räisänen, J. The effect of different strontium concentrations on the efficacy of chlorhexidine-fluoride-strontium gel in preventing enamel softening in vitro. *Arch. Oral Biol.* **1993**, *38*, 107–112. [CrossRef]
44. Spets-Happonen, S.; Luoma, H.; Seppä, L. High strontium addition to chlorhexidine-fluoride gel does not increase its caries-preventive effect in rats. *Acta Odontol. Scand.* **1996**, *54*, 92–95. [CrossRef] [PubMed]
45. Spets-Happonen, S.; Luoma, H.; Forss, H.; Kentala, J.; Alaluusua, S.; Luoma, A.R.; Grönroos, L.; Syvåoja, S.; Tapaninen, H.; Happonen, P. Effects of a chlorhexidine-fluoride-strontium rinsing program on caries, gingivitis and some salivary bacteria among Finnish school children. *Eur. J. Oral Sci.* **1991**, *99*, 130–138. [CrossRef] [PubMed]
46. Kytömaa, I.; Paakkola, O. Strontium-90 in Teeth: A Comparison of Methods. *Acta Odontol. Scand.* **1971**, *29*, 321–326. [CrossRef]
47. Yassen, G.H.; Lippert, F.; Eckert, G.; Eder, J.; Zandoná, A.F. The effect of strontium and combinations of strontium and fluoride on the remineralization of artificial caries lesions in vitro. *Quintessence Int.* **2012**, *43*, e95–e103.
48. Thuy, T.T.; Nakagaki, H.; Inukai, H.; Tsuboi, S.; Robinson, C. Effect of strontium on enamel remineralization in vitro. *Caries Res.* **2006**, *40*, 338.
49. Wang, Y.-L.; Chang, H.-H.; Chiang, Y.-C.; Lin, C.-H.; Lin, C.-P. Strontium ion can significantly decrease enamel demineralization and prevent the enamel surface hardness loss in acidic environment. *J. Formos. Med. Assoc.* **2019**, *118*, 39–49. [CrossRef]
50. Featherstone, J.; Shields, C.; Khademazad, B.; Oldershaw, M. Acid reactivity of carbonated apatites with strontium and fluoride substitutions. *J. Dent. Res.* **1983**, *62*, 1049–1053. [CrossRef] [PubMed]
51. Featherstone, J.; Rodgers, B.; Smith, M. Physicochemical requirements for rapid remineralization of early carious lesions. *Caries Res.* **1981**, *15*, 221–235. [CrossRef] [PubMed]
52. Thuy, T.T.; Nakagaki, H.; Kato, K.; Hung, P.A.; Inukai, J.; Tsuboi, S.; Nakagaki, H.; Hirose, M.N.; Igarashi, S.; Robinson, C. Effect of strontium in combination with fluoride on enamel remineralisation in vitro. *Arch. Oral Biol.* **2008**, *53*, 1017–1022. [CrossRef]
53. Ashrafi, M.; Spector, P.; Curzon, M. Pre-and posteruptive effects of low doses of strontium on dental caries in the rat. *Caries Res.* **1980**, *14*, 341–346. [CrossRef]
54. Krishnan, V.; Bhatia, A.; Varma, H. Development, characterization and comparison of two strontium doped nano hydroxyapatite molecules for enamel repair/regeneration. *Dent. Mater.* **2016**, *32*, 646–659. [CrossRef] [PubMed]
55. Rajendran, R.; Nair, K.R.; Sandhya, R.; Ashik, P.M.; Veedu, R.P.; Saleem, S. Evaluation of remineralization potential and cytotoxicity of a novel strontium-doped nanohydroxyapatite paste: An in vitro study. *J. Conserv. Dent. (JCD)* **2020**, *23*, 330. [CrossRef] [PubMed]
56. Dai, L.L.; Nudelman, F.; Chu, C.H.; Lo, E.C.M.; Mei, M.L. The effects of strontium-doped bioactive glass and fluoride on hydroxyapatite crystallization. *J. Dent.* **2021**, *105*, 103581. [CrossRef]
57. Splieth, C.H.; Tachou, A. Epidemiology of dentin hypersensitivity. *Clin. Oral Investig.* **2013**, *17*, 3–8. [CrossRef]
58. Yoshiyama, M.; Masada, J.; Uchida, A.; Ishida, H. Scanning electron microscopic characterization of sensitive vs. insensitive human radicular dentin. *J. Dent. Res.* **1989**, *68*, 1498–1502. [CrossRef] [PubMed]
59. Pawlowska, J. Strontium chloride—Its importance in dentistry and prophylaxis. *Czas. Stomatol.* **1956**, *9*, 353–357.
60. Griffiths, H.; Morgan, G.; Williams, K.; Addy, M. Dentine hypersensitivity: The measurement in vitro of streaming potentials with fluid flow across dentine and hydroxyapatite. *J. Periodontal Res.* **1993**, *28*, 60–64. [CrossRef] [PubMed]
61. Absi, E.; Addy, M.; Adams, D. Dentine hypersensitivity: A study of the patency of dentinal tubules in sensitive and non-sensitive cervical dentine. *J. Clin. Periodontol.* **1987**, *14*, 280–284. [CrossRef] [PubMed]
62. Gillam, D.; Orchardson, R. Advances in the treatment of root dentine sensitivity: Mechanisms and treatment principles. *Endod. Top.* **2006**, *13*, 13–33. [CrossRef]
63. Pashley, D.H. Dentin permeability, dentin sensitivity, and treatment through tubule occlusion. *J. Endod.* **1986**, *12*, 465–474. [CrossRef] [PubMed]
64. Hodge, H.C.; Gavett, E.; Thomas, I. The adsorption of strontium at forty degrees by enamel, dentin, bone, and hydroxyapatite as shown by the radioactive isotope. *J. Biol. Chem.* **1946**, *163*, 1–6. [CrossRef] [PubMed]
65. Porto, I.C.; Andrade, A.K.; Montes, M.A. Diagnosis and treatment of dentinal hypersensitivity. *J. Oral Sci.* **2009**, *51*, 323–332. [CrossRef] [PubMed]
66. Saeki, K.; Marshall, G.W.; Gansky, S.A.; Parkinson, C.R.; Marshall, S.J. Strontium effects on root dentin tubule occlusion and nanomechanical properties. *Dent. Mater.* **2016**, *32*, 240–251. [CrossRef]
67. Oberg, C.; Pochapski, M.T.; Farago, P.V.; Granado, C.; Pilatti, G.L.; Santos, F.A. Evaluation of desensitizing agents on dentin permeability and dentinal tubule occlusion: An in vitro study. *Gen. Dent.* **2009**, *57*, 496–501.
68. Olley, R.C.; Pilecki, P.; Hughes, N.; Jeffery, P.; Austin, R.S.; Moazzez, R.; Bartlett, D. An in situ study investigating dentine tubule occlusion of dentifrices following acid challenge. *J. Dent.* **2012**, *40*, 585–593. [CrossRef] [PubMed]
69. Kodaka, T.; Kuroiwa, M.; Kuroiwa, M.; Okumura, J.; Mori, R.; Hirasawa, S.; Kobori, M. Effects of brushing with a dentifrice for sensitive teeth on tubule occlusion and abrasion of dentin. *J. Electron Microsc.* **2001**, *50*, 57–64. [CrossRef] [PubMed]
70. Banfield, N.; Addy, M. Dentine hypersensitivity: Development and evaluation ofamodel in situ to study tubulepatency. *J. Clin. Periodontol.* **2004**, *31*, 325–335. [CrossRef] [PubMed]
71. Arnold, W.; Prange, M.; Naumova, E. Effectiveness of various toothpastes on dentine tubule occlusion. *J. Dent.* **2015**, *43*, 440–449. [CrossRef] [PubMed]

72. Arrais, C.A.G.; Micheloni, C.D.; Giannini, M.; Chan, D.C. Occluding effect of dentifrices on dentinal tubules. *J. Dent.* **2003**, *31*, 577–584. [CrossRef] [PubMed]
73. Addy, M.; West, N.X. The role of toothpaste in the aetiology and treatment of dentine hypersensitivity. *Toothpastes* **2013**, *23*, 75–87.
74. Li, Y.; Lee, S.; Zhang, Y.P.; Delgado, E.; DeVizio, W.; Mateo, L.R. Comparison of clinical efficacy of three toothpastes in reducing dentin hypersensitivity. *J. Clin. Dent.* **2011**, *22*, 113.
75. Dotta, T.C.; Hayann, L.; de Padua Andrade Almeida, L.; Nogueira, L.F.B.; Arnez, M.M.; Castelo, R.; Cassiano, A.F.B.; Faria, G.; Martelli-Tosi, M.; Bottini, M.; et al. Strontium Carbonate and Strontium-Substituted Calcium Carbonate Nanoparticles Form Protective Deposits on Dentin Surface and Enhance Human Dental Pulp Stem Cells Mineralization. *J. Funct. Biomater.* **2022**, *13*, 250. [CrossRef] [PubMed]
76. Tirapelli, C.; Panzeri, H.; Soares, R.G.; Peitl, O.; Zanotto, E.D. A novel bioactive glass-ceramic for treating dentin hypersensitivity. *Braz. Oral Res.* **2010**, *24*, 381–387. [CrossRef]
77. Hench, L.L. Bioactive materials: The potential for tissue regeneration. *J. Biomed. Mater. Res. Off. J. Soc. Biomater. Jpn. Soc. Biomater. Aust. Soc. Biomater.* **1998**, *41*, 511–518. [CrossRef]
78. Hench, L.L. Chronology of bioactive glass development and clinical applications. *New J. Glas. Ceram.* **2013**, *3*, 67–73. [CrossRef]
79. Xia, W.; Qin, T.; Suska, F.; Engqvist, H. Bioactive spheres: The way of treating dentin hypersensitivity. *ACS Biomater. Sci. Eng.* **2016**, *2*, 734–740. [CrossRef]
80. Acevedo, L.A.; Campos, L.A.; Dechandt, I.C.; Alegria, G.; Siqueira, R.L.; Zanotto, E.D.; Serbena, F.C.; Santos, F.A. Effect of bioactive glasses containing strontium and potassium on dentin permeability. *J. Biomed. Mater. Res. Part B Appl. Biomater.* **2022**, *110*, 517–526. [CrossRef]
81. Martins, C.; Firmino, R.; Riva, J.; Ge, L.; Carrasco-Labra, A.; Brignardello-Petersen, R.; Colunga-Lozano, L.; Granville-Garcia, A.; Costa, F.; Yepes-Nuñez, J. Desensitizing toothpastes for dentin hypersensitivity: A network meta-analysis. *J. Dent. Res.* **2020**, *99*, 514–522. [CrossRef] [PubMed]
82. Karim, B.; Gillam, D. The efficacy of strontium and potassium toothpastes in treating dentine hypersensitivity: A systematic review. *Int. J. Dent.* **2013**, *2013*, 573258. [CrossRef]
83. West, N.X.; Seong, J.; Davies, M. Management of dentine hypersensitivity: Efficacy of professionally and self-administered agents. *J. Clin. Periodontol.* **2015**, *42*, S256–S302. [CrossRef] [PubMed]
84. Bae, J.H.; Kim, Y.K.; Myung, S.K. Desensitizing toothpaste versus placebo for dentin hypersensitivity: A systematic review and meta-analysis. *J. Clin. Periodontol.* **2015**, *42*, 131–141. [CrossRef]
85. Hu, M.-L.; Zheng, G.; Zhang, Y.-D.; Yan, X.; Li, X.-C.; Lin, H. Effect of desensitizing toothpastes on dentine hypersensitivity: A systematic review and meta-analysis. *J. Dent.* **2018**, *75*, 12–21. [CrossRef]
86. Hu, M.-L.; Zheng, G.; Lin, H.; Yang, M.; Zhang, Y.-D.; Han, J.-M. Network meta-analysis on the effect of desensitizing toothpastes on dentine hypersensitivity. *J. Dent.* **2019**, *88*, 103170. [CrossRef]
87. Cunha-Cruz, J.; Zeola, L.F. Limited evidence suggests that many types of desensitizing toothpaste may reduce dentin hypersensitivity, but not the ones with strontium or amorphous calcium phosphate. *J. Evid. Based Dent. Pract.* **2019**, *19*, 101337. [CrossRef]
88. Barros, A.P.O.; de Melo Alencar, C.; de Melo Pingarilho Carneiro, A.; da Silva Pompeu, D.; Barbosa, G.M.; Araújo, J.L.N.; Silva, C.M. Combination of two desensitizing protocols to control dentin hypersensitivity in non-carious lesions: A randomized, double-blind clinical trial. *Clin. Oral Investig.* **2022**, *26*, 1299–1307. [CrossRef] [PubMed]
89. Pompeu, D.D.S.; de Paula, B.L.F.; Barros, A.P.O.; Nunes, S.C.; Carneiro, A.M.P.; Araújo, J.L.N.; Silva, C.M. Combination of strontium chloride and photobiomodulation in the control of tooth sensitivity post-bleaching: A split-mouth randomized clinical trial. *PLoS ONE* **2021**, *16*, e0250501. [CrossRef] [PubMed]
90. Lee, J.-H.; Mandakhbayar, N.; El-Fiqi, A.; Kim, H.-W. Intracellular co-delivery of Sr ion and phenamil drug through mesoporous bioglass nanocarriers synergizes BMP signaling and tissue mineralization. *Acta Biomater.* **2017**, *60*, 93–108. [CrossRef] [PubMed]
91. Mandakhbayar, N.; El-Fiqi, A.; Lee, J.H.; Kim, H.W. Evaluation of Strontium-Doped Nanobioactive Glass Cement for Dentin-Pulp Complex Regeneration Therapy. *ACS Biomater. Sci. Eng.* **2019**, *5*, 6117–6126. [CrossRef] [PubMed]
92. Basheer, N.; Madhubala, M.M.; Jayasree, R.; Mahalaxmi, S.; Ts, S.K. Effect of Strontium Substituted Tetracalcium Phosphate Cement on Proliferation and Mineralization Potential in Human Dental Pulp Stem Cells. *Eur. Endod. J.* **2021**, *6*, 295. [CrossRef] [PubMed]
93. Mardas, N.; Dereka, X.; Stavropoulos, A.; Patel, M.; Donos, N. The role of strontium ranelate and guided bone regeneration in osteoporotic and healthy conditions. *J. Periodontal Res.* **2021**, *56*, 330–338. [CrossRef]
94. Jia, X.; Long, Q.; Miron, R.J.; Yin, C.; Wei, Y.; Zhang, Y.; Wu, M. Setd2 is associated with strontium-induced bone regeneration. *Acta Biomater.* **2017**, *53*, 495–505. [CrossRef]
95. Jia, X.; Miron, R.J.; Yin, C.; Xu, H.; Luo, T.; Wang, J.; Jia, R.; Wu, M.; Zhang, Y.; Li, Y. HnRNPL inhibits the osteogenic differentiation of PDLCs stimulated by $SrCl_2$ through repressing Setd2. *J. Cell. Mol. Med.* **2019**, *23*, 2667–2677. [CrossRef]
96. Bizelli-Silveira, C.; Pullisaar, H.; Abildtrup, L.A.; Andersen, O.Z.; Spin-Neto, R.; Foss, M.; Kraft, D.C. Strontium enhances proliferation and osteogenic behavior of periodontal ligament cells in vitro. *J. Periodontal Res.* **2018**, *53*, 1020–1028. [CrossRef]
97. Wang, L.; Pathak, J.L.; Liang, D.; Zhong, N.; Guan, H.; Wan, M.; Miao, G.; Li, Z.; Ge, L. Fabrication and characterization of strontium-hydroxyapatite/silk fibroin biocomposite nanospheres for bone-tissue engineering applications. *Int. J. Biol. Macromol.* **2020**, *142*, 366–375. [CrossRef]

98. Jiang, S.; Wang, X.; Ma, Y.; Zhou, Y.; Liu, L.; Yu, F.; Fang, B.; Lin, K.; Xia, L.; Cai, M. Synergistic Effect of Micro-Nano-Hybrid Surfaces and Sr Doping on the Osteogenic and Angiogenic Capacity of Hydroxyapatite Bioceramics Scaffolds. *Int. J. Nanomed.* **2022**, *17*, 783. [CrossRef]
99. Saidak, Z.; Marie, P.J. Strontium signaling: Molecular mechanisms and therapeutic implications in osteoporosis. *Pharmacol. Ther.* **2012**, *136*, 216–226. [CrossRef] [PubMed]
100. Marx, D.; Yazdi, A.R.; Papini, M.; Towler, M. A review of the latest insights into the mechanism of action of strontium in bone. *Bone Rep.* **2020**, *12*, 100273. [CrossRef]
101. Tobeiha, M.; Moghadasian, M.H.; Amin, N.; Jafarnejad, S. RANKL/RANK/OPG pathway: A mechanism involved in exercise-induced bone remodeling. *BioMed Res. Int.* **2020**, *2020*, 6910312. [CrossRef] [PubMed]
102. Lafzi, A.; Nejad, A.E.; Rad, M.R.; Namdari, M.; Sabetmoghaddam, T. In vitro release of silver ions and expression of osteogenic genes by MC3T3-E1 cell line cultured on nano-hydroxyapatite and silver/strontium-coated titanium plates. *Odontology* **2023**, *111*, 33–40. [CrossRef] [PubMed]
103. Kargozar, S.; Montazerian, M.; Fiume, E.; Baino, F. Multiple and promising applications of strontium (Sr)-containing bioactive glasses in bone tissue engineering. *Front. Bioeng. Biotechnol.* **2019**, *7*, 161. [CrossRef] [PubMed]
104. Tsai, S.-W.; Yu, W.-X.; Hwang, P.-A.; Huang, S.-S.; Lin, H.-M.; Hsu, Y.-W.; Hsu, F.-Y. Fabrication and characterization of strontium-substituted hydroxyapatite-CaO-$CaCO_3$ nanofibers with a mesoporous structure as drug delivery carriers. *Pharmaceutics* **2018**, *10*, 179. [CrossRef] [PubMed]
105. Tsai, S.; Yu, W.; Hwang, P.; Hsu, Y.; Hsu, F. Fabrication and characteristics of PCL membranes containing strontium-substituted hydroxyapatite nanofibers for guided bone regeneration. *Polymers* **2019**, *11*, 1761. [CrossRef] [PubMed]
106. Tsai, S.-W.; Hsu, Y.-W.; Pan, W.-L.; Vadivelmurugan, A.; Hwang, P.-A.; Hsu, F.-Y. Influence of the Components and Orientation of Hydroxyapatite Fibrous Substrates on Osteoblast Behavior. *J. Funct. Biomater.* **2022**, *13*, 168. [CrossRef]
107. Osman, R.B.; Swain, M.V. A critical review of dental implant materials with an emphasis on titanium versus zirconia. *Materials* **2015**, *8*, 932–958. [CrossRef]
108. Ting, M.; Jefferies, S.R.; Xia, W.; Engqvist, H.; Suzuki, J.B. Classification and effects of implant surface modification on the bone: Human cell–based in vitro studies. *J. Oral Implantol.* **2017**, *43*, 58–83. [CrossRef]
109. Jemat, A.; Ghazali, M.J.; Razali, M.; Otsuka, Y. Surface modifications and their effects on titanium dental implants. *BioMed Res. Int.* **2015**, *2015*, 791725. [CrossRef] [PubMed]
110. Schünemann, F.H.; Galárraga-Vinueza, M.E.; Magini, R.; Fredel, M.; Silva, F.; Souza, J.C.; Zhang, Y.; Henriques, B. Zirconia surface modifications for implant dentistry. *Mater. Sci. Eng. C* **2019**, *98*, 1294–1305. [CrossRef]
111. Awasthi, S.; Pandey, S.K.; Arunan, E.; Srivastava, C. A review on hydroxyapatite coatings for the biomedical applications: Experimental and theoretical perspectives. *J. Mater. Chem. B* **2021**, *9*, 228–249. [CrossRef]
112. Sergi, R.; Bellucci, D.; Cannillo, V. A comprehensive review of bioactive glass coatings: State of the art, challenges and future perspectives. *Coatings* **2020**, *10*, 757. [CrossRef]
113. Kim, J.; Kang, I.-G.; Cheon, K.-H.; Lee, S.; Park, S.; Kim, H.-E.; Han, C.-M. Stable sol–gel hydroxyapatite coating on zirconia dental implant for improved osseointegration. *J. Mater. Sci. Mater. Med.* **2021**, *32*, 81. [CrossRef] [PubMed]
114. Helen, S.; Kumar, A.R. Electrical, mechanical and surface analysis of ion-doped hydroxyapatite for antibacterial activity. *Appl. Phys. A* **2018**, *124*, 535. [CrossRef]
115. Rabiee, S.M.; Nazparvar, N.; Azizian, M.; Vashaee, D.; Tayebi, L. Effect of ion substitution on properties of bioactive glasses: A review. *Ceram. Int.* **2015**, *41*, 7241–7251. [CrossRef]
116. Geng, Z.; Li, X.; Ji, L.; Li, Z.; Zhu, S.; Cui, Z.; Wang, J.; Cui, J.; Yang, X.; Liu, C. A novel snail-inspired bionic design of titanium with strontium-substituted hydroxyapatite coating for promoting osseointegration. *J. Mater. Sci. Technol.* **2021**, *79*, 35–45. [CrossRef]
117. Newman, S.D.; Lotfibakhshaiesh, N.; O'Donnell, M.; Walboomers, X.F.; Horwood, N.; Jansen, J.A.; Amis, A.A.; Cobb, J.P.; Stevens, M.M. Enhanced osseous implant fixation with strontium-substituted bioactive glass coating. *Tissue Eng. Part A* **2014**, *20*, 1850–1857. [CrossRef]
118. Capuccini, C.; Torricelli, P.; Sima, F.; Boanini, E.; Ristoscu, C.; Bracci, B.; Socol, G.; Fini, M.; Mihailescu, I.; Bigi, A. Strontium-substituted hydroxyapatite coatings synthesized by pulsed-laser deposition: In vitro osteoblast and osteoclast response. *Acta Biomater.* **2008**, *4*, 1885–1893. [CrossRef] [PubMed]
119. Gentleman, E.; Fredholm, Y.C.; Jell, G.; Lotfibakhshaiesh, N.; O'Donnell, M.D.; Hill, R.G.; Stevens, M.M. The effects of strontium-substituted bioactive glasses on osteoblasts and osteoclasts in vitro. *Biomaterials* **2010**, *31*, 3949–3956. [CrossRef] [PubMed]
120. Tao, Z.S.; Zhou, W.-S.; He, X.W.; Liu, W.; Bai, B.L.; Zhou, Q.; Li, H.; Huang, Z.L.; Tu, K.; Hang, L.; et al. A comparative study of strontium-substituted hydroxyapatite coating on implant's osseointegration for osteopenic rats. *Med. Biol. Eng. Comput.* **2016**, *54*, 1959–1968. [CrossRef] [PubMed]
121. Li, Y.; Li, Q.; Zhu, S.; Luo, E.; Li, J.; Feng, G.; Liao, Y.; Hu, J. The effect of strontium-substituted hydroxyapatite coating on implant fixation in ovariectomized rats. *Biomaterials* **2010**, *31*, 9006–9014. [CrossRef]
122. Katunar, M.R.; Pastore, J.I.; Cisilino, A.; Merlo, J.; Alonso, L.S.; Baca, M.; Haddad, K.; Cere, S.; Ballarre, J. Early osseointegration of strontium-doped coatings on titanium implants in an osteoporotic rat model. *Surf. Coat. Technol.* **2022**, *433*, 128159. [CrossRef]
123. Zhang, W.; Cao, H.; Zhang, X.; Li, G.; Chang, Q.; Zhao, J.; Qiao, Y.; Ding, X.; Yang, G.; Liu, X. A strontium-incorporated nanoporous titanium implant surface for rapid osseointegration. *Nanoscale* **2016**, *8*, 5291–5301. [CrossRef] [PubMed]

124. Li, L.; Yao, L.; Wang, H.; Shen, X.; Lou, W.; Huang, C.; Wu, G. Magnetron sputtering of strontium nanolayer on zirconia implant to enhance osteogenesis. *Mater. Sci. Eng. C* **2021**, *127*, 112191. [CrossRef] [PubMed]
125. Marie, P.; Ammann, P.; Boivin, G.; Rey, C. Mechanisms of action and therapeutic potential of strontium in bone. *Calcif. Tissue Int.* **2001**, *69*, 121. [CrossRef] [PubMed]
126. Kołodziejska, B.; Stępień, N.; Kolmas, J. The influence of strontium on bone tissue metabolism and its application in osteoporosis treatment. *Int. J. Mol. Sci.* **2021**, *22*, 6564. [CrossRef]
127. Dai, J.; Fu, Y.; Chen, D.; Sun, Z. A novel and injectable strontium-containing hydroxyapatite bone cement for bone substitution: A systematic evaluation. *Mater. Sci. Eng. C* **2021**, *124*, 112052. [CrossRef]
128. Zhang, Y.; Wei, L.; Wu, C.; Miron, R.J. Periodontal regeneration using strontium-loaded mesoporous bioactive glass scaffolds in osteoporotic rats. *PLoS ONE* **2014**, *9*, e104527. [CrossRef] [PubMed]
129. Zhao, D.-W.; Zuo, K.-Q.; Wang, K.; Sun, Z.-Y.; Lu, Y.-P.; Cheng, L.; Xiao, G.-Y.; Liu, C. Interleukin-4 assisted calcium-strontium-zinc-phosphate coating induces controllable macrophage polarization and promotes osseointegration on titanium implant. *Mater. Sci. Eng. C* **2021**, *118*, 111512. [CrossRef] [PubMed]
130. Zhou, C.; Chen, Y.; Zhu, Y.; Lin, G.; Zhang, L.; Liu, X.; He, F. Antiadipogenesis and osseointegration of strontium-doped implant surfaces. *J. Dent. Res.* **2019**, *98*, 795–802. [CrossRef] [PubMed]
131. Choi, S.M.; Park, J.W. Multifunctional effects of a modification of SLA titanium implant surface with strontium-containing nanostructures on immunoinflammatory and osteogenic cell function. *J. Biomed. Mater. Res. Part A* **2018**, *106*, 3009–3020. [CrossRef] [PubMed]
132. Okuzu, Y.; Fujibayashi, S.; Yamaguchi, S.; Masamoto, K.; Otsuki, B.; Goto, K.; Kawai, T.; Shimizu, T.; Morizane, K.; Kawata, T.; et al. In vitro study of antibacterial and osteogenic activity of titanium metal releasing strontium and silver ions. *J. Biomater. Appl.* **2021**, *35*, 670–680. [CrossRef] [PubMed]
133. Offermanns, V.; Andersen, O.Z.; Riede, G.; Sillassen, M.; Jeppesen, C.S.; Almtoft, K.P.; Talasz, H.; Öhman-Mägi, C.; Lethaus, B.; Tolba, R.; et al. Effect of strontium surface-functionalized implants on early and late osseointegration: A histological, spectrometric and tomographic evaluation. *Acta Biomater.* **2018**, *69*, 385–394. [CrossRef]
134. Wang, S.; Yang, Y.; Li, Y.; Shi, J.; Zhou, J.; Zhang, L.; Deng, Y.; Yang, W. Strontium/adiponectin co-decoration modulates the osteogenic activity of nano-morphologic polyetheretherketone implant. *Colloids Surf. B Biointerfaces* **2019**, *176*, 38–46. [CrossRef] [PubMed]
135. Chang, L.-C.; Chung, C.-Y.; Chiu, C.-H.; Lin, M.H.-C.; Yang, J.-T. The Effect of Polybutylcyanoacrylate Nanoparticles as a Protos Delivery Vehicle on Dental Bone Formation. *Int. J. Mol. Sci.* **2021**, *22*, 4873. [CrossRef] [PubMed]
136. Göker, F.; Ersanlı, S.; Arısan, V.; Cevher, E.; Güzel, E.E.; İşsever, H.; Ömer, B.; Altun, G.D.; Morina, D.; Yılmaz, T.E.; et al. Combined effect of parathyroid hormone and strontium ranelate on bone healing in ovariectomized rats. *Oral Dis.* **2018**, *24*, 1255–1269. [CrossRef]
137. Arepalli, S.K.; Tripathi, H.; Hira, S.K.; Manna, P.P.; Pyare, R.; Singh, S.P. Enhanced bioactivity, biocompatibility and mechanical behavior of strontium substituted bioactive glasses. *Mater. Sci. Eng. C* **2016**, *69*, 108–116. [CrossRef]
138. Zhang, W.; Ma, S.; Wei, Z.; Bai, P. The Relationship between Residual Amount of Sr and Morphology of Eutectic Si Phase in A356 Alloy. *Materials* **2019**, *12*, 3222. [CrossRef]
139. Saxena, K.; Ann, C.M.; Azwar, M.A.B.M.; Banavar, S.R.; Matinlinna, J.; Peters, O.A.; Daood, U. Effect of strontium fluoride on mechanical and remineralization properties of enamel: An in-vitro study on a modified orthodontic adhesive. *Dent. Mater.* **2024**, *40*, 811–823. [CrossRef]
140. Amudha, S.; Ramya, J.R.; Arul, K.T.; Deepika, A.; Sathiamurthi, P.; Mohana, B.; Asokan, K.; Dong, C.-L.; Kalkura, S.N. Enhanced mechanical and biocompatible properties of strontium ions doped mesoporous bioactive glass. *Compos. Part B Eng.* **2020**, *196*, 108099. [CrossRef]
141. Lunawat, K.; Kavitha, S.; Rajkumar, G.; Dhivya, V.; Kumar, N.R.; Mahalaxmi, S.; Shaik, F.A. Influence of strontium containing fluorophosphate glass onto structural and mechanical behavior of MTA network. *J. Mech. Behav. Biomed. Mater.* **2023**, *140*, 105750. [CrossRef]
142. Rezaei, Y.; Moztarzadeh, F.; Shahabi, S.; Tahriri, M. Synthesis, characterization, and in vitro bioactivity of sol-gel-derived SiO_2–CaO–P_2O_5–MgO-SrO bioactive glass. *Synth. React. Inorg. Met.-Org. Nano-Met. Chem.* **2014**, *44*, 692–701. [CrossRef]
143. Arron, J.R.; Choi, Y. Bone versus immune system. *Nature* **2000**, *408*, 535–536. [CrossRef]
144. Zhao, T.; Chu, Z.; Ma, J.; Ouyang, L. Immunomodulation effect of biomaterials on bone formation. *J. Funct. Biomater.* **2022**, *13*, 103. [CrossRef]
145. Wei, Z.; Tian, P.; Liu, X.; Zhou, B. Hemocompatibility and selective cell fate of polydopamine-assisted heparinized PEO/PLLA composite coating on biodegradable AZ31 alloy. *Colloids Surf. B Biointerfaces* **2014**, *121*, 451–460. [CrossRef]
146. Orihuela, R.; McPherson, C.A.; Harry, G.J. Microglial M1/M2 polarization and metabolic states. *Br. J. Pharmacol.* **2016**, *173*, 649–665. [CrossRef]
147. Lee, C.-H.; Kim, Y.-J.; Jang, J.-H.; Park, J.-W. Modulating macrophage polarization with divalent cations in nanostructured titanium implant surfaces. *Nanotechnology* **2016**, *27*, 085101. [CrossRef]
148. Lu, X.; Zhang, W.; Liu, Z.; Ma, S.; Sun, Y.; Wu, X.; Zhang, X.; Gao, P. Application of a strontium-loaded, phase-transited lysozyme coating to a titanium surface to enhance osteogenesis and osteoimmunomodulation. *Med. Sci. Monit. Int. Med. J. Exp. Clin. Res.* **2019**, *25*, 2658. [CrossRef]

149. Martinez, F.O.; Helming, L.; Gordon, S. Alternative activation of macrophages: An immunologic functional perspective. *Annu. Rev. Immunol.* **2009**, *27*, 451–483. [CrossRef]
150. Zhang, W.; Zhao, F.; Huang, D.; Fu, X.; Li, X.; Chen, X. Strontium-substituted submicrometer bioactive glasses modulate macrophage responses for improved bone regeneration. *ACS Appl. Mater. Interfaces* **2016**, *8*, 30747–30758. [CrossRef]
151. Li, S.; Zhang, L.; Liu, C.; Kim, J.; Su, K.; Chen, T.; Zhao, L.; Lu, X.; Zhang, H.; Cui, Y.; et al. Spontaneous immunomodulation and regulation of angiogenesis and osteogenesis by Sr/Cu-borosilicate glass (BSG) bone cement to repair critical bone defects. *Bioact. Mater.* **2023**, *23*, 101–117. [CrossRef]
152. Yuan, X.; Cao, H.; Wang, J.; Tang, K.; Li, B.; Zhao, Y.; Cheng, M.; Qin, H.; Liu, X.; Zhang, X. Immunomodulatory effects of calcium and strontium co-doped titanium oxides on osteogenesis. *Front. Immunol.* **2017**, *8*, 1196. [CrossRef]
153. Cipriano, A.F.; Sallee, A.; Tayoba, M.; Alcaraz, M.C.C.; Lin, A.; Guan, R.-G.; Zhao, Z.-Y.; Liu, H. Cytocompatibility and early inflammatory response of human endothelial cells in direct culture with Mg-Zn-Sr alloys. *Acta Biomater.* **2017**, *48*, 499–520. [CrossRef]
154. Graney, P.L.; Roohani-Esfahani, S.-I.; Zreiqat, H.; Spiller, K.L. In vitro response of macrophages to ceramic scaffolds used for bone regeneration. *J. R. Soc. Interface* **2016**, *13*, 20160346. [CrossRef]
155. Marx, D.; Papini, M.; Towler, M. In vitro immunomodulatory effects of novel strontium and zinc-containing GPCs. *Bio-Med. Mater. Eng.* **2022**, *33*, 377–391. [CrossRef]
156. Lourenço, A.H.; Torres, A.L.; Vasconcelos, D.P.; Ribeiro-Machado, C.; Barbosa, J.N.; Barbosa, M.A.; Barrias, C.C.; Ribeiro, C.C. Osteogenic, anti-osteoclastogenic and immunomodulatory properties of a strontium-releasing hybrid scaffold for bone repair. *Mater. Sci. Eng. C* **2019**, *99*, 1289–1303. [CrossRef]
157. Xu, A.-T.; Xie, Y.-W.; Xu, J.-G.; Li, J.; Wang, H.; He, F.-M. Effects of strontium-incorporated micro/nano rough titanium surfaces on osseointegration via modulating polarization of macrophages. *Colloids Surf. B Biointerfaces* **2021**, *207*, 111992. [CrossRef]
158. Fenbo, M.; Xingyu, X.; Bin, T. Strontium chondroitin sulfate/silk fibroin blend membrane containing microporous structure modulates macrophage responses for guided bone regeneration. *Carbohydr. Polym.* **2019**, *213*, 266–275. [CrossRef]
159. Zhong, Z.; Wu, X.; Wang, Y.; Li, M.; Li, Y.; Liu, X.; Zhang, X.; Lan, Z.; Wang, J.; Du, Y.; et al. Zn/Sr dual ions-collagen co-assembly hydroxyapatite enhances bone regeneration through procedural osteo-immunomodulation and osteogenesis. *Bioact. Mater.* **2022**, *10*, 195–206. [CrossRef]
160. Yu, D.; Guo, S.; Yu, M.; Liu, W.; Li, X.; Chen, D.; Li, B.; Guo, Z.; Han, Y. Immunomodulation and osseointegration activities of Na_2TiO_3 nanorods-arrayed coatings doped with different Sr content. *Bioact. Mater.* **2022**, *10*, 323–334. [CrossRef]
161. Chantelle, L.; Kennedy, B.J.; de Oliveira, C.P.; Gouttefangeas, F.; Siu-Li, M.; Landers, R.; Ciorita, A.; Rostas, A.M.; dos Santos, I.M.G.; De Oliveira, A.L.M. Europium induced point defects in $SrSnO_3$-based perovskites employed as antibacterial agents. *J. Alloys Compd.* **2023**, *956*, 170353. [CrossRef]
162. Wu, T.; Lu, T.; Shi, H.; Wang, J.; Ye, J. Enhanced osteogenesis, angiogenesis and inhibited osteoclastogenesis of a calcium phosphate cement incorporated with strontium doped calcium silicate bioceramic. *Ceram. Int.* **2023**, *49*, 6630–6645. [CrossRef]

MDPI AG
Grosspeteranlage 5
4052 Basel
Switzerland
Tel.: +41 61 683 77 34

Journal of Functional Biomaterials Editorial Office
E-mail: jfb@mdpi.com
www.mdpi.com/journal/jfb

www.ingramcontent.com/pod-product-compliance
Lightning Source LLC
Chambersburg PA
CBHW061546210926
51850CB00009B/2015
* 9 7 8 3 7 2 5 8 2 1 3 3 4 *